D0794152

Water Contamination Emergencies
Enhancing Our Response

Water Contamination Emergencies
Enhancing Our Response

Edited by

K. Clive Thompson
ALcontrol Laboratories, Rotherham, UK

John Gray
Drinking Water Inspectorate, Whitehall, London, UK

RSC Publishing

The proceedings of the Water Contamination Emergencies: Enhancing Our Response conference held on 12–15 June 2005 at Manchester University.

Special Publication No. 302

ISBN-10: 0-85404-658-5
ISBN-13: 978-0-85404-658-4

A catalogue record for this book is available from the British Library

Published by The Royal Society of Chemistry,
Thomas Graham House, Science Park, Milton Road,
Cambridge CB4 0WF, UK

Registered Charity Number 207890

For further information see our web site at www.rsc.org

Printed by Henry Ling Ltd., Dorchester, Dorset, UK

PREFACE

There is a need to react to emergencies involving low probability/high impact contamination events (chemical, biological or radiological) in source waters or treated water, in sufficient time to allow an effective response that will significantly reduce or avoid adverse impacts on consumers or the environment.

The international conference "Water Contamination Emergencies: Enhancing our response" was held in June 2005 and developed the themes established at the first conference held in March 2003 at Kenilworth and which was entitled "Water Contamination Emergencies: Can we cope?" The 2003 conference concluded that we could cope, but had much to learn. The second conference clearly demonstrated that the worldwide water industry had learnt much over the intervening two years and sharing this new knowledge at the Manchester conference should prove invaluable to all those in the water industry and other interested parties who ever have to deal with any aspect of water contamination emergencies.

The book opens with a short scene setting section which identifies the themes and objectives of the conference.

Almost 40 chapters of the book, produced by experts in the field of handling and responding to water contamination incidents and associated issues, consider in detail various aspects of the key features of a water contamination incident, including: awareness of that the incident is occurring; the extent of the incident; accurate and rapid identification and quantification of the contamination (the chemical/microbiological/radiological species present); risks to the consumer and other water users; the origin of the contamination; short and long term remediation; communications with consumers and other users; liaison with public health officials and local authorities; and communications with the media and general public.

It is hoped that a third conference in the series "Water Contamination Emergencies" will be held in 2008. Full details will be available nearer the time at http://www.dwi.gov.uk

Dr John Gray
Prof. K. Clive Thompson
October 2005

Contents

POSTERS

INTRODUCTION: THEMES AND OBJECTIVES

J. Gray

Drinking Water Inspectorate, M08, 55 Whitehall, London, SWIA 2EY

1 INTRODUCTION

In 2001, after the events in New York, the water industry started to look a little harder at the systems it had in place to protect the public water supply. It would be unfair to infer that such matters had never been a priority. Water companies had developed, generally, a strong culture of safety and security which included physical protection of their assets. What 2001 did was to cause the industry to ask the question "If there is an accidental or even an intentional contamination of the public water supply, can we cope?".

This led to the first conference held in 2003 at Warwick which came to the conclusion "Yes – but.....". The "but" identified a number of gaps in activities and responses. I hope that this conference will address those potential gaps and, as a corollorary, identify those areas where individuals or organisations are duplicating others' activities. This wastes resources and diverts attention from areas which are in need.

What the organising committee has done in this conference, as you will see from the programme, is to bring together a wide range of professions, skills, knowledge, organisations and individuals on whom the water industry relies and with whom the industry will need to work in the event of a contamination incident. The committee has worked hard to bring together such a diverse and powerful group and I am grateful to them.

We have established the environment. It is now up to you as individuals to develop and expand your contacts and understanding of others' roles and responsibilities and to make it all worthwhile.

2. THEMES AND OBJECTIVES

A number of words and phrases recur throughout the programme for the next few days - safety, security and water safety plans. These latter are nothing new. Every company has a safety plan in some form or other, whether they recognise it as such or not. What may be missing are some individual components of a comprehensive plan or the organisational and management structure for that plan.

Water safety plans consider the whole process of protecting the quality of the water supply from source to tap. They do not rely solely on monitoring the quality of the supply as it leaves the control of the supplier. This monitoring apparently gives comfort about the quality of the supply at the supposed time of supply and only gives information after the water has been drunk.

It should not be forgotten that all stages of the water treatment process reinforce the effectiveness of the other individual components. None, with the possible exception of disinfection, work in isolation.

Security is considered within the WSP approach and whilst some parts of the process chain are protected, for example padlocked and alarmed hatches, others are open and almost inviting, for example access ladders on aerators. Have companies assessed each and every part of their chains of supply from source to tap? There is a need for constant vigilance, continued effort and continual awareness. This conference hopes to provide the opportunities to share experiences, knowledge and expertise. Have a good conference and make the most of the opportunities that arise.

SAFETY, SECURITY, (UN)CERTAINTY

J Colbourne

Drinking Water Inspectorate, M08, 55 Whitehall, London, SWIA 2EY

1 INTRODUCTION

"The most effective means of consistently ensuring the safety of a drinking water supply is through the use of a comprehensive risk assessment and risk management approach that encompasses all steps in water supply from catchment to consumer"

So begins Chapter 4 of the third edition of "Guidelines for Drinking Water Quality"[1]. Water Safety Plans (WSPs) are the foundation of the measures water suppliers need to take to preserve drinking water quality and protect public health. Just as water treatment is a multilayered approach, with each state of treatment tailored to the specific needs represented by the quality of the water source, so water safety plans are multilayered and so too are water security requirements.

2 WATER SAFETY PLANS

There are three key components of any WSP, each guided by health considerations:

- System assessment
- Effective operational performance
- Management arrangements

An holistic assessment of a water supply system will quickly identify a number of vulnerabilities including those in the water catchments, sources, abstraction points, storage (impounding) reservoirs, treatment processes and within the distribution system.

Any assessment of risk must consider the likelihood of unauthorised access at any of these points and the potential impact of the introduction of any deliberate contaminant on consumers' health or general wellbeing.

3 SECURITY CONSIDERATIONS

Security requirements are an integral part of a water company's WSP. WSPs are nothing new – water supply management plans were the way of working for the industry for many years but in recent times this holistic, integrated approach has been less to the fore. There is thus a need to educate a new generation of water supply managers using WSPs as the tool.

It has been said that a Water Safety Plan would alarm consumers. This ignores the potentially disastrous consequences of failure for consumers individually as well as for business. The way to gain the confidence and trust of consumers is to demonstrate the existence of robust, well thought through plans for safeguarding drinking water quality and public health against the likely (and the unlikely) risk scenarios.

4 UNCERTAINTY

Water Safety Plans provide for uncertainty by the setting of health based criteria for each water supply system. From these criteria it is then possible to identify the critical control points where checks can be carried out at an appropriate frequency to determine, with reasonable confidence, that all is well. In choosing a critical control point it is important to ensure that the monitoring is simple, easy and effective – giving immediate information that, if need be, can be acted upon to bring the system back into control. If measurements (as opposed to human observation) are to be relied on, then these should be in-line or on-line. Information should be displayed and archived automatically. Measurement points should be early enough in the system to facilitate corrective action – reliance should not be place on end point monitoring, particularly if this involves samples being conveyed for analysis elsewhere. The purpose of end point monitoring is to provide overall reassurance, after the fact, that safe water has been supplied.

5 CONCLUSIONS

There is nothing special, sacred or scary about water supply security arrangements. Security risk is but one component of the overall risk matrix inherent in a WSP. The greatest risk is arguably that of managing security risks outside of the normal water supply management framework.

References

1 World Health Organisation, *Drinking Water Guidelines*, 3rd Edn. 2003

THE WATER INDUSTRY'S PERSPECTIVE OF WATER CONTAMINATION EMERGENCIES

J. Dennis
Wessex Water, Claverton Down, Bath, BA2 7WW

1 INTRODUCTION

Since the days of the industrial revolution and the ravages that cholera and typhoid brought on the population as a result of contaminated water supplies, water treatment has improved greatly. However, since the middle of the twentieth century, the increased use and disposal of chemicals into our environment, together with the increased pressure of population growth in our towns and cities, allied with a better understanding of the health issues, has led to a greater appreciation of the threat that our modern way of life has to water supplies.

2 SECURITY AND PUBLIC HEALTH

The desire to provide safe and wholesome drinking water for all is now enshrined in the Bonn Charter (2004). However, water Utilities and society itself face additional difficulties with the rise of fundamentalism and the terrorist who see society and the utilities that support it as a legitimate target in emphasizing or furthering their cause.

These developments have led to the re-examination of how we measure risk and how we use risk assessment to formulate security and response plans to protect public health regardless of how the threat may manifest itself.

Modern security and treatment systems can be expensive and a particular type of treatment, on its own, may not meet the standards needed for treatment of water for supply to the public or protect it from accidental or deliberate contamination. However, the multi-barrier approach to water treatment and supply protection is now accepted as standard, but the selection of the right barriers in relation to population need, expectation, vulnerability, economics and source is fundamental to providing a safe water supply and securing customer confidence.

It is therefore essential that the risks are understood and treatment and security are properly tested and applied through good risk assessment. This risk assessment will also inform the development of *water safety plans* that will not only enshrine an understanding of the system from source to tap, but will also support the development of event management plans.

These event management plans set out the measures and actions to be undertaken to protect the public in the event that a supply is lost or contaminated. They have to be robust, rehearsed and tested, ensuring that they are fit for purpose so that safe and wholesome water is produced and delivered to where it is needed.

3 CONCLUSION

In summary, water contamination events on a scale that will cause the public to loose confidence in the water supply, are rare.

Properly applied risk assessment, integrated management from source to tap and emergency planning will limit the impact of such events.

And as we live in a free and vibrant society, the risk of major contamination through deliberate or accidental contamination remains a reality. We must, therefore, never be complacent.

THE CUSTOMER'S VIEW ON WATER CONTAMINATION

R. STURT
WaterVoice Southern, 4th Floor (South), High Holborn House, 52-54 High Holborn, London, WC1V 6RL

1 INTRODUCTION TO ROLE OF WATERVOICE

WaterVoice is the statutory body which represents water customers throughout England and Wales. It consists of a national council and nine regional committees for England, plus one for Wales. The customers whom it represents are the customers, or bill-payers, of the appointed water undertakers. The existing WaterVoice structure, which is part of Ofwat, is due to be abolished and replaced on the 1 October 2005 by a new statutory body, the Consumer Council for Water ("the CCW"), a central corporate body. The new CCW will be completely independent of Ofwat, but will continue to represent consumers through a number of regional committees in England, plus one for Wales, as well as through its central organisation. It will have a wider remit than WaterVoice, in that it will represent consumers (as opposed to customers), including future consumers, and will monitor the activities not only of the appointed undertakers, but of the new licensed suppliers established under the Water Act 2003. It will have an explicit duty to:—

> "exercise and perform its powers and duties in the manner which it considers is best calculated to contribute to the achievement of sustainable development."

It will also continue to have a statutory duty to have regard to the interests of the disabled and vulnerable consumers.

In the context of this presentation, the CCW will have a duty to try and ensure that the environmental consequences of development are such that the development can be sustained. This would include ensuring that there are plans for preventing contamination of the public supply and for mitigating and isolating its effects if it occurs.

2 THE CUSTOMER BASE

As at 31st March, 2004, there were about 21.3 million households in England and Wales connected to a mains water supply, plus about 1.6 million non-household customers. The population, or consumer base, behind these households and non-households was about 53 million people.

3 CUSTOMER PRIORITIES

During the PR04 process, WaterVoice was involved in a detailed opinion survey carried out on behalf of various stakeholders, including WaterVoice itself, Ofwat and the DWI. It was obvious from that survey that water customers place the highest priority on safety of supply:—

> "maintaining the quality and safety of the drinking water and ensuring a reliable continuous supply of water were reported as significantly more important that any other service area listed."

WaterVoice relies on the DWI to police the requirements about quality and safety but there is a limit to the amount which either DWI or the companies, or the EA for that matter, can do to cope with the unexpected.

4 CUSTOMER PERSPECTIVE ON THE RISKS

So I am pleased that the water customers' representative should be here today to put the point of view of customers about contamination of water supply. Whilst the threat of widespread contamination of supplies may be more apparent than real, there are nevertheless considerable risks. Contamination can take many forms and although not all forms would necessarily be injurious to health, they may have other adverse consequences for consumers, such as interruption of supplies.

The commonest forms of contamination are, perhaps:—

- industrial discharges
- sewage in watercourses
- nitrates and pesticides from agricultural activities
- cryptosporidium
- accidental spillages

Rarer and riskier ones might be the result of terrorist activity, such as:—

- explosives damaging mains
- deadly poisons, such as arsenic, being introduced into the supply chain
- disease bearing organisms, such as anthrax

5 "ROUTINE CONTAMINATION"

5.1 Naturally occurring substances

In this list there are various sources of contamination which occur routinely. For instance, fluoride and benzene can occur naturally in groundwater. The companies carry out continuous sampling of their water and the DWI of course polices the results which show the levels of these "contaminants" which are not generally a problem for consumers. These do not generally impose any significant cost on water customers.

5.2 Contamination by excessive concentrations

Of more concern are the other pollutants, such as:—

- lead
- nitrates
- pesticides
- phosphates
- cryptosporidium

5.3 Deliberate introductions

Some contaminants are deliberately introduced into the water supply for a specific purpose, for example chlorine as a disinfectant and fluoride for the protection of teeth, as well as orthophosphate treatment to reduce the effects of lead contamination.

5.4 Safety and environmental purity come at a cost

The combined effect of these forms of treatment, some of which affect water and some form part of the treatment of wastewater, are largely routine; but very expensive. The capital cost of the quality programme in PR04 is upwards of £2 billion, with concomitant operating cost increases of £29 million, in each case subject to efficiency reductions laid down by Ofwat. This cost has a potent adverse effect on the affordability of the bills which customers are required to pay. When added to the environmental programme required of companies in the sewerage sector, the combined effect is a gross increase of £30 per domestic customer, of which about £9 is attributable to the removal of contaminants. These figures are of course in addition to the sums customers are already paying for the quality programme.

5.5 What do customers think of the costs?

Customers are very unhappy at the steep increase in costs, particularly in the first year of the 5 year AMP4 period. Whilst they approve of capital expenditure on infrastructure maintenance and on securing future supplies, they do not like paying for the removal of nitrates (£288) or pesticides (£73 million) as they believe that these forms of contamination should be paid for by the polluter. The orthophosphate lead treatment is something which they welcome. The jury is still out on fluoridation, as there is a conflict between those who feel that it confers an important public health benefit and those who do not wish to have potentially unsafe additions made to their supplies, although in this case the cost of the introduction of fluoride would have to be borne by the health authorities, so there would be no financial impact on water customers.

6 PROTECTION OF SOURCES AGAINST LIFE-THREATENING CONTAMINATION

The first essential is that the most vulnerable areas of a water company's supply are adequately protected. This is easier said than done.

6.1 Open reservoirs at risk

Open reservoirs are obviously at risk, and so are water mains and service reservoirs, although in the latter case, the contamination of any one reservoir would affect only a limited section of the population. In the case of large impounding reservoirs, very large quantities of toxic material would be needed to cause significant damage. This picture of part of Bewl Water, one of the biggest reservoirs in the South East, demonstrates the point.

Figure 1. Bewl Water

Bewl Water, like many of the reservoirs in the South, not only serves the population directly, but acts as a "header tank" for the River Medway (from which abstractions intended for the public supply are made), and more recently as a resource for replenishing Darwell Reservoir, which in turn acts partly as a header tank for the River Rother.

6.2 Advantage where reservoir used as a header tank

This means of storage has advantages in the context of contamination. First it enables a contaminated reservoir to be isolated from the river which it replenishes; and secondly it enables a contaminated river to be left to recover while supplies are taken from the unaffected reservoir. Reservoirs also hold large amounts of fresh water which could be released into a contaminated river, for dilution.

7 IMPORTANCE OF ISOLATION AND BY-PASSING CONTAMINATION

The area of the damage caused by deliberate or accidental contamination should be confined. In reality, there is very little the company can do to deter a well-organised attacker. What is probably more important is that the company should be able to respond quickly when contamination occurs and to have a degree of "redundancy" in its supply and distribution network. This would enable a company to isolate any contaminated source, service reservoir or distribution area and bypass the isolated area, or in an extreme case moving supplies overland or even by sea. It is also obvious that a company should have a sufficient degree of surplus capacity to cope with a short term interruption of supply in any particular zone. Companies already attempt to meet these requirements, by keeping some surplus capacity, creating links between zones and by cross-border bulk supply agreements. These measures give a supply company "headroom".

8 SECURITY OF SUPPLIES

Ofwat monitors companies' compliance with a Security of Supply Index, which is at least partly designed to ensure that companies have sufficient headroom and distribution flexibility to cope with a contamination incident. The index enables Ofwat to assess water resource availability and leakage issues within a wider security of supply context. Ofwat defines "headroom" as the difference between the amount of water a company has available to supply (or water available for use) under specified planning conditions and service assumptions, and the volume of water it will need to introduce into its network ("distribution input") under the same conditions.

8.1 Target headroom

Target headroom is the difference between water available for use and the distribution input that companies need in each of their resource zones, to take account of future supply and demand uncertainties. Achieving target headroom shows that a company can deliver its planned level of service in a variety of conditions. However, the assumptions about interruptions on which Ofwat bases its judgments are quite limited. Its reference levels of service require a company to be able to cope without imposing a three month hosepipe ban (meaning a reduction in demand of 5% during dry weather) more often than once in ten years and a further 5% reduction for three months once every 40 years.

8.2 Typical supply/demand from Chalky's borehole

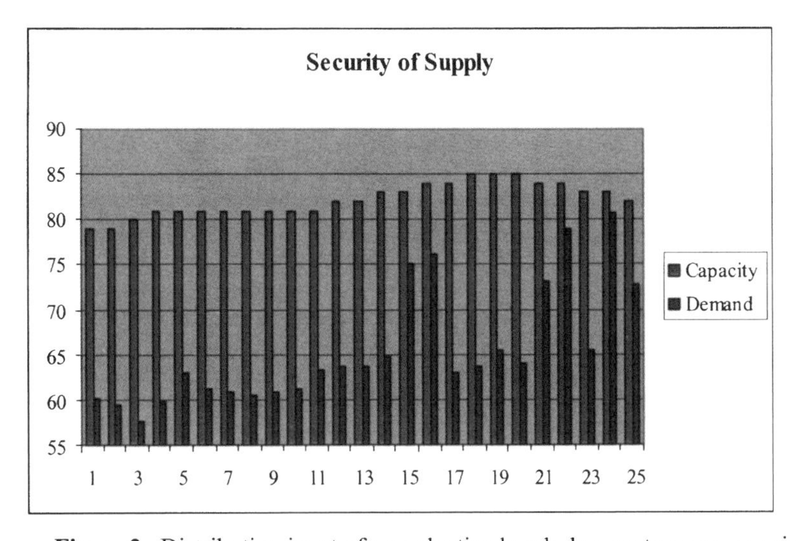

Figure 2. Distribution input of a productive borehole over two years against adjusted demand.

Figure 2 shows the actual distribution input of a particular (very productive) borehole over two years against adjusted demand. A typical year in the supply/demand balance of one of the groundwater sources is shown of a hypothetical company which I shall call "Chalky Water".

The graph shows clearly that whilst at most times there is ample headroom, there are times when any threat to supply could cause serious problems. Clearly any company needs a choice of sources if it is to be capable of coping with contamination, even where the normal headroom is very high. It also needs some storage capacity, either for raw water or for treated water at strategically important points in its network.

8.3 Comparison of the appointed companies

In fact many companies have much less headroom than Chalky Water. The Ofwat Security of Supply Index shows that only about half the regulated companies, including (I shall assume) Chalky, have A Grade ratings. Some have very serious deficiencies and would be in great difficulties if a major contamination incident coincided with a drought. During PR04 most companies with deficits are planning major improvements but I wonder whether sufficient account has been taken of the risks of this combination of events.

Figure 3 sets out, by company, the index results for both the planned and reference service assumption. Companies with higher index score bands have better security of supply.

Company	Planned service	Reference service
Northumbrian North	A	A
Wessex	A	A
Yorkshire	A	A
Bournemouth & W Hampshire	A	A
Bristol	A	A
Cambridge	A	A
Dee Valley	A	A
Portsmouth	A	A
South Staffordshire	A	A
Sutton & East Surrey	A	A
Tendring Hundred	A	A
Anglian	B	B
United Utilities	B	B
Mid Kent	B	B
South West	B	B
Three Valleys	B	A
Severn Trent	C	C
South East	C	C
Dŵr Cymru	C	B
Northumbrian South	C	C
Southern	D	D
Folkestone & Dover	D	C
Thames	D	D

Figure 3. Index results for both the planned and reference service assumption
A: No deficit against target headroom in any resource zone.
B: Marginal deficit against target headroom.
C: Significant deficit against target headroom.
D: Large deficit against target headroom.

9 WHAT WOULD HAPPEN IN EXTREMES

Under Section 37 of the Water Industry Act 1991:—

"37.—(1) It shall be the duty of every water undertaker to develop and maintain an efficient and economical system of water supply within its area and to ensure that all such arrangements have been made—

(a) for providing supplies of water to premises in that area and for making such supplies available to persons who demand them; and
(b) for maintaining, improving and extending the water undertaker's water mains and other pipes,

as are necessary for securing that the undertaker is and continues to be able to meet its obligations under this Part."

Under extreme conditions the company could be forced into honouring its commitment to maintain supplies by physically moving water into the contaminated zone by ship or lorry. This expedient is so expensive that most companies could only contemplate it for a very short time. If the situation continued for a number of weeks, say, a company could easily run out of funds and have to be rescued by Ofwat.

10 BOTTLED SUPPLIES

There have been a number of small-scale incidents in the last few years where drinking water has become contaminated and the companies have demonstrated a fast-footed ability to deliver bottled supplies to the doorsteps of their customers and some of these incidents have caused major ecological damage, but in recent times there has been no generalised contamination over a wide area.

11 ACTUAL SECURITY AGAINST THREATS

Ofwat also requires companies to spend a limited amount of resource on physical security of their undertakings, designed to prevent intrusion. In the Final Price Determinations announced in December 2004, Ofwat has approved plans by companies to spend over £300 million on improving this sort of security and authorised extra revenue expenditure of about £4 million. It is difficult for customers to work out whether this expenditure is justified and equally difficult to know whether existing security levels are adequate. Companies have to have their emergency plans audited by Defra, but the problem is twofold: how to prevent disastrous contamination occurring; and how to mitigate the consequences if it does. All good emergency plans have a strong focus on business continuity, which is especially important with a vital commodity like drinking water. The consequences for a major hospital of a major interruption or, worse, major contamination of the supply, can be readily imagined.

12 CIVIL CONTINGENCIES ACT 2004

12.1 Section 2 of the Civil Contingencies Act

Under Section 2 of the Civil Contingencies Act 2004, water undertakers are treated as "Category 2 responders" and as such must:—

> "(a) from time to time assess the risk of an emergency occurring,
> (b) ... assess the risk of an emergency making it necessary or expedient for the company to perform any of its functions,
> (c) maintain plans for the purpose of ensuring, so far as is reasonably practicable, that if an emergency occurs the company is able to continue to perform its functions,
> (d) maintain plans for the purpose of ensuring that if an emergency occurs or is likely to occur the company is able to perform its functions for the purpose of—
> (i) preventing the emergency,
> (ii) reducing, controlling or mitigating its effects, or
> (iii) taking other action in connection with it ..."

12.2 Mass fatalities

Customers will expect companies to take seriously their obligations under the Civil Contingencies Act. Under this Act, water companies have to collaborate with other agencies, and also liaise with the blue light emergency services, carrying out regular exercises, exchanging plans, etc. They are also required to have plans for coping with mass fatalities. Most agencies, including, I suspect the water companies, have rather shied away from planning for such an Armageddon. If large numbers of fatalities occurred as a result of the contamination of the public supply, it is unlikely that the water consumers would die simultaneously and even more unlikely that they would be concentrated all in the same place.

12.3 Limited but vital role of water companies in a disaster

Hence the job of the water companies would be confined to preventing the problem spreading, warning the public of the risk, and restoring supplies at the earliest possible moment while maintaining supplies from some uncontaminated source. Public confidence would have to be regained and the companies would have a major public relations job on their hands. These functions could and should be planned for, but experience shows that emergency planning is the Cinderella of public administration and tends to be under-resourced until a calamity occurs.

13 SECURITY AND EMERGENCY MEASURES DIRECTION

The Security and Emergency Measures Direction predates the Civil Contingencies Act. It is concerned with the measures which companies should take to ensure that, even in a major calamity, companies can deliver a minimum amount of 10 litres of potable water per person per day.

14 ACTUAL CONTAMINATION EXAMPLES

Restoration of damaged installations could well take a great deal of time. Some forms of groundwater contamination may be impossible to rectify as the contamination may have originated in diffuse pollution, which may have taken a long time to percolate down into the public supply bore hole. An actual example of this has occurred recently in an area served by a company which I shall call "Chalky Water". The company has kindly allowed me to use the facts of this case as an example, but anonymised.

15 CHALKY WATER'S SAMPLE BOREHOLE WTW

A recent example of accidental contamination can be used to illustrate the risks and consequences of contamination. It occurred in the area of one of Chalky Water's main chalk aquifers.

15.1 The event

On the afternoon of 5th October 2004, the Environment Agency informed the company's Supply Manager that they had been notified by a householder that approximately 2,000 litres of heating oil had leaked from a central heating storage tank at a local farm, "Greenacre". The property was approximately 1.2km 'up catchment' of "Sample Borehole" WTW in Zone 1 of the Agency's Groundwater Protection Zone. The Company and its predecessors have been abstracting groundwater from Sample Borehole for over 100 years. Groundwater quality at Sample Borehole is excellent and no physical treatment is required to ensure that the raw water meets drinking water quality standards. After some minimal chlorination it is pumped to "Regional Reservoir".

15.2 Importance of the source

As a result of its reliable water quality and low cost, Sample Borehole is used as a 'baseload' source by Chalky Water and abstraction averages 18 Ml/d in a normal year. It is of great importance, not only as a source for local supplies amounting to about 5 Ml/d but also provides another 13 Ml/d for other parts of Chalky Water's network. Chalky is also licensed to abstract another 10 Ml/d but keeps this as a vital strategic reserve and as a header tank for summer peaks.

15.3 Environment Agency investigations

The Environment Agency has so far been unable to determine the full extent of pollution and investigations continue. The level of water in the borehole is low at the moment as a result of the dry autumn and winter, and it will not be possible to say how serious the contamination is until the source has recharged.

15.4 Steps taken

As a precautionary measure abstraction at Sample Borehole ceased within 35 minutes of the report of pollution being received. Supplies from elsewhere in the network were brought in to the area served by Sample Borehole. Lost resource capacity was made up by operating three other sources, including the River "Watercourse", at higher output than would normally be expected at this time of year. This has incurred significant additional operating costs.

15.5 Fears for the future

No heating oil had been detected in the catchment of the Sample Borehole boreholes up to the middle of March 2005. However, it is quite possible that the oil may still be located within the unsaturated zone of the chalk aquifer; so heavy rainfall and/or high levels of groundwater recharge in future years might mobilise the oil at some later date.

15.6 Recommissioning

Chalky Water is constructing a Granular Activated Carbon (GAC) plant, which is due to be completed by mid-May and this will enable the Company to recommission Sample Borehole, although at a lower throughput. Additional 'on-line' water quality monitoring will be installed to ensure that in the event that the oil is identified in the raw water at Sample Borehole, the Company is aware of this eventuality and action taken to ensure that drinking water quality is not compromised. The expense of this action would be for the company to bear, although if it were very substantial, customers would eventually see the cost reflected in higher prices.

15.7 What the customer feels

Customers of Chalky Water would feel justifiably concerned that such a trivial incident could cause such a major interruption to vital supplies. The fact that the company was able to replace the loss from other sources is of some comfort, but over-abstraction from those sources cannot be allowed to continue indefinitely. It is lucky for customers that the company, as a result of prudent investment over many years, is one of those which have an adequate security of supply and also that no contamination of customers' drinking water has emerged so far.

15.8 PR aspects

It is also important that there is an effective and pro-active campaign by the EA and the company to educate the public about the desirability of avoiding contamination. Chalky Water has probably the purest groundwater of any undertaker and all possible steps must be taken to preserve it. In this particular case there has been no danger to the public as a result of the incident and Chalky Water has been able to make good the supply from other sources, so there has been no need for the company to make any contact with customers.

16 CAMELFORD

Whilst the Chalky Water incident required no customer handling from the company, this is not always the case. A very serious contamination incident, again with an accidental origin, occurred on 6th July 1988.

16.1 The event

The South West Water Authority failed to warn the public of the very real danger to public health, when a contractor dumped 20 tons of liquid aluminium sulphate into the wrong tank at the Lowermoor treatment plant operated by South West Water at Camelford, a small town of about 2,500 people in Cornwall. The plant was an unmanned installation and the contractor was a relief driver unfamiliar with the plant layout and delivery procedures. The resultant acidic water entered the water supply directly, causing public complaints about the taste, skin irritation and corrosive effects on plumbing and fixtures. The cause of the problem was not determined for two days and even after it had been, the public were assured by a spokesman for the water authority that the water, while tasting slightly acidic, was safe to drink.

16.2 Dangerously high levels of contamination

It was estimated that consumers were exposed for up to three days to water with pH as low as 3.9 to 5.0. An aluminium content of up to 620 milligram per litre and a sulphate concentration of up to 4,500 milligram per litre were recorded in the water supply. Once the cause of the problem was determined, a program of flushing reduced levels rapidly to 1 milligram per litre.

16.3 Failure of public relations

No public warning was issued to prevent the ingestion of the water. At no time did South West Water Authority inform their customers that the water was unsafe to drink. None of the health authorities, hospitals or general practitioners in the local area was informed of the incident. The public and public authorities were not informed about the true nature of the incident until 17 days later on the 23rd July 1988, when it was reported on the bottom of the sports page at the back of the Western Morning News, a local newspaper.

16.4 Immediate after-effects

The water was consumed by the local population who as a consequence suffered injuries to the mouth, stomach, and skin and other disorders. These effects lasted in some cases for weeks in the first instance. Other long term adverse health effects have been slowly emerging since the incident.

16.5 Long term brain damage

In 1991 a study was conducted Dr Paul Altmann into these effects by those tested "suffered considerable damage to cerebral function." The researchers found that transmission of signals from the eye to the optic cortex of the patients – an objective measure of brain function - was significantly delayed relative to a control group. The results also correlated with poor scores on other more subjective measurements of brain function such as hand-eye co-ordination.

16.6 Lessons to be learned

This case illustrates how vital it is for water companies:—

to react immediately to any contamination threat; and where it is found to exist;

to keep the public and the health institutions promptly and fully informed of the event, the dangers to public health, and any steps that need to be taken to safeguard their welfare.

17 CONCLUSION

There are many forms of contamination, and many sources from which it may be introduced into the supply. Some forms are removed from the public supply and others added. The DWI does a very good job of policing the purity of water available for public consumption and the companies themselves routinely achieve very high quality standards.

The new measures required of companies to plan for emergencies give added protection. The risks of a major catastrophe caused by widespread contamination are slight but nevertheless real.

ACHIEVING AN APPROPRIATE BALANCE? – AN OFWAT PERSPECTIVE

R. J. Tye[1] and W.H. Emery
Office of Water Services, Centre City Tower, 7 Hill Street, Birmingham, B5 4UA

1 INTRODUCTION

Ofwat is the economic regulator of the licensed water and sewerage companies in England and Wales. Our key role is setting the maximum amount companies can charge for providing water and sewerage services for domestic and most business customers.

This apparently simple statement hides some complexities and this paper cannot address them. However, there is more detailed information on the Ofwat website (www.ofwat.gov.uk).

In December 2004, the Director General of Water Services announced maximum price limits, companies can charge for water and sewerage services for the five years from April 2005 to March 2010, the PR04 or AMP4 programme.

Customers receiving water bills in England and Wales in spring 2005 have realised that charges for water have gone up. Water charges are increasing on average by 8.5% in real terms this year, and will be on average 18% higher by the end of the five year period in March 2010. For this conference, it will be useful to explain the basis of making decisions and the impact on water customers. This will provide a context for aspirations and potential standards proposed by other presenters.

2 THE PRICE SETTING PROCESS

The vast majority of the resources employed by the Water and Sewerage companies are needed to meet the current standards for quality, customer performance and other legal requirements that affect all businesses, such as employment or Health and Safety legislation. We assume that within this base service of business as normal each company manages its own affairs. Companies have reported summary financial and non-financial information to Ofwat since 1990 and we have an understanding of the overall costs of providing services to customers and we can and do make comparisons across companies. This also includes seeking to understand why individual company costs may be different.

[1] Correspondence should be addressed to Dr Rowena J Tye
Tel 0121 625 1364 email rowena.tye@ofwat.gsi.gov.uk

Each company makes the decisions on using its revenue from water charges to deliver these services in compliance with regulations and its licence. The incentive based price cap regulation allows the regulator to stay at arms' length. Each company makes its own decisions on how to run its business, such as when assets need maintaining and how to deal with its customers.

As the economic regulator, we compare the historical and forecast costs and then set price limits for the next pricing period (5 years) based on our judgement of the efficient costs of continuing to deliver water and sewerage services. It is then for the company management to deliver the services. Both the strategy and the day to day management are for the company.

We have explained the process we used to reach the 2004 price limits, publishing the assumptions we used.(Ofwat 2004) This is our basis for price setting but no company is tied to the operating and capital expenditure levels we assumed. We expect them all to deliver the specified outputs, both those which are legally binding and others that are set as part of the periodic review. Regulation is not based on either an activity or investment basis, but on outputs, performance measures, targets and the delivery of legally required quality standards. We do monitor activities and indeed scrutinise reported expenditure, chiefly to inform year on year assessment of relative efficiency and future decisions on price limits as well as giving comfort that each company is progressing and maintaining its assets.

3 ENHANCEMENT TO WATER COMPANY ASSETS

During price setting, we also consider any enhancements a company needs to deliver either over the next five years or occasionally for a longer-term project. Not surprisingly, there are a number of criteria any proposals need to satisfy before they can be considered for inclusion in price limits. There needs to be firm justification of such enhancements before water customers can be expected to pay more in real terms. A company can, of course, carry out whatever improvement it wishes as long as it is financed by the base revenue it receives from customers.

There are five criteria each proposal must meet before it can be included:

- it is required by the quality regulators, and confirmed by Ministers, or are new obligations under current legislation;
- it delivers a measurable defined output, which is enforceable;
- it has a clearly defined timetable and due date for delivery in line with regulations or other legislation;
- it has defined asset improvements or changes to operational procedures to deliver the output; and
- it has identified costs for the proposed solution which have been challenged and validated by the company's reporter.

We also expect that proposals are cost-effective and offer value for money for both the customer and the environment. Costs of the proposals must be in line or below assessments of the benefits they deliver. Once accepted and incorporated into price limits, a company must deliver the specified outputs as part of the price limit package.

These outputs include:

- the specified quality programme;
- supply/demand enhancement projects;
- other enhancements; and

- of course maintenance of the current asset base

At the last price setting we assumed that overall the industry would be likely to need to carry out £16.7bn of capital investment (in 2002-03 prices).

Final Determinations	***Water***	Sewerage	Total
	£bn	£bn	£bn
Capital expenditure (AMP4 – five year total – April 2005 – March 2010)			
Capital Maintenance	4.2	4.2	8.4
Supply/demand balance	1.7	0.6	2.3
Quality enhancements	2.1	3.4	5.5
Enhanced service levels	0.0	0.6	0.6
Total	8.0	8.8	16.8
£ per property	341	391	732

Figure 1. 2004 Periodic Review – Ofwat capex assumptions (Ofwat 2004)

2002-03 prices

4 PLANNING AND DEALING WITH EMERGENCY SITUATIONS

Within the price limits set for the base service we expect each company to manage and deal with all situations that may arise during the normal course of providing water and sewerage services. This includes managing all incidents including emergencies, the threat of terrorist acts, loss of supply, severe weather conditions, breakdown of plant etc.

We have always assumed that planning and having the resources to deal with these situations is part of the base service. It affects all companies and is a continuing obligation that should be constantly reviewed and refined. It is a company responsibility and requires detailed knowledge of company assets. We cannot and should not manage it. However, we must have comfort that the company is carrying out necessary planning and its systems are adequate.

The attributes of this base service are not static, as over time, customer expectations will change as will general legislation. All such changes should be accommodated within the base service, many of them being reflected when price limits are raised by the rate of inflation. There is an opportunity for a company to explain how any changes have affected its forecast costs when it submits its annual return and also its business plan during the process of price setting. Generally, we do not classify company proposals as enhancement, which attracts an extra provision, unless there has been a definite step change in legal requirements specific to the water industry, or other justification by previously agreed routes – such as improvements and customer service.

5 WATER TREATMENT AND THE DISTRIBUTION SYSTEM

In England and Wales there are over 1,300 works treating water prior to entering the public supply. The distribution network comprises over 4,600 service reservoirs as well as 330,000 km of water distribution mains. These assets are needed to ensure the public water supply distributed to customers' taps is wholesome. All water treatment works reduce the level of contamination, chemical or microbiological, generally by coagulation, filtration, adsorption and disinfection. The distribution system not only transports the public water supply to point of use but should also be of sufficient integrity to protect the quality of the water it transports. We do not seek to split the expenditure to identify that required to deal with contamination.

6 OFWAT EXPECTATION FROM ALL COMPANIES

We expect every company to manage its business to prevent emergencies and when incidents happen to minimise their effect both in degree of impact and length of time. Companies should minimise customers' exposure to such emergencies and incidents.

Within this context, customers should not have a greater burden in paying for company plans than is necessary. No company should respond to these expectations by taking a risk averse approach and gold plating its proposals. A thorough examination of the systems employed, rather than just building extra assets, is a more efficient route. Crucially, each company must balance the costs of its proposals against the risks of failing to provide wholesome water, with the impact this has on both service to customers and public health.

A risk based approach does not mean delivering a zero risk at any cost. It does mean making difficult decisions on the appropriate level of risk, to balance the perceived and actual risks for the public water supply with the likely costs and willingness to pay.

6.1 Some indications of the extent of the risk

The World Health Organisation encourages the use of health-based targets when considering the risks of public water distribution systems. This type of analysis is difficult to carry out in developed countries where the incidence of mortality or illness linked to the public water supply is very low indeed and indeed in most countries cannot be reliably measured. Therefore, it is necessary to look at secondary indicators to identify potential risks. The Drinking Water Inspectorate reports yearly on the number of incidents reported to them. (DWI 2004) This has fluctuated since it was first reported in 1997 and after peaking in 1999, now appears to be on a downward trend. The incidents in 2003 indicated the annual risk of an incident was : 1 per 130 treatment works, 230 service reservoirs, or 4,200km of water mains.

However, not all incidents are an immediate or a long term threat to public health. They may just be a warning. Another indication is the number of prosecutions brought by the Drinking Water Inspectorate under Section70 of the Water Industry Act for water 'unfit for human consumption'. Since 1997 there have generally been fewer than ten per year and are frequently lower. None was reported in 2002 and 2003, although incidents may have occurred then and not been prosecuted until later years.

England and Wales can be justifiably proud of the very high quality of the public water supply. There have been very few instances of illness affecting communities being attributed to the public water supply. There have been a number of high profile incidents which are unfortunate not only for those affected but also for their impact on public confidence in tap water.

The Communicable Disease Record reported two outbreaks with possible links to the public water supply in 2002 reported as affecting 52 people, none was reported in 2003. This is a good record for the water industry and it is improving. However, there are no grounds for complacency. This very high standard cannot be maintained without constant vigilance, appraisal and refining of strategies. The majority of contamination is controlled by the operation of treatment works. The day to day running of these works, particularly those with variable raw water quality, is a major responsibility.

6.2 Customer perception of the public water supply

Sometimes the actual risk of unwholesome supplies is secondary. Public perception of any risk, although it may be misdirected, must be respected, understood and as far as possible managed. Public confidence can be very fragile and those involved in running and providing water services, as well as regulators and other commentators must be responsible when both reporting and communicating water-related issues. It would be easy to undermine public confidence and it is not easily regained. The joint customer survey carried out in preparing for the periodic review 2004 did indicate that customers placed a safe reliable water supply top of the list of priorities. It is highly valued and 87% of those interviewed were satisfied with the water supply service, and 6% dissatisfied. (MORI 2002)

7 THE COST OF MAINTAINING A SAFE SUPPLY OF DRINKING WATER

From 1989 to 2005 the water industry has invested £12 billion to maintain the works and pipes used to supply water and an additional £8 billion to improve them.

High investment levels will continue for the next five years with £4 billion to maintain an increasingly large and complex asset base, £2 billion to improve drinking water quality as well as an additional £2 billion of capital investment to deal with the increasing demand for water both from new houses and also from increasing use of water in existing houses. The 2004 price review assumed £8 billion of capital investment for 2005-10. Overall, approximately £340 extra investment is planned on average for each property connected to the public water supply.

7.1 Quantity of water supplied

Ensuring that there is a sufficient volume of water for the public water supply is particularly important. We assumed the level of metering will go up from 24% to 36% which would help to control demand and increase security. Most importantly there will be significant improvements in the security of a continuing adequate supply. Instead of 12 companies and 75% of the population having a marginal to large deficit in 2003-04, we expect that by 2008-09, 95% of the population served by 22 companies will have no deficit.

7.2 Dealing with pollution of raw water supplies

Planned investment is necessary at water treatment works to contend with increasing pollution of the raw water sources used for public water supplies. Standards for the maximum levels of key pollutants have been in place for some time, yet still more work is needed to maintain compliance. In 2005-10 when setting prices we assumed that nearly £300m will be needed to carry on reducing nitrate levels, and pesticide reduction needs a further £73m of capital investment. This is on top of almost £1 billion capital investment since 1989 on reducing levels of pesticides and pesticide metabolites to produce wholesome drinking water.

Reducing the risk from cryptosporidium in 2005-10 attracted planned investment of over £100m. This is in addition to planned investment in the 2000-05 quality programme of £500m. Also, of course, there are the operating costs for carrying out several daily assays at some works to determine the number of cryptosporidium oocysts in treated water, to criminal evidence standards. This is a substantial burden on companies and their customers.

7.3 What is an acceptable risk?

This level of investment prompts the question what is an acceptable level of risk to the population from the public water supply? We are now moving quite correctly to a risk based approach for dealing with these 'natural' contaminants but it should be just that. It cannot realistically be a zero risk approach. To give two examples:

a) Cryptosporidium risk reduction.

There has been significant investment planned and delivered over the period 2000-10 to reduce the risk from cryptosporidium. This will amount to over £600m of capital investment and considerable operating costs, all paid for by customers' bills.

Have there been any studies on the valuation of the benefits arising from this investment? For example how much is it costing for each case of illness averted or premature mortality? Conversely how much are customers 'willing to pay' to reduce further their risk from cryptosporidium, having been informed of the reported levels of illness before and after this investment.

b) The pesticides standard

There has been very considerable investment to reduce levels of pesticides and pesticide residues in drinking water. This level of drinking water is set at 0.1µg/l, very considerably less than the pesticide residues permitted in some fruit and vegetables. Most pesticides are very extensively tested before registration and the levels found in either food or tap water are unlikely to be harmful. Has it been a good use of the country's resources to invest over £1 billion in additional assets to reduce pesticides levels to such low levels? What has been the impact on human public health? Operating these assets has also used energy and contributed to the CO_2 burden as well as producing chemical by-products.

Similar arguments could also be raised for the nitrate parameter.

To place these water-related issues into context, on 9 June 2005 the Environment Agency published a report on the State of the Environment. (EA 2005) This reported on the wide spectrum of issues affecting the environment, air, land water with flooding and climate change. It is interesting that the EA reported that poor air quality is thought to contribute to between 12,000 and 24,000 premature deaths per year, and in the 2003 heatwave, 800 people may have died from air pollution caused by traffic. There has also been the recent debate on deaths and infections from MRSA, and other hospital acquired infections. This leads to very important questions:

- what is the right balance of risk for society?
- where should finite resources, indeed household finances be focused?
- would these be better spent on more efficient vehicles or cleaner fuels?
- or on further significant investment to reduce risk levels for the public water supply?

These are not questions anyone at this conference can answer. Decisions must be made with the help of sound, well-balanced information to allow informed decisions by those responsible that are understood by the public.

8 THE ADVANCE OF TECHNOLOGY

In the past it has not been necessary to deal with risk as only evident problems were addressed. However, improvements in detecting potential contaminants have advanced alongside improvements in water treatment technology. Almost any standard could be met, if enough is invested in, for example, membrane technology. Such processes can produce drinking water to very exacting standards, but are expensive, both in financial terms and in potential impacts on the environment. All of these processes are energy intensive and resource intensive.

The easy option is to carry out rigorous analytical sampling, find a detection limit, set this as a standard, then fit the highest specification plant available to meet such a standard. This only takes account of the financial costs, for example in higher water bills, it does not address the impact on household budgets, business finance, or the wider issues surrounding sustainable development. The goals do not always pull in the same direction, be it the environmental impact of use of resources, such as energy and chemicals, alongside meeting risk to health. All should be reviewed together by those with responsibility.

9 PROTECTING THE INTEGRITY OF ASSETS DELIVERING THE PUBLIC WATER SUPPLY

Companies have responsibility for the integrity of their assets and the distribution system. This is an integral part of providing water and sewerage services. This should be dealt with by good practice and systems, rather than designing redundancy in the system or assuming major capital investment is needed. A well managed company who considers the integrity of its assets should have few water contamination emergencies. Exceptionally, the need to improve security may change due to external circumstances. There has been a step change in the estimation of risk to these assets in recent years and the final determination in 2004 did recognise that some companies needed to reinforce their assets over and above that which a prudent company would previously have done. Therefore, an additional 82 projects were included in price limits at a cost of £340m.

Capital investment alone does not guarantee security. Security may be improved at the planning stage, for example by adoption of a Pokayoke approach to 'mistake proofing'. Would padlocks with unique locks prevent Camelford type incidents? Unique keys for padlocks would mean it was less likely for chemicals to be pumped into the wrong tanks, and cost a few hundred pounds. Whilst we can be sure that each company has learnt these simple lessons, there will be others.

10 THE WAY AHEAD

Dealing with the risk of contamination and other emergencies is just like capital maintenance, apparently very simple. Capital maintenance and security should be viewed alongside all other aspects of the public water supply in an integrated way, the right investments, at the right time at the right place.

However, it is much easier said than done. There should be sufficient information on the risks and benefits, alongside the costs, to make informed decisions. These may be at a political or policy level for national standards, or high level risks; or at company level when dealing with the day to day maintenance of assets, or protecting the integrity of assets from emergencies and incidents.

This conference is discussing the Drinking Water Safety Plans and the contribution that they can make. Security issues may also be discussed and can be reviewed alongside each company's responsibilities to maintain its assets. This conference is also dealing with more detailed examination of certain risks. However, this information is only useful if it is reviewed, scrutinised and applied prudently, balancing all the areas so that the outcomes are in line with the rest of society and public expectations.

References

Drinking Water Inspectorate *Drinking Water 2003- a Report by the Chief Inspector* July 2004

Environment Agency *State of the Environment Report 2005 – a better place?* June 2005

MORI *The 2004 Periodic Review: Research into Customers' Views,* August 2002

Office of Water Services (Ofwat*) Final determinations: future water and sewerage charges 2000-05,* November 1999

Office of Water Services (Ofwat) *Future water and sewerage charges 2005-10 - final determinations*, December 2004

WATER CONTAMINATION: CASE SCENARIOS

D. Russell

Centre for Radiation, Chemicals and Environmental Hazards, Chemical Hazards & Poisons Division (Cardiff), The Health Protection Agency, UWIC, Cardiff CF23 9XR, UK.

1 INTRODUCTION

An adequate and clean supply of drinking water is long recognised as being mandatory for healthy living and sustainable communities. This is well established, with the link between contaminated water and public health noted by ancient society. Ancient scribes note that Cyrus the Great (5th.Century BC) carried boiled water in silver flasks, a combined strategy of bacteriocidal and bacteriostatic practice respectively.[1]

With the advent of the Roman Empire, rain water was initially collected in cisterns, reservoirs and alluvia.[2] However, with the development of large cities, it was necessary to transport sufficient quantities of clean water to a given populus. Consequently, bronze, lead and clay pipes were constructed to gravitationally transport water from clean sources to the developing cities, whilst aqueducts transported water across valleys.[1]

Further problems arose with continuing with greater urbanisation of society and the contamination of potable water supplies with raw sewerage in the 19th. Century. As a consequence, cholera and typhus were a major public health concern resulting in morbidity and mortality. John Snow identified the Broadwick Street pump as the source of cholera in the London outbreak of 1854. This led to a major change in practice to protect public health, with abstraction of water upstream of sewerage and the introduction of sand and gravel filtration.[3] During the course of the 20th.Century, chemical treatment of water supplies commenced, within the introduction of the first chlorination plants.[1]

Today, the UK is recognised as having a sophisticated water industry that provides safe potable water to more than 20 million properties i.e. the majority of the UK's population. The industry consists of approximately 1,000 reservoirs, more than 2,500 water treatment works, ·9,000 sewage treatment works and more than 700,000 kilometres of mains & sewerage pipes.[4] This infrastructure dates back to Victorian times (1837-1901) when a considerable proportion of today's existing network was constructed.

Most of the water supplied in the UK is from upland resources that are remote from sources of industrial pollution, whilst the potential for contamination of groundwater sources has been dramatically reduced in the UK by the introduction of the Groundwater Directive (EC 80/68/EEC) and the designation of groundwater protection zones *[NCSA 2003]*. This legislation, the Groundwater Regulations 1998 (SI 1998 No. 2746), protects aquifers from sources of industrial development, which have the potential to release aqueous discharges. In addition, cleanliness is assured by extensive coagulation and flocculation, filtration and oxidation. The benchmark for the quality of the final product is determined by the European Directive on the Quality of Water for Human Consumption, itself in many instances based upon WHO Guidelines.[5]

Despite this extensive mains supply, many rural and remote communities receive a private water supply derived from wells and other abstraction points that are not subject to the extensive purification measures described above. Indeed the Drinking Water Inspectorate (DWI) describes approximately 50,000 such supplies in England and Wales alone, supplying 300,000 people and derived from boreholes, springs, streams, rivers, lakes or ponds [6].

This paper describes some water pollution incidents relevant to such private water supplies. The incidents are of public health concern and reflect examples that the Chemical Hazards and Poisons Division (CHaPD) of the Health Protection Agency has provided advice and support in 2004. These will be discussed in turn.

2 CASE SCENARIOS

2.1 Lead Contamination

In March 2004, 5 members of a small rural community in Wales reported symptoms of lethargy and malaise (age range 13-51 years). It had previously been noted that all received a common private water supply delivered from a lead tank and transported through lead pipes. Lead contamination is subsequent to the dissolution of lead in water in contact with such plumbing materials [1].

Environmental samples taken at the site revealed water lead levels of greater than 1290ug l^{-1} (WHO Guidelines,10ug l^{-1}). Subsequent measurement of whole blood lead (WBL) by atomic absorption spectroscopy revealed individual levels ranging from 8.3 to 44 $\mu g\ dl^{-1}$ (reference range, 10$\mu g\ dl^{-1}$). As all were clinically asymptomatic, chelation therapy was not considered appropriate.

As a consequence an enforcement order was served, an alternative source of potable water made available and a recommendation made that new casing and piping should be installed, whilst children should not bathe in water from the contaminated source. All individuals were followed up clinically, with sequential measurement of WBL and haematological parameters.

It is long recognised that lead is detrimental to health. Thus, early Romans recognised its neurotoxicity.[1] It is now recognised that acute exposure induces a range of clinical symptoms including, lethargy, hypertension, weakness, anaemia and encephalopathy. Chronic exposure is associated with anorexia, constipation, abdominal pain, peripheral neuropathy and classical gingival "lead lines". [7] The most widely recognised effect, however, is the effect of lead exposure on neurological development in children, with a reduction of intelligence quotient with increased exposure.[8]

2.2 Nitrate Contamination

In June 2004, the unit was notified of an incident in the Republic of Ireland involving contamination with nitrates. Measured water levels to a rural community revealed a concentration of 88mg L^{-1} (WHO Guideline <50 mgL^{-1}). The acceptable daily intake (ADI) for nitrate is 3.65 $mg^{-1}kg^{-1}day^{-1}$. [9] Assuming that exposure from food is at the upper estimate value of 180mg, then an adult of 70kg consuming 2 L day^{-1} will consume a total of 365 mg day^{-1}, equal to 5.08 mg^{-1} kg-[1] day^{-1}. Similarly, a bottle fed infant of 10kg will consume 66mg at this concentration from 750 ml of water and assuming a contribution from food of 54 mg, a total daily intake of 12mg kg^{-1} day^{-1}.

Neonates and infants in particular are at risk of developing metheamoglobinaemia as a consequence of nitrate conversion nitrite in the gastrointestinal tract and subsequent oxidation of haem Fe (II) to Fe(II), resulting in reduced oxygen carriage. Clinically this presents as cyanosis [7].

It was therefore advised that the water supply should be replaced with an alternative potable supply and the cause of the contamination explored. It was postulated that this was likely to be application of agricultural fertiliser, although the source was not identified.

2.3 Total Petroleum Hydrocarbons

In August 2004, a member of the public in a rural part of the Republic of Ireland complained of a "petrol-like" odour in tap water that originated from a mountain spring, supplying approximately 70 houses. Initial water samples revealed that Polyaromatic Hydrocarbon concentration were in the range of 15-18mg L^{-1} (2001 Water Supply Water Quality Regulation guideline <0.1 μgL^{-1}), whilst total petroleum hydrocarbon (TPH) concentrations ranged from 46.0 - 68.0 ug L^{-1} (Water Quality Regulations 1989 quote a taste and odour threshold of 10 ug L^{-1}) . Based on these values, it was recommended by the unit that alternative water supplies should be used for consumption and for bathing, based on the possible dermal, ingestion and inhalation uptake of these hydrocarbons.

It is interesting that subsequent analysis of the water supply by an accredited laboratory could not verify these results, subsequent analysis being unable to detect elevated levels of PAHs or TPH. Consequently, the general public were advised to continue using the original supply. No further symptoms were reported.

3 CONCLUSION

The case studies described exemplify the on-going significance of private, potable water supplies and the chemically distinct nature of potential contaminants, whether heavy metals, inorganics or organics. The public implications are of significance and acute and chronic health effects are recognised. The cases further illustrate the need for a holistic, multi-disciplinary, multi-agency approach to the on-going issue of water pollution.

References

1 J. Fawell and G. Stanfield, *Pollution, Causes, Effects and Control*, 4th Edition; Editor R.M.Harrison, RSC, 2001.

2 www.crystalinks.com/romanempire5.htm

3 Snow J, Frost WH and Richardson BW., *Snow on Cholera,* Commonwealth Fund, New York, 1936..

4 Water UK. 2004. www.water.org.uk

5 World Health Organisation, *Guidelines for Drinking water Quality*, Volume 1, WHO Geneva, 1984.

6 www.dwi.gov.uk

7 *Tietz Textbook of Clinical Chemistry,* Editors C.A. Burtis and E.R. Ashwood, 3rd.Edition, Saunders ,1999.

8 Pocock, S.J., Smith, M. and Baghurst, P., *Environmental Lead and Children's intelligence: a systematic review of the epidemiological evidence. Br Med J* **309**:1189-1197, 1994.
9 European Commission Scientific Committee for Food, *Opinion on Nitrate and Nitrite*, Annex 4 Document III/5611/95, 1995.

CHEMICAL CONTAMINATION OF WATER – TOXIC EFFECTS

P.C. Rumsby, W.F. Young, N. Sorokin, C.L. Atkinson and R. Harrison.

National Centre for Environmental Toxicology, WRc plc. Frankland Road, Blagrove, Swindon, Wilts, SN5 8YF, UK

1 INTRODUCTION

National Centre for Environmental Toxicology (NCET) is the environmental consultancy arm of WRc plc (formally known as the Water Research Centre) and has expertise in chemical risk assessment, the setting of environmental and human health standards and guidelines, plus work on fate and behaviour of chemicals and microbiology in the environment. One of the main roles of NCET is the provision of enquiry services for a number of clients including the Environment Agency and the Scottish Environment Protection Agency (SEPA). NCET has, for the last fifteen years, run a comprehensive toxicity advisory service for the members of UKWIR (UK Water Industry Research) which includes all the UK water companies. This service consists of three parts: quarterly current awareness reviews on scientific and regulatory developments; an enquiry and emergency service; and the preparation, review and revision of datasheets on chemicals and processes of importance to the water industry. The toxicological advisory service and the ways in which they are used to deal with incidents of chemical contamination of the water supply will be outlined in greater detail below.

The working of the emergency and enquiry service and the preparation and use of datasheets are important in the handling of chemical contamination by the UK water industry. The procedures involved in the processing of enquiries and the preparation and review of datasheets have ISO9001 accreditation.

2 QUARTERLY TOXICITY UPDATE REPORT

This update report ensures that the water industry is kept aware of the current issues and topics worldwide as well as legislative advances with respect to water. It aims to alert the water industry to toxicological issues with regard to water supply and treatment; reporting on the latest scientific literature. Chemicals of interest which are covered regularly include arsenic, fluoride, disinfectant by-products and endocrine disrupting chemicals.

3 UKWIR ENQUIRY AND EMERGENCY SERVICE

This consists of a 24 hours a day, seven days a week service to the members of UKWIR. Indeed, use of the enquiry and emergency service together with the datasheets is written into the procedures of many water companies. The scope of the enquiries is wide and covers all aspects of the presence of chemicals in both clean and dirty water:

- Potential health effects associated with the intentional or unintentional contamination of drinking water
- Taste and odour problems
- Effects of chemicals on sewage treatment
- Biodegradation and removal in sewage treatment
- Ecotoxicity of chemicals
- Environmental fate and behaviour of chemicals
- Pollution Prevention and Control (PPC) and other regulatory drivers
- Intentional contamination of the water supply
- Public and media interest e.g. fluoride, disinfectant by-products and bracken carcinogens

The emergency enquiries comprise about 10% of all enquiries with the majority being on human health effects. The use of the word, emergency, rather than incident, is an operational term and signifies that a response is required within an hour of receiving the enquiry. These again cover a wide range of problems involving threats to the continuation of drinking water supply, efficient sewage treatment and environmental pollution. Below is a list of those encountered in the last twelve months (2004-2005):

- Spillage of diaryl phenylenediamine
- Sewage toxicity of two isothiozolines
- Pollen in a reservoir
- Botulism in cattle
- Methoxypentanone in supply
- High pH in drinking water
- Ecotoxicity of bromide in a river
- Effect of forest fire on a water reservoir
- Effect of copper on a sewage treatment works
- Toluene, ethylbenzene and xylene in drinking water
- Guidelines for phenol
- Guidelines for total petroleum hydrocarbons
- Kerosene in drinking water
- Effects of pharmaceuticals on sewage treatment (on two separate occasions)
- The presence of human pharmaceuticals in waste water
- Contamination of the disinfection process
- Metal contamination of drinking water

Figure 1 shows the use of the enquiry service by UKWIR members in the year from July 2004 to July 2005 divided into enquiries where effects on human health are the main concern and others where non-human health effects on sewage treatment or the environment are more important and indicating the proportion of emergencies.

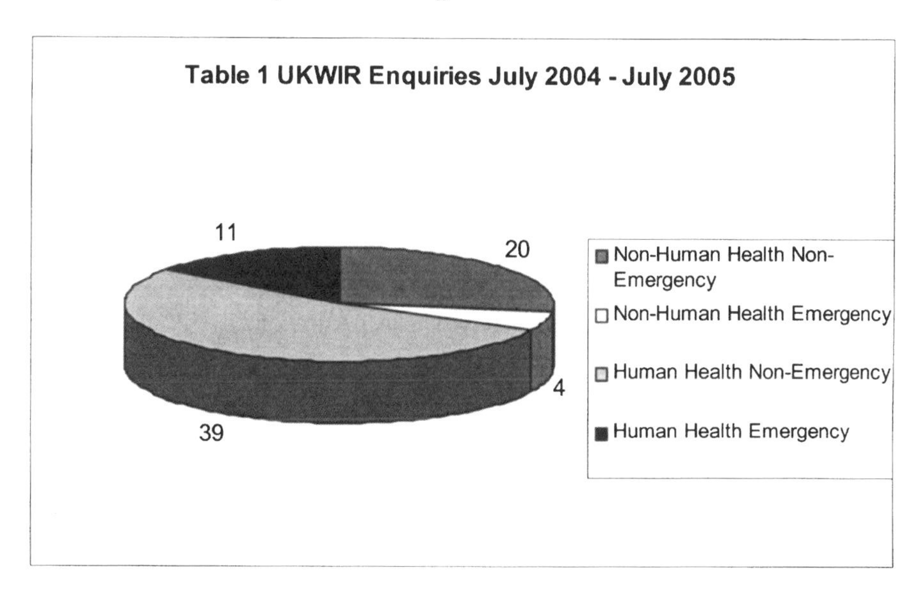

Figure 1 Use of the enquiry service by UKWIR members July 2004 to July 2005

4 DATASHEETS

Five hundred and twenty datasheets have so far been prepared on chemicals of importance to the water industry. As the datasheets have evolved and increased, the number of enquiries and emergencies have decreased somewhat, as more information and answers have been sought for and found in the datasheets. Often enquiries are now on more unusual chemicals for which there are no datasheets or where further confirmation or checking that there is no very recent information is required. The evolution of the datasheets is an ongoing process with new chemicals being added at the rate of about five a year following discussion with the Steering Group for the project, which consists of representatives of the water industry. The preparation and revision of the datasheets includes five stages of review including external peer-review.

These datasheets are of two main types, full and cursory. The latter are of a non-standard format usually for chemicals where there is limited information, a briefing note style is more appropriate or where materials are grouped by function or class. In this last case, reference is then made to datasheets available for individual substances (e.g. membrane treatment chemicals). Full datasheets are specific and tailored to a particular chemical with respect to its use by the water industry. The content of these datasheets are regularly reviewed and updated with a maximum of a four-year update cycle with the date of the last update recorded in the sheet. These datasheets are available to water quality professionals via CD-ROM (updated every six months), company intranets and UKWIR/AWWARF websites. The contents page of the datasheet gives some indication of the depth of information present:

- Synopsis
- Occurrence and use
- Mammalian toxicology
    - Pharmacokinetics
    - Human toxicity
    - Experimental animal toxicity (acute, irritation & sensitisation, repeat dose, genotoxicity, carcinogenicity, reproductive toxicity)
- Evaluations by authoritative bodies
- Emergencies and SNARLs
- Standards (drinking and environmental water and occupational)
- Ecotoxicity (freshwater, marine and terrestrial)
- Bioaccumulation
- Biodegradation and removal
- Toxicity to sewage treatment organisms
- Taste and odour
- Reactivity with chlorine and information on by-products
- Reaction with other disinfectants
- Removal during drinking water treatment
- Chemical properties
- Analytical chemistry
- References

4.1 SNARLS

While much of this information may be of importance in an emergency, attention is particularly drawn to the Emergencies and SNARLs section of the datasheet. A SNARL is a Suggested No Adverse Response Level. These values are specifically designed for human exposure through drinking water in emergency situations, i.e. short-term exposure. This makes them unique to datasheets and the experience of both NCET and the water industry suggest that they are invaluable in the management of pollution incidents. They are designed to give a 'safe' value to compare with levels in both raw water and supply. A brief description of the procedures involved in the derivation of SNARLs is outlined below and it can be seen that while being derived as short-term emergency levels, there is considerable conservatism built into the values.

For the derivation of SNARLs, there is a need to derive an acceptable or tolerable daily intake (ADI/TDI), a concentration considered 'safe' and expressed as mg/kg body weight/day. These acceptable daily values are preferably taken from an established hierarchy of toxicological sources comprising of UK regulatory bodies such as the expert committees of the UK Department of Health, Defra and Food Standards Agency, EU committees on toxicity, food additives and pesticides and, of course, the World Health Organization (WHO) Guidelines on drinking water quality are invaluable together with other international and national authoritative bodies.

When there are no relevant values available for the estimation of SNARLs, conservative values are derived using suitable study results. Although SNARLs are short-term, a No Observed Adverse Effect Level (NOAEL) from long-term study is used where possible to increase the margin of safety and a NOAEL/L(lowest)OAEL from a chronic study (>90 days) preferred to a subchronic (<90 days) value. Uncertainty Factors, which are widely used by regulatory authorities, are then used to account for unknowns in the study

NOAEL: division by 10 for interspecies variation and division by 10 for intraspecies variation. A number of other possible uncertainty factors can be also be applied to take into account the use of short-term studies, LOAEL to NOAEL extrapolation, whether the effect is severe/quick acting and/or irreversible or whether the dataset is poor. These deductions inevitably involve case-by-case expert assessment. When the dataset is very poor, for example, when the only study value available is an LD_{50}, a Tentative 24-hour SNARL may be set. This very conservative value includes the use of an Uncertainty Factor of 10 000 and the use of the most vulnerable receptor, a 10 kg child plus a 50% allocation to drinking water.

The usual default allocation of the TDI to drinking water is 100% for a 24-hour SNARL reducing to 50% for 7 days. This is only altered when there is evidence of additional exposure through other routes, primarily through the diet, when the substance is specifically present in foods/crops although this change is never carried out for pesticides. The default subject and water intake is a 60 kg adult drinking 2 litres water/day. If evidence suggests a vulnerable subpopulation (e.g. the NOAEL is derived from a developmental study), there may be additional margins of safety: for example, 10 kg child drinking 1 litre of water or a 5kg child drinking 0.75 litre. There is always a reasoned argument for the SNARLs in the datasheet.

In the next section, two case studies will be outlined. Although the incidents proved to be not a significant risk to human health, they illustrate the way in which the water companies use the enquiry service during incidents and how SNARLs are used in the management of these incidents.

5 CASE STUDIES

5.1 Case 1 - Back-siphonage incident

5.1.1 Operational details This first example, concerned a direct contamination of supply, and the investigation was initiated following consumer complaints. Initially, the source and the substances involved were unknown. A similar incident had occurred elsewhere previously for which NCET had also been contacted.

This incident took place in Brindley, Zone 155 (population 9000), a zone fed from Hurleston Water Treatment Works. This is dairy farming land in rural Cheshire where the water pressure was poor. A complaint of intermittent blue water was received from a member of the public through a call centre.

Analysis of this blue water yielded the following details:

- An unknown peak was detected at a wavelength 632 nm (copper sulphate was detected at 745 nm and methylene blue at 666 nm).
- The pH at Hurleston and water in distribution was pH 7.9.
- The levels of chloride were 352 mg/l (the Prescribed Concentration Value, PCV, is 400 mg/l but the normal concentration about 40 mg/l) and;
- The magnesium level was 236 mg/l (the PCV for magnesium is 50 mg/l and the local level normally about 5 mg/l).

Further investigation of the site revealed byelaws contraventions and a linking of a feed system to the local water supply in order to facilitate getting feed to the cattle. Also found on the site was a drum labelled 'Rumag Aqua'. This industrial farming product consisted of a 5% solution of magnesium chloride, which was given as a prophylactic feed supplement to guard against hypomagnesaemia ("grass staggers"). This product also contained 0.007% Dalfcol Brilliant Blue Food Dye E133 as a marker. Bacteriological samples were also taken at the same time.

5.1.2 Use of enquiry service Initial telephone enquiry 16th May. One property had complained of 'blue water', but as this was also found in the main, it was therefore unlikely to be due to copper/copper plumbing and did NCET have any ideas?

5.1.3 Initial advice By use of our considerable database of prior incidents, emergencies and enquiries (over 1200), it was noted that perhaps this could be a similar type of incident to a previous situation in which 'blueish' water had been detected. This proved to be back siphonage involving ethylene glycol containing a blue dye. Could a similar situation i.e. back siphonage with a blue dye be involved in this incident? A further enquiry, two days later, gave the information that the blue water had been tracked down to a back siphonage on a diary farm, involved magnesium chloride (a dietary supplement) and that this blue colour was probably due to a food colouring 'Brilliant Blue', however, there was no quantification of the Brilliant Blue available. The concentration of the magnesium was 236 mg/l and toxicological advice was requested.

Examination of our extensive databases of the hierarchy of toxicological sources, identified that for Brilliant Blue, an ADI of 12.5 mg/kg body weight/day had been set by WHO/FAO Joint Expert Committee for Food Additives (JECFA) (this equates to 375 mg/l for an adult drinking two litres of water/day). This being the case, it was likely that the concentration of Brilliant Blue being used as a marker was low and so aesthetic and taste problems were likely to be a key issue rather than any risk to human health.

The other ingredient of 'Rumag Aqua', magnesium, is of low oral toxicity and used in laxatives/antacids. Thus it was not considered to be a significant risk to human health in the short-term at these concentrations and taste was likely to become unpleasant before toxic concentrations could be reached. In fact, magnesium is an essential element (with 270-350 mg/day required by an adult) with more published on its deficiency than its toxicity. The adult daily intake is about 500 mg/day with 200-400 mg/day being taken in from food. JECFA considers it unnecessary to specify an ADI for man for many magnesium salts. Assuming 2 litres of water were consumed, the level in the water of 236 mg/l equates to a dose of 472 mg/day which is in the region of normal daily intake from food.

In summary, this was an unusual incident that involved the direct contamination of the water supply. However, it is not a one-off incident and there have been other examples of such contamination of the water supply. In this case, the enquiry service was able to provide reassuring advice to the water company about the nature and severity of the incident.

5.2 Case study 2 - 'Oil' contamination.

5.2.1 Introduction 'Oil' contamination incidents are probably the most common incidents with which the enquiry service has to deal involving, for example, diesel, petrol, kerosene and various other petroleum products. These incidents are interesting toxicologically as they often involve a mixture of compounds of different toxicities depending on the 'oil' involved. Incidents have been reported in both public and private supplies and involve consultation between water companies, local authorities and the relevant health agencies often originally alerted by customer complaints.

These incidents many of which are emergencies may involve the contamination of raw water or via the distribution (e.g. permeation of MDPE pipes, leaking petrol tanks either industrial or domestic) and may involve large populations served by a water supply through to residences such as blocks of flats or individual properties. Other incidents have included backsiphonage, for example, following fire-fighting procedures.

Two major examples of diesel contamination have occurred in Scotland in the last decade. In the East of Scotland, diesel fuel contaminated an aqueduct and passed through the treatment works (slow sand filtration) serving a population of 140 000 in Edinburgh. 'Do not drink ' notices were issued and the water supply was disrupted for two days. At the start of the incident (Friday night), concentrations of hydrocarbons leaving the works were 500 μg/l, dropping to 200 μg/l at the consumers' taps. By the Sunday morning, the maximum concentration of hydrocarbons at the taps was 30 μg/l. In a similar incident in the West of Scotland, reports of severe 'diesel' tastes from consumers were being received from the Glasgow area (population served 60 000). The origin of the contamination was traced back to a spill of diesel from a generator near an upland treatment works outside Glasgow. Similarly, 'Do not drink' notices were issued. The concentration of total hydrocarbons leaving the works was 250 μg/l, dropping to 80-90 μg/l along the supply. Analysis by Gas chromatography- Mass Spectrometry (GC-MS) indicated the presence of C8-C12 hydrocarbons.

5.2.2 Operational details The background to this incident was that 140m length of 75mm pipe providing water to 16 residences had been scraped and lined with polyurethane. During this process, the pipe was contaminated with 40 litres of heat transfer mineral oil (Crownlube 2118) from a spraying device. The contractors had then attempted to clean it out from the system but had not informed the water company. Upon re-connection of supply, a customer had complained of a film of oil on his bathwater. This led to an 'apocryphal' tale of a customer leaping from his bath and running down the street in true 'Archimedean' fashion.

Concentrations of the oil up to 1.3 mg/l were detected by infrared methodology. Customers were advised to use tapwater only for flushing the toilet and were then provided with a temporary overland supply. A programme of serial cleaning and flushing with detergent was then started. By GC-MS analysis, it took 3 days for the cold water system to clear while taking 20 days for the hot water system to be decontaminated. In this incident the use of a hand-held photo-ionisation detector (PID) proved invaluable in assessing the effectiveness of the flushing regime, although this instrument may not be adequately sensitive in many situations.

5.2.3 Use of the Enquiry Service WRc were informed that a petroleum-derived substance had entered the public supply and may have been present since the previous day. There had been consumer complaints and analysis was currently being undertaken. A written response to the enquiry was given on the same day.

A WRc chemist contacted chemists at the manufacturing company to gain information on the composition of the product (Crownlube 2118). It proved to be a recovered, distilled mineral oil, with a boiling point of 260-330°C which contained a complex mixture of hydrocarbons

The composition of the oil was as follows:

- 3% aromatics, mostly alkylnaphthalenes (C1 to C6)
- Smaller amounts of fluorene, acenaphthene, anthracene, phenanthrene
- 42% aliphatics - mostly nC14 - nC19 alkanes, smaller amounts of branched alkanes
- 55% naphthenics - equal amounts of tri- and tetra-cyclic alkanes and their alkyl substituted derivatives

Key issues relating to petroleum hydrocarbons are associated with a) the presence of the so-called BTEX chemicals: benzene, toluene, ethylbenzene, and xylene; b) specific polycyclic aromatic hydrocarbons (PAHs) such as benzo(a)pyrene. However, information on the boiling range and percentage of each class in the mixture indicated that the BTEX chemicals were unlikely to be present.

Short-term SNARLs needed to be derived for each fraction and for Total Petroleum Hydrocarbons to give some indication of 'safe levels' of these compounds. These SNARLs were based on oral reference doses (oral RfDs) developed by the US Total Petroleum Hydrocarbons Working Group.

	24-hour SNARL	**7-dav SNARL**
3% aromatics	0.9 mg/l	0.45 mg/l
42% paraffinics	3 mg/l	1.5 mg/l
55% naphthenics	3 mg/l	1.5 mg/l

Table 1. *SNARLS derived for fraction of mineral oil*

Fractions were considered additive in their toxicity and so the SNARLs were combined to give a Total Petroleum Hydrocarbon 24 hour SNARL = 3 mg/l.

The toxicological advice given was that skin irritation during washing was possible, but this toxic effect was only observed at relatively high concentrations and therefore this was considered unlikely. Taste/odour effects would be observed at concentrations in the region of 10 μg/l. It was suggested that the water company contacted an NCET toxicologist if analysis indicated that the levels in supply were approaching the SNARLs. For information on specific components of the oil, individual datasheets available on many of the chemical constituents were recommended to the water company.

In summary, oil contamination is a common incident which may involve direct contamination of a water supply. In this case, the enquiry service was used and was able to provide reassuring toxicological advice with the key issue being taste and odour effects due to the contamination of the water supply.

6 CONCLUSION

In conclusion, during incidents involving the contamination of water supplies with chemicals, the NCET Toxicity Advisory Service offers UKWIR members an integrated system based on the specific needs of the water industry. This service includes the use of specially prepared datasheets on over 520 chemicals with the derivation of SNARLs, short-term emergency guidance levels for chemicals in the water supply, and a 24/7 emergency and enquiry service offering toxicological advice.

7 ACKNOWLEDGEMENTS

The authors would like to thank Alan Godfree of United Utilities and Gary O'Neill of Yorkshire Water for their help in preparing the case studies and UKWIR for their funding and support of the Toxicity Advisory Service.

HPA ROLE ON HEALTH RISK ADVICE TO PUBLIC HEALTH TEAMS

P. Saunders

HPA, Chemical Hazards and Poisons Division

1 BACKGROUND

Chemicals play a major role in human societies and the speed of growth of the global market has dramatically accelerated the range of chemicals and the ways in which the population can be exposed to chemicals. More than 600 new chemicals enter the market place each month adding to the more than 11 million already known and 70,000 in regular use (International Labour Organisation. Safework. 13th June 2005. http://www.ilo.org/public/english/protection/safework/papers/smechem/ch2.htm). In addition to occupational exposures, the population can be exposed to these chemicals through inadvertent or deliberate contamination in food, water, commercial products, air or soil. Up to 25% of the disease burden in industrialised countries has been attributed to environmental factors (Smith, K.R., Corvalan, C.F., Kjellstrom, T. How much ill health is attributable to environmental factors? Epidemiology 2003, 10(5), 573-584). A study of the contribution of environmental pollutants to the incidence, prevalence, mortality, and costs of four categories of paediatric disease in American children estimated total annual costs to be $54.9 billion (range $48.8-64.8 billion) comprising $43.4 billion for lead poisoning, $2.0 billion for asthma, $0.3 billion for childhood cancer, and $9.2 billion for neurobehavioral disorders; 2.8 percent of total U.S. health care costs. (Environmental Pollutants and Disease in American Children: Estimates of Morbidity, Mortality, and Costs for Lead Poisoning, Asthma, Cancer, and Developmental Disabilities Philip J. Landrigan, Clyde B. Schechter, Jeffrey M. Lipton, Marianne C. Fahs, and Joel Schwartz *Environ Health Perspect.*, 110:721-728 (2002)). Improving environmental quality is a key element of European and UK strategies to improve health.

In parallel with these trends has been a growing awareness and concern by the public about potential damaging health and environmental consequences of exposures, especially perceived risks of cancers and reproductive health. Concerns are fuelled by the unpredictable nature of chemical incidents and the often poor risk communication during and after such crises. This increased awareness of potential health effects has contributed to around 80% of all European legislation having an environmental focus and the Health Service being given a statutory duty to identify and respond to the public health impact of most industrial processes through the Integrated Pollution Prevention Control regime.

Major chemical incidents such as the river Severn pollution in 1994, the Sea Empress oil spill in 1996 or the Distellex Tyneside incident in 2002 are relatively rare in the UK; Government and Industry expend considerable effort to prevent incidents occurring. However, national surveillance has shown that less serious incidents are common, with over 1,000 identified each year and the incidence appears to be increasing (Kibble, A., Dyer, J., Wheeldon, C, Saunders, P..J. Public-health surveillance for chemical incidents. Lancet 2003;357(9265):1365). The number of people potentially exposed is measured in hundreds of thousands. Acute health effects are common, as in the Sea Empress Oil Spill, where one third of the population of the affected towns reported symptoms within days of exposure (Lyons, R.A., Temple, J.M., Evans D., Fone, D.L. and Palmer, S.R. Acute health effects of the Sea Empress oil spill. Journal of Epidemiology and Community Health, Vol 53, 306-310). But longer-term effects are less well defined. In the Sea Empress incident the exposed population was significantly more anxious and had lower mental health scores a year later. In the Lowermore incident, chronic neurological effects have been claimed ten years later. A new area of concern is the psychosocial impact of incidents, especially following deliberate release.

2 THE ROLE OF THE HPA

The HPA Chemical Hazards and Poisons Division covers health aspects of chemical contamination of the environment, poisons information and clinical toxicological advice. [Food contamination, workplace incidents, unless the latter results in exposure of the wider population beyond the workforce, pesticide and pharmaceutical safety issues are not covered. In particular the Division will

- anticipate and prevent the adverse effects of acute and chronic exposure to hazardous chemicals and other poisons
- identify, prepare and respond to new and emerging diseases and health threats
- identify and develop appropriate responses to childhood diseases associated with biological, chemical or radiation hazards
- improve preparedness of responses to health protection emergencies, including those caused by deliberate release
- strengthen information and communications systems for identifying and tracking diseases and exposures to infectious and chemical and radiological hazards

The Division provides a 24-hour, 365-day of expert support and advice on chemical incident management to local NHS bodies and to other local agencies, e.g. local authorities, police and fire services. This includes environmental, toxicological, scientific and managerial support.

The Division manages an active chemical incident surveillance system This involves activity data from the Division, together with some routine data from key agencies such as the Fire Service and the Environment Agency in some regions. This has developed into the existing National Chemical Incident Surveillance Programme (established April 1998), which includes data from CHaPD units and Scottish Centre for Infection and Environmental Health.

3 SURVEILLANCE AND EPIDEMIOLOGY

Underpinning the Alert and Response System is a developing public health surveillance system for chemical hazards to provide the essential context for risk assessment.

The Division is a central source of authoritative scientific/medical advice on both the acute and chronic health effects of environmental chemicals/poisons. This covers public health concerns relating to exposure to chemicals via environmental pathways and from, non-food related, consumer products. The HPA advice includes matters such as treatment of patients, personal protective equipment, decontamination and evacuation, to toxicological and epidemiological advice on likely health effects for the public and the appropriateness of industrial operational conditions in protecting public health, clinical advice on antidotes and medical treatment, to the advisability and methods for follow up of exposed people. It also includes advice to the public and the media using sound risk communication strategies.

PREVENTING DRINKING WATER EMERGENCIES – WATER QUALITY MONITORING LESSONS FROM RECENT OUTBREAK EXPERIENCE

S.E. Hrudey[1] and S. Rizak[2]

[1]Department of Public Health Sciences, 10-102 Clinical Sciences Building, University of Alberta, Edmonton, Alberta, Canada T6G 2G3
[2]Department of Epidemiology & Preventive Medicine, Monash University, Prahran (Melbourne), Vic, 3181, Australia

1 INTRODUCTION

A recent analysis of documented drinking water disease outbreaks over the past 30 years in affluent nations reveals some recurring themes about the underlying causes of outbreak failures and some broader issues about the role of drinking water quality monitoring for the protection of public health.[1] While this analysis of over 70 outbreak case studies acknowledges the relatively high level of drinking water safety that is experienced in wealthy countries, these case studies almost universally demonstrate that the outbreaks involved were eminently preventable and in most circumstances, the solutions for assuring safety from the risks are not complex and rely on improved system management and operation. Yet, when outbreaks do occur, a common regulatory response is to propose more intensified monitoring and/or more stringent numerical water quality standards to assure the safety of drinking water. While intuition may suggest that we should devote most of our monitoring resources to checking treated water, the inherent limitations in monitoring treated drinking water quality and the difficulties arising in interpretation of results for informing risk management decisions make this an ineffective strategy for assuring the safety of water supplies and protecting public health. Some recently documented risk management approaches are more inherently preventive.[2-4]

2 MONITORING TREATED WATER DOES NOT ASSURE SAFETY

2.1 Evidence of Monitoring Limitations for Outbreak Detection

Compliance monitoring as a primary means of public health protection suffers from many limitations.[5] Firstly, monitoring is relatively infrequent and only a very small proportion of total water production is sampled; hence the detection of intermittent contamination by routine monitoring would be fortuitous. The scope of monitoring is also limited, it is neither technically feasible nor economically sustainable to monitor for every possible parameter. Indicator organisms must generally be used as surrogates for pathogens but these poorly represent viruses and protozoa; and, trihalomethanes are used as surrogates for as yet unidentified disinfection by products that might cause health effects. Furthermore water quality parameters are presumed to offer useful insights about health-related water quality, but the insights they offer may be tenuous.

Perhaps the most significant limitation however is the fundamentally reactive nature of monitoring treated water. For the vast majority of parameters, reliable results are only available a substantial time (in many cases, days) after a sample is taken. Thus if adverse monitoring results indicate potential contamination, often customers are likely to have already consumed the water. There are recent examples that highlight the reactive nature of monitoring data in relation to recognizing outbreaks and taking steps to intercede to reduce adverse health consequences, even after the opportunity to prevent an outbreak has been forsaken.

The Walkerton, Ontario outbreak in May 2000 is infamous for causing 7 deaths along with over 2300 cases of gastroenteritis, 65 hospitalizations and 27 cases of haemolytic uremic syndrome (median age of 4), a serious kidney condition with potential lifelong consequences. The pathogens (*Escherichia coli* O157:H7 and *Campylobacter jejuni*) causing the Walkerton outbreak were attributed to contamination arising from cattle manure from a local farm. These pathogens contaminated a shallow production zone (only 5 to 8m depth) of Walkerton Well 5. The contamination was believed to have occurred following a period of exceptionally heavy spring rainfall, approximately 134 mm over 5 days, estimated to be a 1 in 60 year occurrence.[6] Although there were too many failures involved in this outbreak to properly address in this paper, the weekly samples from the distribution system only revealed contamination 5 days after the contamination entered the system. Unfortunately even this belated warning was not conveyed to the local health unit, who eventually called a boil water advisory when illness was already widespread in the community about nine days after the contamination occurred (Figure 1).

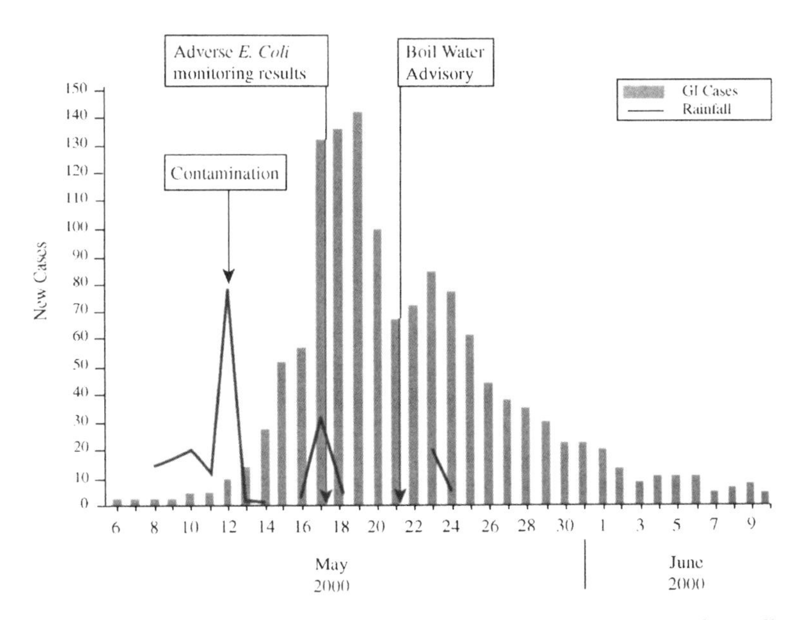

Figure 1. *Chronology of events and gastrointestinal disease cases in the Walkerton outbreak (based on data from BGOSHU 2000)*[7]

Another Canadian outbreak occurred in North Battleford, Saskatchewan only 11 months after Walkerton. This outbreak, caused by *Cryptosporidium*, produced between 5,800 and 7,100 cases when maintenance performed on the solids contact clarifier at the water treatment plant led to negligible turbidity removal for several weeks. This plant was vulnerable to contamination, being only 3.5 km downstream from the community's sewage effluent outfall. The treated water showed no evidence of exceeding regulated requirements and the process monitoring revealing negligible turbidity removal was not recognized as a basis to question water safety.

The maintenance took place on March 20, 2001, with contamination likely entering the system shortly thereafter (Figure 2). The first confirmed case of cryptosporidiosis was not detected until April 4. By April 24, there were still only 8 laboratory-confirmed cases of cryptosporidiosis in the community, although gastroenteritis was now widespread. The drinking water outbreak was not recognized until April 25 when a regulatory official who understood the implications of negligible turbidity removal in these circumstances was informed of the problem.

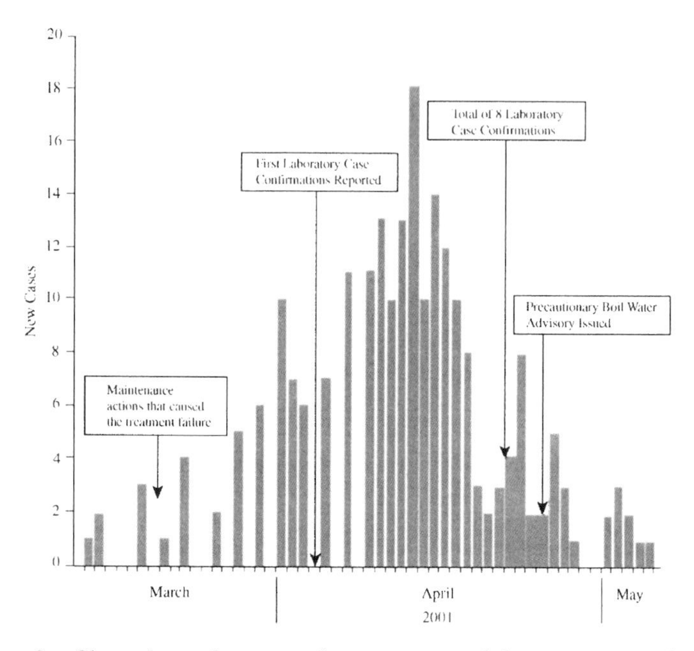

Figure 2. *Chronology of events and gastrointestinal disease cases in the North Battleford outbreak (based on data from Stirling et al. 2001)*[8]

The limitations of monitoring treated drinking water quality as a primary means for preventing outbreaks are evident from these two examples. Rather than by means of any routine monitoring efforts, outbreaks generally become known when epidemiologic disease surveillance is able to detect increases in cases of gastrointestinal illness or by other means such as hospital admissions, diarrhea in nursing homes or increasing absenteeism in schools. The real missed opportunity for any monitoring to have revealed and prevented or reduced the scope of these outbreaks was the failure of the operators to understand their water supply systems and perform adequate process control monitoring. In the case of Walkerton, where manure contamination evidently overwhelmed the fixed chlorine dose, had the operators measured the chlorine residual, monitoring they were supposed to conduct daily, they should have been able to detect this problem within 24 hours of the contamination, shut the contaminated well down and transfer to another safer well that was capable of supplying the entire demand for the town. In North Battleford, had the operators understood the contamination challenge of the source water and the risk of poor turbidity removal performance they could have assured that treatment was either returned to required levels or warned community health officials that treatment was inadequate for a source under the influence of a sewage effluent outfall.

To further highlight the limitations of monitoring data in protecting public health, outbreaks can still occur when performance criteria and conventional water quality targets are met. For example, the infamous Milwaukee outbreak of cryptosporidiosis in 1993, infecting an estimate of more than 400,000 consumers coincided with a spike in filtered water turbidity that still met their regulatory targets of that time.[1] Even when continuous sampling for pathogens has been used, interpretation of the significance of the results can prove difficult. A striking illustration is continuous sampling data for *Cryptosporidium* oocysts that were collected in Belfast during an outbreak in 2001 involving at least 191 laboratory-confirmed cases of cryptosporidiosis. Monitoring data were collected for treated water consistent with the monitoring requirements developed by the Drinking Water Inspectorate (DWI) for England and Wales.[9]

The data shown in Figure 3 were obtained by this monitoring scheme during an extended outbreak of cryptosporidiosis that was eventually tracked back to one of Belfast's drinking water plants. Evidence of the outbreak was provided by epidemiology, signs of sewage contamination of the treated water at the plant site and finding *Cryptosporidium* contamination of sewage in the onsite septic system. The interesting feature of these data is that they never exceed the regulatory standard of 1 oocyst per 10 L, a standard thought to preclude the possibility of outbreaks. The reasons for this discrepancy are not clear, but they range across various explanations from inefficient sampling, to presence of a particularly virulent strain, to further contamination occurring downstream of the sampling location.

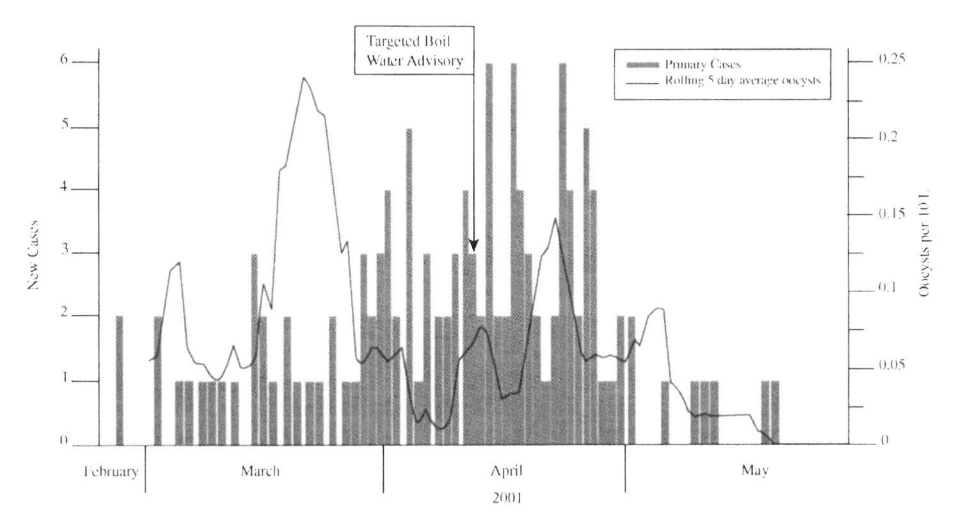

Figure 3 *Combined primary cases of laboratory-confirmed cryptosporidiosis in relation to rolling 5 day average continuous oocyst monitoring of treated water – Belfast.* ***Note****: all oocyst numbers are less than 1 per 10 L (Based on data from Smyth et al., 2002)*[10]

Another illustration of difficulties arising with interpretation of treated water monitoring for pathogens occurred with the Sydney water crisis in 1998 in which *Giardia* cysts and *Cryptosporidium* oocysts were detected in treated water in numerous samples at concentrations of several hundreds per 10 L less than two years after the opening of Sydney's new filtration plant.[1] This incident led to a series of boil water notices being issued and withdrawn for a service population of about 4 million consumers over a period of almost 3 months.

A commission of inquiry was called.[11] There was substantial controversy, yet, the one feature of this affair that almost everyone agreed upon was that there was no detectable increase in giardiasis or cryptosporidiosis among consumers.

There have been divergent views on what happened in Sydney from attributing the entire affair to an enormous false positive monitoring mistake[12] to the water contamination being authentic even though no illness was caused among consumers.[13] On balance, there were major anomalies associated with the simultaneous appearance of huge numbers of pathogens across a drinking water distribution system with water travel times of up to 3 weeks. These were certainly consistent with lab contamination causing false positives.[12]

2.2 Fundamental Limitations of Treated Water Monitoring

The false positive dilemma described above highlights another difficulty with compliance monitoring and interpretation of data from treated water. A recent analysis of screening tests for detecting unusual drinking water contaminants such as might arise in a terrorist attack provides an illustration of this more fundamental problem with compliance monitoring.[14] States et al.[15] highlighted the importance of using screening assays that have adequate analytical sensitivity and specificity along with low rates of false positive (***FP*** rate = α) and false negative (***FN*** rate = β) responses.

There are some vitally important quantitative insights on this topic that need to be considered in relation to any monitoring program that seeks to detect rare contamination episodes. The following insights derive from decades of rigorous development in the field of medical diagnostic screening, but they are fully applicable to this problem, provided we can overcome some potential confusion in terminology.

States et al.[15] referred to the importance of false positive rates noting the best claim was for a false-positive rate of only 3% for *Bacillus anthracis* with some assays having false-positive rate up to 83%. Suppose the best screening technology (3% false-positive rate) was applied to circumstances of rare contamination where, for the sake of illustration, only 1 out of 300 samples is truly hazardous. Given the specified conditions, we can ask: *If we get a "positive" result (contaminant is detectable or exceeds the standard) from an analytical test, how likely is that positive result to be correct?* (i.e. Is it a true positive?).

On average, we will need to screen 299 samples free of detectable levels of the contaminant to find 1 sample that contains detectable levels. With a false positive rate of 3%, we will detect approximately 9 false positives (299 • 0.03) in our search for the 1 true positive. Consequently, the answer about how likely is a positive result from this analytical test to correctly reflect the presence of a hazard turns out to be only 10% (1 true positive out of 10 total positive responses). This discouraging finding is an inescapable reality (as a function of the false positive rate and frequency of true hazards) for any analytical screening test. As the hazard that we are searching for becomes rarer, we can expect the number of false positive results to exceed true positive results.

To generalize this analysis, it is necessary to systematically evaluate the possible outcomes of screening evidence. In so doing, we need to confront some key differences in terminology across disciplines. In water quality, we are used to thinking of the analytical sensitivity as being the lowest concentration of a substance that we are able to detect reliably. We think of analytical specificity as the ability of the method to discriminate our analyte of interest from among a universe of interfering substances.

The field of medical diagnostics uses these same words (sensitivity, specificity) to mean different quantities,[16] where both terms are unit-less, conditional probabilities. To avoid confusion for this commentary, we will refer to the medical usage as being diagnostic sensitivity (***DSe***) and diagnostic specificity (***DSp***), but we must realize that the "diagnostic" qualifier is not attached to these terms in the medical literature. These terms are best understood according to a 2x2 table, as shown in Figure 4.

Hazard Reality

		Truly a **Hazard**, *H*	Truly a **non-Hazard**, *nH*	
Screening Evidence	Positive analytical result indicates a **Hazard**, *EH*	True Positives (TP) occurs with rate = ***DSe***	False Positives (FP) occurs with rate = α	[TP+FP]
	Negative analytical result indicates a **non-Hazard**, *EnH*	False Negatives (FN) occurs with rate = β	True Negatives (TN) occurs with rate = ***DSp***	[FN+TN]
		Cases of Hazard *P[H]* • n = [TP+FN]	Cases of non-Hazard *P[nH]* • n = (1 – *P[H]*) • n = [FP+TN]	n = [TP+FP+FN+TN]

Figure 4 *Framework for hazard detection and judgment of evidence*

This shows the decision problem dichotomized (even when working with continuous concentration data) into screening evidence indicating evidence of a hazard (***EH***), such as detection of a pathogen or toxic contaminant, or evidence of a non-hazard (***EnH***), shown here as the rows of the table. The reality that we seek to know, whether there is either truly a hazard (***H***) present or truly no hazard present (*nH*), is shown by the columns of the 2x2 table in Figure 4.

Diagnostic Sensitivity, *DSe*, is the conditional probability ***P[EH /H]***: given that there is truly a hazard, the chance that the screening evidence will correctly identify it as a hazard. This is calculated from the true positives (***TP***) and false negatives (***FN***) and is equal to the complement of the false negative rate (β):

$$DSe = \frac{TP}{TP + FN} = 1 - \beta \qquad [1]$$

Diagnostic Specificity, *DSp*, is the conditional probability ***P[EnH /nH]***: given that there is truly a non-hazard, the chance that the screening evidence will correctly identify it as a non-hazard. This is calculated from the true negatives (***TN***) and the false positives (***FP***) and is equal to the complement of the false positive rate (α):

$$DSp = \frac{TN}{FP + TN} = 1 - \alpha \qquad [2]$$

Although the false positive and false negative rates of screening assays may be estimated using laboratory studies with authentic standards, the effect of the likely hazard frequency, ***P[H]***, for the situation being assessed, is an important consideration because the decision about whether any positive result is a true positive in a specific scenario depends very strongly on ***P[H]***. The interpretation of a positive test is informed by the positive predictive value, ***PPV***.

PPV is the conditional probability ***P[H /EH]***: given that the evidence identifies the presence of a hazard, the chance that a hazard truly is present.

$$PPV = \frac{TP}{TP + FP} \qquad [3]$$

Figure 5, which is derived from spreadsheet calculations using the table in Figure 4 and the designated values for α, β and ***P[H]***, shows that ***PPV*** becomes increasingly dependent on ***P[H]*** for low values, ultimately becoming linearly related for true hazard frequencies below about 1 in a 100 to 1 in a 1,000.

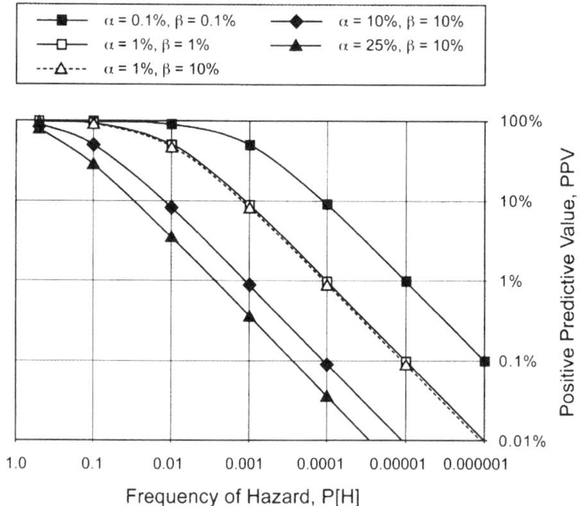

Figure 5 *Quantitative dominance of Positive Predictive Value (PPV) by low frequency of hazard occurrence, P[H]*

In practical terms, this shows that the ***PPV***, our ability to correctly predict danger with a single source of screening evidence, inevitably will be poor for rare hazards, i.e. small ***P[H]***, caused by the overwhelming number of false positive results.

The essential insight from this analysis is that unless we have a very effective, targeted sampling scheme that will ensure that the hazard we seek is likely to be present in any given sample, we can expect to face many more false positives than true positives when screening for rare hazards, even if the methods are much better than those currently available. This insight is not limited to field screening programs. Any isolated evidence, even if produced by a very sophisticated laboratory with the best analytical support, will encounter the same problem if the probability of a given sample containing the hazard is low relative to the overall false positive rate (including sampling and analytical errors).

Hrudey and Leiss[17] noted that although the dependency of ***PPV*** on ***P[H]*** has been recognized for medical diagnostics (***PPV*** will be inherently poor for rare diseases), this insight has not been commonly recognized in the environmental sciences where very low values of ***P[H]*** are the norm rather than the exception unless we are dealing with polluted environments. Certainly for hazards in treated drinking waters, we would normally expect ***P[H]*** to be low after treatment if water treatment processes are working effectively.

To further illustrate the important effects of prevalence on the predictive value, Rizak and Hrudey[18] have considered a hypothetical screening test for some pathogenic hazard in drinking water, say *Giardia*, with a diagnostic sensitivity of 90% and a diagnostic specificity of 85%. The same test is applied in three different settings where the prevalence or likelihood of *Giardia* in water samples is estimated to be 0.1%, 5% and 50% (Table 1). From the analysis presented by the 2x2 table, a positive test result yields predictive values of 0.60%, 24% and 86% respectively (Table 1). It can be seen that at higher prevalence of the hazard, the test performs relatively well, but at lower prevalence, the *PPV* drops to nearly zero and our ability to confidently predict a true hazard with this single source of monitoring evidence is greatly compromised.

Setting	Estimated Prevalence of *Giardia* (% of all water samples truly positive)	Positive Predictive Value, *PPV* (% of positive results truly positive)
finished (treated) water at treatment plant	0.1	0.60
raw water, protected catchment	5	24
raw water, polluted river source	50	86

Table 1 *Effect of prevalence on Positive Predictive Value (PPV) of Giardia in various settings using a testing method with DSe = 90%, DSp=85%*

This illustration demonstrates the decreasing information value that goes with testing treated water where the prevalence of contaminants should be very low, compared with sampling further upstream, where contaminants are likely to be found with greater frequency. Even higher prevalence can be achieved by targeting the sampling effort on circumstances or events that are known to contribute to source contamination. For example, during spring runoff in cold climates, or after storm events that may increase combined sewer overflows or increased contaminated runoff.

Of course a logical response to all of this is – we do not drink raw water so what good does it do to focus our monitoring there? The answer is that effective use of monitoring relies on knowing the capability of the treatment system combined with developing effective surrogates that may be used to monitor treatment performance continuously, in real time. Turbidity and chlorine residual are the classical parameters that are typically used in this way. For even more sensitive monitoring of filtration performance, particle counters have come into use. In reality, the combination of knowing the contaminant challenge to the system, the treatment capability of the system and real time monitors of performance provide much greater information value than intermittent monitoring of treated water quality aimed at the target contaminant.

2.3 Understanding of the Limitations of Monitoring

Rizak and Hrudey[18] have also found that water quality professionals do not necessarily recognize these fundamental limitations for analytical results screening for rare hazards. In a survey of 21 experienced water quality personnel more than 80% of respondents were 80 to 100% confident that a hypothetical water quality monitoring result (with the characteristics ***DSe*** = 95%, ***DSp*** = 95%, ***P[H]*** = 0.001) was correct even though these hypothetical characteristics correspond to a ***PPV*** of less than 2% because of the low value specified for ***P[H]***. Such over-confident responses are consistent with medical students given similar evidence, notwithstanding the widespread incorporation of these concepts into current medical education.[19]

A full understanding of this decision framework (Figure 4) can provide a mathematical justification for intuitions normally derived from extensive experience and informed judgement. Clearly, response actions with serious consequences (e.g., total interruption of water supply) need to be validated by other evidence or by further testing. The inevitable balancing between making false positive and false negative decision errors needs to be informed by an understanding of these relationships. Hrudey and Leiss[13] have elaborated on these balancing considerations including the effect of these parameters on negative predictive value (***NPV***) and the resulting implications for precaution.

3 CONCLUSIONS

A review of drinking water outbreaks in affluent countries over the past 30 years provides a rich body of evidence for drinking water providers who are committed to ensuring the delivery of safe drinking water.[1] A key finding is that monitoring treated water quality is not the primary protection for consumers for a number of demonstrable reasons. Case studies illustrate situations where compliance monitoring failed to prevent outbreaks thereby confirming the importance of a preventive risk management, rather than strictly reactive, compliance monitoring approach to assuring drinking water safety. In particular, greater attention to watershed protection and to real-time process control, rather than reliance solely on treated water monitoring is clearly more effective in assuring drinking water safety.

Furthermore, the fundamental nature of monitoring evidence is such that when source protection and treatment systems perform better, to make the occurrence of contamination in finished water rarer, the information value of monitoring that higher quality water for compliance also declines. This insight is derived from basic statistical concepts proven from decades of application in medical diagnosis, yet it is not commonly reflected in regulatory philosophy or practice. The combination of knowing the contaminant challenge to the system, the treatment capability of the system and real time monitors of performance will provide much greater information value than intermittent monitoring of treated water quality aimed at compliance for specific target contaminants.

Assuring drinking water safety requires a commitment to a comprehensive approach to risk management. There is a growing international consensus moving towards this approach to assuring safe drinking water, which provides the prospects of making water even safer most places in the developed world.

4. ACKNOWLEDGEMENTS

Financial support from the Australian Cooperative Research Centre for Water Quality and Treatment, the Canadian Water Network, Health Canada, Alberta Health and Wellness and the Natural Sciences and Engineering Research Council of Canada are gratefully noted.

References

1. Hrudey, S.E. and E.J. Hrudey (2004). *Safe Drinking Water – Lessons from Recent Outbreaks in Affluent Nations*. London. IWA Publishing: 514 pp.
2. NHMRC (2004). Australian Drinking Water Guidelines. National Health and Medical Research Council. Canberra, ACT. http://www.nhmrc.gov.au/publications/synopses/eh19syn.htm
3. WHO (2004). Water Safety Plans, Chapter 4. *WHO Guidelines for Drinking Water Quality 3rd Edition*. Geneva, World Health Organization: 54-88. http://www.who.int/water_sanitation_health/dwq/gdwq3/en/
4. IWA (2004). The Bonn Charter for Safe Drinking Water. International Water Association, London http://www.iwahq.org.uk/pdf/Bon_Charter_Document.pdf
5. Sinclair, M. and S. Rizak. (2004). Drinking water quality management: the Australian Framework. *J. Toxicol. Environ. Health*, Part A. 67: 1567-1579.
6. O'Connor, D. R. (2002). *Report of the Walkerton Inquiry. Part 1. The Events of May 2000 and Related Issues.* Toronto, The Walkerton Inquiry: 504 pp. http://www.attorneygeneral.jus.gov.on.ca/english/about/pubs/walkerton
7. BGOSHU (2000). *The Investigative Report of the Walkerton Outbreak of Waterborne Gastroenteritis May-June 2000.* Owen Sound, Ontario, Bruce-Grey-Owen Sound Health Unit: 58 pp. plus appendices.
8. Stirling, R., J. Aramini, A. Ellis, G. Lim, R. Meyers, M. Fleury and D. Werker (2001). "Waterborne cryptosporidiosis outbreak, North Battleford, Saskatchewan, Spring 2001." *Can. Commun. Dis. Rep.* **27**(22).
9. Waite, M. and P. Jiggins (2003). Cryptosporidium in England and Wales. *Drinking Water and Infectious Disease — Establishing the Links.* P. R. Hunter, M. Waite and E. Ronchi. Boca Raton, CRC Press and IWA Publishing: 119-126.
10. Smyth, B. and OCT (2002). *The Report of the Outbreak Control Team of an outbreak of Cryptosporidiosis during February / April 2001.* Belfast, Communicable Disease Surveillance Center (NI): 41.
11. McClellan, P. (1998). *Sydney Water Inquiry.* Sydney, New South Wales Premier's Department. http://www.premiers.nsw.gov.au/our_library/archives/sydwater/5threp/r5vol1.htm
12. Clancy, J. (2000). Sydney's 1998 water quality crisis. *J. Am. Water Works Assoc.* **92**(3): 55-66.
13. Cox, P., I. Fisher, G. Kastl, V. Jegatheesan, M. Warnecke, M. Angles, H. Bustamante, T. Chiffings and P. R. Hawkins (2003). "Sydney 1998 — lessons from a drinking water crisis." *J. Am. Water Works Assoc.* **95**(5): 147-161.
14. Hrudey, S.E. and S. Rizak. 2004. Discussion of rapid analytical techniques for drinking water security investigations. *J. Am. Water Works Assoc.* 96(9):110-113
15. States, S., J. Newberry, J. Wichterman, J. Kuchita, M. Scheuring and L. Casson. 2004. *J. Am. Water Works Assoc.*, 96(1): 52-64.
16. Gordis L. 2000. *Epidemiology*. 2nd ed. Philadelphia:WB Saunders.
17. Hrudey, S. E. and W. Leiss (2003). "Risk management and precaution: insights on the cautious use of evidence." *Environ. Health Perspect.* **111**: 1577-1581.
18. Rizak, S. and S.E. Hrudey. 2004. Improved understanding of water quality monitoring evidence for risk management decision-making. *Proc. 11th Canadian National Conference on Drinking Water.* Calgary. Canadian Water and Wastewater Association.

19. Hoffrage U., S. Lindsay, R. Hertwig and G. Gigerenzer. 2000. Communicating statistical information. *Science*, 290: 2261-2262.

WATER SAFETY PLANS AND THEIR ROLE IN PREVENTING AND MANAGING CONTAMINATION OF THE WATER SUPPLY

R.Aertgeerts

WHO European Centre for Environment and Health, Rome, WHO Regional Office for Europe, via F.Crispi, 10, I-00187 Rome, Italy

1 INTRODUCTION

The European Union (EU) Water Framework Directive[2] (WFD), adopted on 23 October 2000, is to provide a basic strategy for the sustainable and integrated management of water resources, meeting the need of different users including the production of safe drinking-water. On 19 September 2003 the European Commission (EC) adopted a proposal for a new directive[3] to protect groundwater from pollution, the groundwater directive (GWD). This daughter directive to the WFD ensures that groundwater is monitored and evaluated across Europe in a harmonized way. It responds to the requirements of the WFD related to the assessment of the chemical status of groundwater and the identification and reversal of significant and sustained upward trends in pollutant concentrations. Groundwater is the main source for drinking-water production in the EU; the GWD therefore acts in a preventive manner to protect human health.

Directives concerning the quality of drinking-water have existed since 1980. The European Council on 3 November 1998 adopted a new Drinking-water Directive (DWD)[4] on the quality of water intended for human consumption. The Directive entered into force on 25 December 1998 and Member States had 2 years, i.e. until 25 December 2000 to transpose the Directive into national legislation. Member States had 5 years, i.e. until 25 December 2003 to ensure that drinking-water complies with the standards (with some exceptions). This directive is currently under revision, with one of the main areas of interest the greater emphasis on risk assessment and risk management.

---

[2] "Directive 2000/60/EC of the European Parliament and of the Council establishing a framework for the Community action in the field of water policy" available from For further information, see URL:
http://europa.eu.int/smartapi/cgi/sga_doc?smartapi!celexapi!prod!CELEXnumdoc&lg=en&numdoc=32000L0060&model=guichett

[3] Commission proposal COM (2003)50: Proposal for a Directive of the European Parliament and of the Council on the protection of groundwater against pollution – the Groundwater Directive available from URL: http://europa.eu.int/eur-lex/en/com/pdf/2003/com2003_0550en01.pdf

[4] Council Directive 98/83/EC of 3 November 1998 on the quality of water intended for human consumption – the Drinking-water Directive available from :
http://europa.eu.int/comm/environment/water/water-drink/index_en.html

This paper recalls the historic basis of risk assessment and risk management in the framework of the hazard analysis and critical control point (HACCP) approach in the food industry, leading to the formulation of water safety plans (WSP) in the drinking-water service, and discusses the role of such WSPs in the protection of public health in case of malevolent introduction of contaminants in the water system.

2 RISK ASSESSMENT AND RISK MANAGEMENT

2.1 HACCP in the Food Industry

The HACCP system, defined as "a system which identifies, evaluates, and controls hazards which are significant for food safety", grew from the conceptual development of the total quality control in the 1950s, and the application thereof during the early years of the US space program. The twentieth session of the Codex Alimentarius Commission (Geneva, Switzerland, 28 June – 7 July 1993) adopted *Guidelines for the application of the Hazard Analysis Critical Control Point (HACCP) system* (ALINORM 93/13A Appendix II).

The HACCP plan is defined as "a document prepared in accordance with the principles of HACCP to ensure control of hazards which are significant for food safety in the segment of the food chain under consideration." A HACCP system has two main characteristics:

- an assessment of hazards
- critical control points, which are "steps at which control can be applied and which are essential to prevent or eliminate a food safety hazard".

This basic concept was later translated into the drinking-water safety plan approach.

2.2 WSPs in the Drinking-water Industry

2.2.1 WHO Framework for Water Safety – Water Safety Plans. The WHO guidelines for drinking-water quality (GDWQ) use the concept of a framework for safe drinking-water production and distribution. The framework has five components:

1. health-based targets derived from a critical consideration of health concerns;
2. system assessment to determine whether the water supply chain (from source through treatment to the point of consumption) as a whole can deliver water of a quality that meets the above targets;
3. operational monitoring of the control measures in the supply chain which are of particular importance in securing drinking-water safety;
4. management plans documenting the system assessment and monitoring, and describing actions to be taken in normal operation and incident conditions, including upgrade and improvement documentation and communication; and
5. a system of independent surveillance that verifies that the above are operating properly.

Items 2–5 together form a Water Safety Plan (WSP). WSPs are comprehensive risk assessment and risk management approaches that encompass all steps in water supply from catchment to consumer. Principles of WSPs are presented in Chapter 4 of the 3rd edition of the WHO guidelines for drinking-water quality.

2.2.2 Experiences with Risk Management Approaches in the Drinking-water Sector

2.2.2.1 Legislators. The risk assessment and risk management principles that guide WSPs are considered important by legislators because they:

- are proactive and present a "push towards innovation" as systematic and integral analysis of water supplies become an obligation;
- provide a consistent, objective and science-based approach that can be applied as an overall framework to all water supply systems irrespective of size, source waters and technologies employed;
- are an instrument for proving that state-of-the-art principles are followed and barriers are intact;
- introduce a sound basis for cost-benefit analysis;
- stimulate sector cooperation between water suppliers and (environmental) health authorities; and
- provide a platform that formally can integrate catchment protection aspects.

2.2.2.2 Water Suppliers. Additional advantages to water suppliers of implementing a WSP include the following.

- For larger supplies, existing procedures can be integrated into a systematic and accessible package that can be easily communicated or integrated into existing quality management schemes (ISO 9001).
- For smaller supplies, WSPs provide a tool for basic assessment of needs and help promote good asset management.
- WSPs target the resources and attention towards the critical issues of drinking-water quality and help focus water supplier's monitoring activities on a smaller number of critical parameters.
- WSPs document that due diligence is in place and thus protect against allegations of negligence.
- WSPs increase transparency towards consumers and surveillance agencies;
- WSPs provide a platform for the management of existing knowledge, competence, and experience within the water supply.
- WSPs provide an opportunity to increase knowledge of personnel.

2.2.2.3 Authorities. Authorities will find WSPs advantageous because they:

- provide a framework that can be audited and assessed in a standardized way;
- provide improved understanding of health risks and assessment of priorities as part of overall public health policies;
- provide greater confidence in the continuous management of drinking-water quality, particularly in small supplies;
- provide improved understanding of operational aspects of supplying drinking-water, the range of hazards and what can go wrong; and
- trigger more intensive cooperation and communication with water suppliers and with environmental authorities.

2.2.3 Differences between HACCP and WSP. The WSP concept is a systematization of procedures used in water management, some of which predate HACCP. In a WSP, there are few really critical control points comparable to interventions in the food industry, such as pasteurization. Rather, a large number of interventions collectively leads to safety. These tend to be of two types: measures against contamination, for example in protecting the source environment, and measures that contribute to progressive reduction in contaminant levels. WSP puts great emphasis on the multibarrier principle, which is not the case in HACCP.

Water science is moving quickly and is now in an era in which quantified description of the efficiency of treatment procedures is possible. Thus with the multi-barrier principle, a quantitative approach can be applied to understanding the progressive achievement of adequate quality through sequential processes. This facilitates the adequacy of a system (i.e. its ability to achieve safe drinking-water); and also an understanding of the significance of failures in single processes in terms of public health significance.

2.3 WSP in Relation to Food Preparation

The water used for food processing is an important consideration, especially in the case of water used for foods that require only minimal processing such as salads, fruits, and some kinds of vegetables. Water produced in compliance with a WSP will also pass the HACCP requirements for food production.

3 PRECAUTIONS AGAINST MAN-MADE CONTAMINATION OF DRINKING-WATER

3.1 Drinking-water Services as a Target

Different reasons may prompt malevolent individuals to target water supply enterprises. Extortion for financial gain is unfortunately not an uncommon occurrence. Today, political or religious ideologies have also prompted attacks by organizations or individuals. It is important to bear in mind that the main aim may not only be to cause death or injury – the mere denial of service can cause significant disruption in the social and economic life of the targeted community, and cause widespread panic. Contrary to food supply, water services have the additional characteristic that alternatives are not readily available.

It will be important to proactively assess the risk associated with the deliberate introduction of microbial or chemical contaminants, to take preventive actions, and to prepare for remedial measures.

3.2 Analysis of systems components

A WSP includes a step whereby a flowsheet of the complete system is developed, and the interrelationships between the different components are clearly identified. By providing a clear insight in the actual integrated operation of a water supply system, a WSP is an important instrument for the identification of potentially weak areas.

3.2.1 Water Sources. Surface water sources are seldom protected; however, dilution effects any contaminant introduced upstream can be so important that massive amounts would need to be introduced to produce any effect. Groundwater aquifers, especially those benefiting from assisted recharge, may not be totally immune from the malevolent introduction of contaminants. Security will depend on the ease of access to the source and the possibility to deliver to it quantities of chemical or biological agents sufficient to cause injury or illness in end-users. All installations associated with the primary water source such as water capture areas should be monitored. Such monitoring activity can be aided by modern technology, while local citizens also have a major role to play. Physical structures need to be secured through basic precautionary measures such as fences, intrusion-resistant building materials, intrusion detectors, and alarm systems.

3.2.2 Raw Water Mains. Raw water mains may be the target of intrusion, but the following treatment unit operations and repeated laboratory analyses militate against this component as a priority target. Nevertheless, it should be recognized that certain microbiological agents can resist subsequent treatment steps, and that most chemicals, including nuclear materials, will not be fully removed or inactivated by conventional treatment.

3.2.3 Treatment Plants. Treatment plants in most major agglomerations are conceived as self-contained installations, with closely controlled access. Existing protection systems can be made more robust by implementing a graded access system, with access to areas of greater concern being more tightly controlled than others. Again, the careful analysis of the operation of a water supply system such as foreseen under a WSP would lead to the grading of installations within the perimeter of a treatment plant and form a basis for a precautionary revision of access.

Responsibility for monitoring access should be accompanied with authority to take immediate action in case breaches of procedure have been observed. The introduction of patrols, closed-circuit television, and installing safety locking devices allowing access to sensitive areas would increase security.

Treatment plants use significant amounts of chemicals, and proper procedures should be put in place to control both quality and quantity. Correct identification of chemicals should be guaranteed at all times, while amounts used should be reconciled with remaining stock at least weekly. Access to areas where toxic chemicals are stored, especially in bulk quantity, should be particularly controlled.

Disinfection is a particularly sensitive area for two reasons.

- Chlorination, a classic disinfection method, is effective against many, but not all, pathogenic biological agents and can be overwhelmed if operational parameters are not adjusted from standard operations in the event of contamination. Ozonization is a more expensive form of disinfection, but is generally more effective against contaminating agents, pathogens, and toxins. Chlorination does provide residual protection during the transportation phase between production and arrival at the consumer, whereas ozonization alone does not provide this additional protection.
- Chlorination processes often require the storage of significant volumes of chlorine gas in pressurized containers. The storage areas where these containers are being kept may, of and by themselves, become targets for theft of the pressurised containers and all precautions possible ought to be taken to prevent such an event.

3.2.4 Distribution Systems. Main distribution systems are usually pressurized, making the deliberate introduction of contaminants difficult but, given a suitable pumping installation in a remote location, not impossible to achieve.

Intermediate locations on water mains such as equalizing towers, water towers, and tanks are usually built in a time of less pressing security concern. Reviewing access and hardening the sites by sealing entry points and erecting multiple barriers to entry are recommended. It must also be realized that the water inside the reservoirs is no longer under pressure and feeds under gravity to the local area network. Introduction of contaminants in these points is therefore significantly easier than in pressurized primary distribution mains.

Local, or secondary, distribution networks are weak points in the water-supply network because they offer many access points and require relatively unsophisticated means of introduction. It can be argued that the risk is limited since the population exposed is more restricted. This argument neglects the psychological impact from seeing highly visible parts of the system collapse under an attack, and the potential for resulting widespread panic. In assessing critical points in a water distribution system, it is therefore important to include local distribution systems serving highly concentrated and visible populations, such as those living or working in highly visible high-rise buildings. Water lines and meters feeding into these buildings or being component parts of the internal distribution system are to be secured, and a control system for access to the system should be developed and implemented in consultation with the building manager.

Local distribution systems should also be reviewed to identify politically sensitive targets such as key government buildings, foreign missions, financial institutions, seats of international organizations etc. whose collapse would have significant political and psychological impact.

For the same reason, analysis of local systems should also include identification of particularly sensitive facilities essential to public health. These include not only hospitals and other public health services, but also law enforcement, civil protection, and bottled water plants. In critical points, installation of additional water-treatment processes may be considered.

3.2.5 Monitoring. Economic and practical considerations preclude a continuous monitoring of all critical control points against all possible contaminants. Continuous monitoring of certain parameters such as pH, conductivity, and turbidity, may provide some non-specific indication of change of water quality and prompt more detailed analysis. In vitro and in vivo tests, particularly biomonitoring, are also of potential use as first-line detection systems. Regular controls of the water quality within the distribution system and follow-up of customer complaints are considered appropriate tools. In particular, complaints related to changes in water quality in the same section of the distribution system within a short time interval should lead to an immediate reaction by the water supplier.

The detection of health effects of malevolent contamination of water supply networks is likely to come from a number of epidemiological clues such as:

- large number of people ill with similar disease or syndromes;
- large number of cases of unexplained disease or syndrome, or deaths;
- higher morbidity and mortality than usual with a common disease, or failure of a common disease to respond to treatment; and
- no illness in people not exposed to common systems.

Such drastic health outcomes need to be communicated to the water supplier, and it is therefore important that a good communication system exists between the public health departments responsible for outbreak detection and the water services.

Even when surveillance has indicated the likely presence of a contaminant, detailed analysis, especially when highly toxic chemicals and/or microbial pathogens are suspected may quickly overwhelm the capacity of standard laboratories operated by water supply enterprises. It is therefore important that water suppliers contact highly specialized laboratories, including those operated by civil protection and military services, and negotiate mutual support arrangements.

Once the nature of the contaminant has been identified, a WSP will allow the identification of potential control points to deal with the intrusion, the change in operational procedures to counter any residual threat, and the development of corrective actions including monitoring of water quality during the recovery period.

4 RECOVERY

4.1 Why to Plan for Recovery

The concept of planning for after an emergency does not imply acknowledgement of defeat. Rather, it is a wise precaution to prevent waste of resources, poor service delivery for extended periods of time, and a prolonged negative impact on the health of the population.

In a society where water not only serves the direct health needs of the community but is also an essential vehicle for the safe elimination of waste, post-emergency planning will need to deal with both supply of drinking-water and planning for the maintenance of sanitary services.

4.2 Rehabilitation

4.2.1 Preventive Planning. Modern water services are highly complex undertakings whose operation depends on the interaction of a variety of skills. During the period of planning for emergency, an analysis should be made of those skills that are essential for the functioning of the system. Staff possessing the necessary skills should be identified and informed of their selection in an emergency management team. Such teams should be composed of staff at all levels of the organization, from management to local operators. Information on the working of the system should be reviewed and updated.

In larger systems that grew historically, it is not uncommon to find unit operations managed by experience rather than by clearly understood scientific principles. In sewerage especially, it is not uncommon for the exact placement of the pipes to be buried in the mist of time. A thorough understanding of all unit operations and knowing the exact geographic location of each component of the system is essential to the planning of any emergency recovery effort.

In addition to the major components of the system, a preventive analysis needs to be made of laboratory capacity. Dealing with an emergency will cause a very significant increase both in the number of samples as well as in the number of parameters. This in turn will create additional pressure on equipment, staff, and supply of spare parts and expendable materials which may lead to breakdowns in the analytical capacity. Correct assessment of the company's analytical capacity and the preparation of fall-back positions needs to be undertaken in a preventive manner.

The possibility to isolate sections of the water service, so as to minimize the affected area, identification of sensitive locations in this area and alternatives to service them, are part of precautionary planning and will allow staff to concentrate on affected areas during the actual emergency.

4.2.2 Distribution First. Priorities in dealing with an emergency will be: to isolate the contamination; to maintain at least a basic supply to unaffected areas; and to improve quality in stages. Isolating affected sections while speedily restoring a continuous supply of water that is not harmful, even if it is not drinkable, is important since interruption of supply on a system-wide basis can lead to generalized contamination and deprive people of water required for sanitary needs. In cases where distributed water would fail to meet quality criteria due to accidental contamination, it may be better to supply sensitive locations such as hospitals with water that can be disinfected and whereby quality can be maintained, for example through tanker trucks. Simple treatment can be provided at local level, such as chlorination of local distribution tanks.

5 COMMUNICATIONS

Because psychological impact is often as important as actual harm when it comes to disrupting society, it is important that all practical measures be taken to build up public trust before any event takes place. Water services should inform the public that:

- plans to ensure water safety are in place;
- coordination between the water service and all other support services is established and tested;
- reviewing these plans and coordination mechanisms is standard procedure and they are being kept up to date; and
- all potential contamination has been scrutinized and appropriate responses have been devised.

Speed is crucial during an emergency, and this is not the time to expose staff who may be expert in technical matters to a first contact with the media. Communication plans need to be developed, and a single spokesperson needs to be identified. This individual should establish cordial professional relationships with the members of the media, and receive specialized training on managing media in case of emergencies.

6 CONCLUSIONS

Risk assessment and risk management principles derived from the HACCP concept developed in the food industry are codified in WSPs whose introduction can be expected to be part of the revision of the current EU Drinking-water Directive.

The detailed analysis of the water supply network, as well as the identification of critical control points and the definition of operational measures to control risks at these control points will be an added value to ensure public safety under normal operational procedures compared to the current parametric approach.

Enhanced safety measures, including security of the installation, monitoring particularly of the (secondary) distribution system, and immediate intervention upon receipt of customer complaints, can harden water services against malevolent introduction of chemicals or pathogenic organisms. Nevertheless, close cooperative agreements will need to be in place with the public health service for the detection of epidemiological clues that may signal such an event, and with specialized laboratories for the identification of the causative agent. Once these steps completed, however, the WSP will greatly facilitate the location within the overall system of those critical points where the new threat may be controlled, and the operational interventions to do so.

Publicity given to actual events of malevolent introduction of contaminants, or to the threat thereof, can be as effective as an actual attack in terms of the psychological impact to the public at large. Furthermore, it can create hoaxes and copycat actions by less organized members of the community which, in turn, can overwhelm emergency-response systems by their mere volume. Professional communication skills are essential from the side of the water industry.

During the rehabilitation phase, water safety plans will allow the determination of the best way in which unaffected zones can continue to be supplied, and by which affected zones can be restored to normal service.

References

1 World Health Organization (2004) "Guidelines for Drinking-water Quality" Third edition Volume 1 Recommendations.

2 World Health Organization (2002) 'Terrorist threats to food – Guidance for establishing and strengthening prevention and response systems', *Food safety issues*, WHO Geneva.

3 World Health Organization (2005) 'WHO Technical notes for emergencies', accessed 14 April 2005, http://www.who.int/water_sanitation_health/hygiene/envsan/technotes/en/ .

4 'CCP decision tree', accessed 14 April 2005, http://www.fao.org/docrep/W8088E/w8088e03.jpg

5 Anon, 'Scientific synthesis report' *Drinking-water seminar* (Brussels, Belgium, 27–28 October 2003).

THE USE OF COMPUTATIONAL TOXICOLOGY FOR EMERGENCY RESPONSE ASSESSMENT

T. White

Marquis & Lord – Consulting Scientists, Albion House, 13 John Street, Stratford upon Avon, Warwickshire CV37 6UB

1 INTRODUCTION

This paper focuses on the use of computational tools in the development of toxicological information not available from literature sources. The development of the information depends on the primary identification, and semi-quantification of the substances concerned by conventional analytical techniques. The techniques advocated here are not proposed as a replacement for the application of robust knowledge published in the literature or available as a result of conventional experimentation. In many circumstances the development of computer generated toxicological data may be a powerful tool that allows a decision to be made in advance of the delivery of literature or conventionally derived experimental data.

The history and development of the ongoing problem and hence the need for computational toxicology revolves around some simple ratios which illustrate the approximate availability of knowledge about organic substances. Depending on the source of information, and its date of publication, it is possible to derive the following approximations. The very nature of the subject matter dictates that these ratios, even as approximations, will never correlate particularly well to the state of current knowledge. However, they do serve to illustrate the problem.

- the ratio of unknown (those for which the structure only is known compared to other data) to known organic compounds is approximately 18:1;

- for those where there is published toxicological information the ratio is closer to 1800:1; and

- in addition there are countless, as yet, undiscovered compounds generated in fires and biologically productive environments such as sediments, decay pathways, and plants.

Clearly, the occasion when mankind is exposed to toxicologically significant concentrations of any one, or a mixture, of such compounds is relatively small. Civilisations have developed, over time, to limit the exposure to harmful concentrations of substances using varying default techniques. The obvious example is the administration of medicinal substances by only experienced or qualified practitioners.

The commercial sector which has the most to gain from the immediate knowledge of substances' toxicity because it sells chemicals which are deliberately administered to the end user is the pharmaceutical industry. In nearly all cases of modern drug design hundreds, if not thousands, of compounds with related drug like performance need to be screened before efficacy trials can begin. For many years the pharmaceutical industry has used sophisticated computational techniques to screen compounds for their toxicological and other fate and behaviour characteristics before expensive bench trials are embarked upon. There are many drivers for the use of this technology. These include the need for a reduction in animal testing, as well as the significant overall cost of bringing a desired therapeutic compound to market.

2 DEVELOPMENT OF THE TECHNOLOGY

There are two types of tool available for the evaluation of substances using this technology. The easiest to consider is the use of proprietary models developed by respected software houses, for widely accepted industry needs such as the evaluation of potential carcinogenicity in a drug candidate compound. The second is the development of in-house models designed for the prediction of a specific property required for a relatively small market sector. An indirect example of this is the forecasting of auto ignition temperatures of compounds to aid the development of safety protocols and fire investigations. In both cases the principles involved in the development of the required models are:

- collecting uniformly produced data (in the case of toxicology data the same test species, and conditions);
- vetting the data and rejecting inconsistency (method variation); and
- determine the most appropriate quantum mechanical and/or atomistic parameters to establish correlations with test data (There are over 1600 molecular descriptors [1] available for correlation assessment, some estimates put the number closer to 2500).

This collection and evaluation of data is followed by the derivation of a multiple regression equation, or the application of neural net technology using a proportion of the overall data as a 'training set'. This is accomplished using established statistical techniques. Once a mathematical model has been identified, it is evaluated against the remainder of the data set to assess its selectivity and the level of uncertainty within any prediction. The more sophisticated software tools are developed with an additional level of selectivity based upon the broad classifications of molecules in sub divisions of a training set, for example chlorinated solvents, phenolics or, in the case of pharmaceutically active materials, more complex compound groupings based upon multiple functional group similarity are often used.

Collectively the technique is termed Quantitative Structural Activity Relationship modelling or QSAR. Modern applications are very robust and the pharmaceutical industry relies heavily on it, on a routine basis, to provide indicative toxicology screening. Over recent years providers of the technology have had to ensure its robustness because of the claimed offset of development cost and improved safety levels required when investigating new drug candidate compounds. This is a very necessary marketing tool when various estimates put cost barriers for drug development in the order of $800m to bring a product to market, over a period 15 years. This can often result in a 10 year patent for which a company will have heavily defended market rights to recoup its investment and derive a profit.

The pharmaceutical industry's use of QSAR has not stopped at the development of preliminary toxicological knowledge. The industry also uses the technology to establish other information relative to a compounds use as a therapeutic material before any kind of animal testing is carried out. These applications are often referred to as Absorption, Distribution, Metabolism and Excretion (ADME) outcomes.

3 THE OPERATIONAL PROCESS

In order for the processes discussed above to be undertaken on large batches of compounds, these software packages process up to 10,000 compounds per hour. As a matter of necessity, computational chemists have developed several data file formats. These allow a fixed description of a molecule to be stored as a binary file which can be translated by many proprietary packages as two dimensional structures familiar to all chemists. One of the most commonly used file types are MOL files (this is a proprietary file type developed by software authors MDL, Molecular Design Ltd [2]). Another and very common structure transfer code is that of the industry standard SMILES string [3]. This is the Simplified Molecular Input Line Entry System which uses a standard set of descriptive rules in much the same way as the IUPAC naming convention, an example of this is given for DDT below together with its corresponding two dimensional structural representations.

SMILES string: Clc1ccc(C(C(Cl)(Cl)Cl)c2ccc(Cl)cc2)cc1

Systematic Name: Benzene, 1,1'-(2,2,2-trichloroethylidene)bis 4-chloro-

Structure:

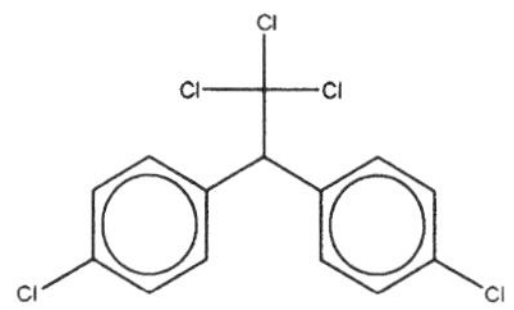

4 LIMITATIONS

As with any determination, be it in a laboratory or 'in silico', there are limits to the technique. Two major technical boundaries to QSAR techniques are those of:

- compatibility of test molecule with training set; and
- test result outside the training range.

The former issue occurs when the structure of the substance of interest does not conform to a broad target group which was used to generate the original calibration. This means that there could be a greater degree of uncertainty associated with such a determination compared to one for a substance within the training boundary. For example in the simplest case if a test substance contains nitrogen and the training set did not there is an obvious difference that could introduce an increased level of uncertainty.

The second boundary is analogous to the limits of a linear calibration; if a result is outside the base data range of the training set then the same cautions apply to using the result as if it had been determined using a laboratory method. For example if the upper boundary of the calibration is 10 units and the model returns a result of 20 then the output should be treated with caution and further validation of the determination should be investigated. With any model there will be clearly defined statistical limits which define this decision need.

There is therefore an inevitable consequence of these limitations in that the use of proprietary software should only be used by experienced practitioners. This cautionary practice is even more appropriate when dealing with the application of custom developed models.

4.1 Issues for the Adoption of the Technology by Different Industries

In many commercial and industrial sectors there is a great deal of focus only on substances used within those wealth generating industries. In addition little attention is paid to what in general terms it is deemed unnecessary to know. This has lead to a lack of awareness in advances in QSAR applications outside the pharmaceutical industry. Consequently this also applies to the lack of understanding of the significant advantages adoption could bring. Furthermore, there is a lot of scepticism by those outside the pharmaceutical industry about the scientific merits of the technology. This is not helped by the industry focused inward marketing which targets the pharmaceutical companies as the biggest users which represent the most lucrative market. The software tools are undoubtedly a high priced product as far as off the shelf applications are concerned. This reflects the high development costs and the perceived benefits purchase can bring. Declines in the fortunes of pharmaceutical companies may cause the software houses to widen their marketing in an attempt to seek new outlets but as always they will tend to concentrate on willing and established markets as low marketing cost sales generators.

In many other markets, the relatively high cost of model development or proprietary software purchase and licensing, makes use of QSAR outside the pharmaceutical industry prohibitive. This is often because of a perceived need for its use only on an 'occasional' basis and an assumption that it is not likely to be cost-effective.

5 USE OF THE TECHNOLOGY IN THE EMERGENCY INCIDENT TIME LINE

In most cases, when a contamination emergency occurs, there will be some immediate knowledge of the substances involved and their most dominant effect. These impacts may present themselves as taste, odour, or rapidly induced illness such as vomiting or skin and eye irritation. In addition there are perceptual impacts such as discolouration or foaming which will also lead to consumer rejection of water.

In all these cases there will be a rapid response to an identification need followed by a summary appraisal of available literature about the substances involved. In any case action will have been taken to isolate affected water from the consumer to allow remedial action to be taken at a measured pace along with medical assessment.

The major benefit of in-silico technology is in the phase of an incident management immediately post the identification of the substances of concern. Typically the maximum benefits are likely to be derived when evaluations of the medium to long term consequences need to be made. It is on these occasions that gaps in the available literature knowledge can be filled using computational techniques. The second major benefit is in assisting with the decision process defining when a point of cleanliness has been attained! The technology allows a rapid development of information across a wide range of parameters in fractions of the time it would take to get results from conventional animal toxicology testing. The very least that can be achieved is to allow a prioritisation of laboratory based testing. This is achieved by highlighting those substances which may present a more insidious exposure risk because of possible positive responses to tests such as mutagenic activity or carcinogenicity.

6 CASE STUDY

Having described the concept of the technology and the range of knowledge gaps that can be filled it is important to illustrate the benefits in the form of a case study. In this case the techniques were applied in order to deliver information absent from the literature to toxicologists dealing with a case of accidental exposure to closed water treatment chemicals. These had potentially been consumed at a potable outlet as a result of a plumbing cross connection. The cross connection had not been noticed for a significant period of time before the matter was investigated in terms of the chemical composition of the water. In summary the toxicology of the inorganic components concerned was understood but that of the organic components was not.

In order for the toxicologists to do their job they needed the following information from the literature or other sources for each of the organic substances identified, and quantified, in the water:

- Lowest Observed Adverse Effect Level (LOAL) data;
- Bioconcentration Factors (BCF) for each substance; and
- Estimates of cumulative body burdens;

In addition it was deemed prudent to obtain indications of any propensity towards:

- Carcinogenicity;
- Mutagenicity; and
- Birth Defects.

In the first instance literature searches were carried out by two sets of professionals, chemists and clinical toxicologists, at different locations using different resources. The information obtained by each search did not substantially improve on that produced by the other party. All of the organic substances concerned were located in the literature. However, it was found only 20% of the compounds had adequate toxicology information available. Table 1 summarises that information:

Substance	Mammalian Lethality			Teratogenic	Carcinogenic	Fatal Dose
	Acute LC_{50}	Chronic LC_{50}	Other	Indication	Indication	mg/kg
Isocyanato-cyclohexane	X	X	X			
N-butylbenzenesulphonamide	X	X		X		
Benzthiazole	X	X	X	X	X	
4-Methyl-1H-benzotriazole	X		X			
N-formyl morpholine	X	X	X	X	X	X
hexanoic acid	X	X	X			
2-(2-Butoxyethoxy)ethanol	X	X	X	X	X	X
Acetophenone	X	X	X	X	X	
1 H-benztriazol, 5-methyl	X				X	
1,3-Dicyclohexylurea					X	
Isothiocyanato-cyclohexane						
N-ethyl-2-benzothiazolamine						
1-[4-(1-methylethenyl)phenyl]-ethanone						
α, α, α, α-Tetramethyl-1,4-benzenedimethanol						
2(3H)-benzothiazolone					X	
1,1'-(1,3-phenylene)bis-ethanone						

Table 1. *Summary of literature derived information for substances identified in water.*

It can be seen from this summary table that no data relative to LOAL was identified readily in the literature. Some of the substances listed are thought to be present as contaminants of the major treatment component N-butylbenzenesulphonamide. In addition there were also references to toxicological responses not directly relating to oral consumption, these were often as a result of peritoneal injected challenges or dermatological evaluations. There was also a distinct lack of information on mutagenic activity. Using QSAR packages the following parameters were derived for all the substances concerned:

- BCF;
- LOAL (Rodent);
- LC_{50} (Rodent); and
- information on critical non-reversible effects, cancer, mutagenic and birth defects.

The data gaps were all filled using two established software packages. The toxicological components were determined using TOPKAT™ produced by the Accelrys Software Inc. BCF values were determined using BCFWIN™ produced for the United States Environmental Protection Agency (US EPA).

The BCF values determined using this software were subsequently applied to an in-house model designed to approximately simulate the cumulative body burdens for each substance. LOAL's (rodent) were determined to allow theoretical water quality limit values to be calculated for lifetime exposure.

The above combination of output was used by toxicologists to evaluate the predicted effects and exposure scenarios against the risks posed by the toxicology. It was concluded that there had been no exposure scenario that represented a risk to human health.

6.1 The Exposure Data in Context

It is often difficult for those who are not used to dealing with therapeutic or harmful exposures to chemicals from any source, to view in context numerical expressions of body burdens. In order to do this reference is made to some commonly applied pharmaceutical products available both over the counter and for use under medical supervision.

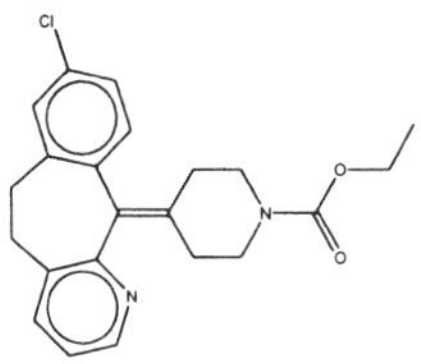

Loratadine is a commonly used over the counter antihistamine with a typical recommended dose of 10 mg/day.

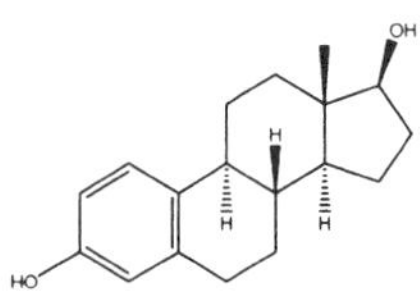

Oestradiol is commonly prescribed as a hormonal therapy and is effective at doses of 1 or 2 mg/day.

It can be seen therefore that intakes of fractions of milligrams per day could prove problematic in the short to medium term if bioconcentration becomes an issue, i.e. an organism cannot excrete or otherwise safely metabolise the entire uptake before the next ingestion occurs.

It is at this point that toxicologists must be engaged to place the results of modelled exposure scenarios into a medically significant context. However, it is relatively easy to see that if a substance is of physiological significance then chronic exposure risks need to be evaluated even after any major impact has been isolated or removed. A high BCF value is not necessarily indicative of a high risk factor. If exposure is low level and over long periods the overriding control factors are likely to be metabolism and excretion rates.

It is important, when considering the overall benefits of the technology described, to view the data acquisition in the context of the time taken to procure the information. The following illustrations place the time scales in context for the case described:

- the toxicology summaries for all the compounds was rationalised in a matter of days which incorporated cross referencing both literature searches;
- the modelling of the 'missing' information such as BCF's and LOAL's took three to four hours; and

- potential persistent body burdens were calculated within hours of the release of BCF data.

In reality the modelling can be completed before the literature search results, thus allowing a preliminary evaluation to be carried out. This has the advantage that the literature data, where available, can be used to supplement, or replace the modelled information, where appropriate.

7 CONCLUSIONS

While not being advocated as a panacea tool there are immense advantages in using the technology as a frontline source of information about compounds in an emergency, or as part of a post incident review. It is an important caveat that some chemical groups may not be amenable to modelling. However, sufficient information can be derived to more comprehensively inform the precautionary principle approach to exposure management.

The technology has the potential to provide a broad range of information reliably about a large range of substances which may be encountered in an emergency situation. In contrast there is a prospect that the alternative could be attempting to cope using a little knowledge about a few substances. This may be accompanied in some cases by having no information in the short term about a large number of substances. There is, therefore, scope to develop extensive qualified knowledge about the majority or all of the substances identified in the evaluation of an emergency incident?

It is very important not to allow over interpretation of the information produced and recognise the limitations of the tools deployed. No single company provides a suite of applications to cover all the determinations needed for every conceivable emergency situation. However, with the correct knowledge and evaluation a very powerful collection of tools can be assembled, and configured, to provide a vital, and powerful, information resource.

References

1 R. Todeschini and V. Consonni: "Handbook of Molecular Descriptors", WILEY-VCH, 2000.

2 CTFile Formats: MDL Information Systems, Inc. October 2003

3 D. Weininger, SMILES, a Chemical and Information System. 1. Introduction to Methodology and Encoding Rules. J. Chem. Inf. Comput. Sci. 28(1): 31-6. 1988.

RISK MANAGEMENT CAPABILITIES – TOWARDS 'MINDFULNESS' FOR THE INTERNATIONAL WATER UTILITY SECTOR

S.J.T. Pollard[1*], J.E. Strutt[1], B.H. MacGillivray[1], J.V. Sharp[1], S.E. Hrudey[2] and P.D. Hamilton[1]

[1]School of Industrial and Manufacturing Science, Cranfield University, Cranfield MK43 0AL, UK
[2]Department of Public Health Sciences, 10-102 Clinical Sciences Building, University of Alberta, Edmonton, Alberta, Canada T6G 2G3

ABSTRACT

Public health protection must be the primary goal of a drinking water utility; delivered through supplying safe drinking water. For complex multi-utilities, this goal may come under pressure from the need to manage a plethora of business risks. We describe a risk management maturity model for assessing the capacity of utilities to manage business risks and comment on the importance of 'mindfulness' as a prerequisite for effective risk management.

1 INTRODUCTION

1.1 A risk management imperative

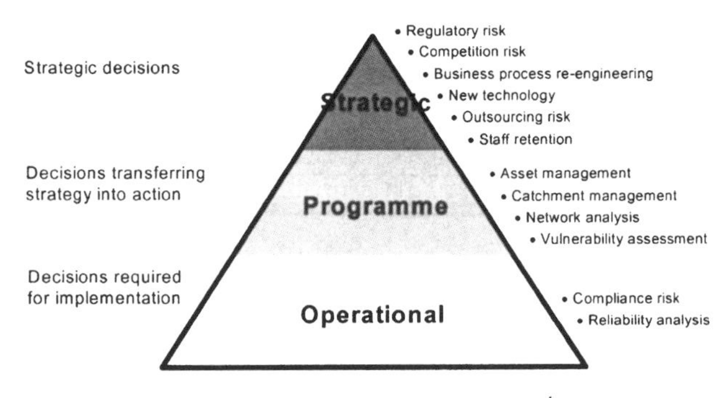

Figure 1 *The risk hierarchy applied to the water utility sector[1].*

*To whom correspondence may be addressed
(s.pollard@cranfield.ac.uk) S.J.T. Pollard; Tel: +44(0)1234 754 101; Fax: +44 (0)1234 751 671

From embedding corporate governance, through to the management of individual assets, the ability to understand, communicate, assess and manage risk has become a mainstream business activity. Many of the larger water utilities have begun integrating their responsibilities for financial control alongside their risk management programmes, including those that exist for asset management and regulatory compliance (Figure 1).

The water industry is witnessing a significant shift in the approach to risk management to one that is increasingly explicit and better integrated with other business processes. This is clearly, in part, a response to the asset management (financial and environmental regulation), public health (drinking water safety) and environmental protection (*e.g.* catchment management) agendas but may also represent a growing recognition that the provision of safe drinking water deserves to be treated as a 'high reliability' service within society and subject to the sectoral and organisational rigours and controls inherent to operations in the nuclear, offshore and aerospace industries. These sectors have learned important lessons and developed significant literatures on the implementation of safety cultures, much of which is transferable directly to the water sector as it progresses with the implementation of risk management

This paper deals with the application of maturity models within the international water utility sector. The research is relevant to the subject of water emergencies because it explores the preparedness, or resilience, of organisations to foresee, prevent, manage and withstand adverse risks. Our research has been conducted as part of a larger study for the American Water Works Association Research Foundation (AwwaRF) on risk analysis strategies for better and more credible utility decision-making[2]

1.2 Risk analysis in the water sector

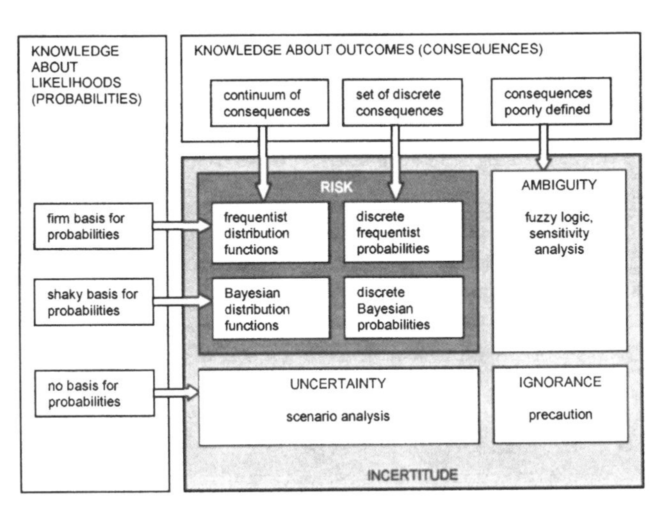

Figure 2 *Approaches to addressing uncertainty, risk and ignorance in decision-making (redrawn from Stirling[6])*

Notwithstanding the increasing application of risk assessment tools in the water utility sector[3, 4] there are many approaches to managing uncertainty in organisations[5] ranging from the use of standards, engineering judgement and good practice through to embedding company values and corporate cultures to safety and risk. Risk analysis plays an important function in decision-making where the probability of a hazard being realised is significant and uncertain and where the outcomes, or consequences are reasonably well understood (Figure 2).

The water sector has made excellent progress towards setting its goal of "providing wholesome, safe drinking water that has the trust of customers" within a risk management context, most recently through the preparation of revised WHO drinking water guidelines that promote a risk management approach[7]. There remain challenging implementation issues to be addressed, however. For example, following the types of incident reviewed by Hrudey and Hrudey[8] it is becoming accepted that risk analyses need to extend their reach beyond engineered systems and view management (system) and human (people) factors as equally central to effective risk management [9,10].

1 CAPABILITIES IN RISK MANAGEMENT

2.1 Risk assessment is not enough – developing capabilities in risk management

Risk assessments do not guarantee risk reduction. Left with their recommendations not implemented, they are a hollow gesture that may only serve to increase legal liability after failures occur. Managing risk competently, wisely and by targeting the risk critical elements of a system for maximum risk reduction is what counts. To understand the organisational competency in risk management, we must look to the *capability* a utility possesses in risk management. Because most water companies manage risk by virtue of the routine provision of safe drinking water, we are generally concerned then with the relative *maturity* of their capability[11] in risk management, rather than its presence or absence *per se*. Practically, we are concerned with their ability to act wisely and to anticipate when things might go wrong and act quickly in a preventative fashion.

A capability maturity model (CMM) is a management tool used to assess the degree of wisdom with which an organisation competently performs the key processes required to deliver a product or a service (Table 1).

N	maturity	mode / style	process characteristic and effect
5	Optimised	Adaptive, double loop learning	The organisation is 'best practice', capable of learning and adapting itself. It not only uses experience to correct any problems, but also to change the nature of the way it operates.
4	Managed	Quantified, single loop learning	The organisation can control what it does in the way of processes. It lays down requirements and ensures that these are met through feedback.
3	Define	Measured, open loop	The organisation can say what it does and how it goes about it but not necessarily act on its analyses
2	Repeatable	Prescriptive	The organisation can repeat what it has done before, but not necessarily define what it does.
1	Ad hoc	re-active	Characterises a learner organisation with complete processes that are not standardised and are largely uncontrolled
0	Incomplete	Violation	Incomplete processes, criminal or deliberate violation tendencies

Table 1 *Interpretation of maturity levels*

The degree of wisdom is represented by levels of maturity. Level 5 (high) organisations exhibit 'best practice'. They are capable of learning and adapting and they use experiences to correct problems and change the nature of the way they operate. Level 1 (low) organisations are learner organisations with non-standard and largely uncontrolled processes.

CMM has its roots in the field of performance measurements[12, 13] and quality management developed in the 1970s[14, 15]. The most widely referenced CMM is that developed by the US Software Engineering Institute to assess the software design capability of software houses[16]. One of the strengths of the CMM approach is its broad applicability and this is leading to increasing numbers of CMM models in other sectors[17, 18]. Capability maturity models can be used both as an *assessment* tool and as an *improvement* tool. Both approaches are used in practice.

Risk management (there are various paradigms; Figure 3 indicates one such approach for environmental risks) can be viewed as a 'service' most organisations undertake on behalf of their internal and external stakeholders to ensure business continuity and the delivery of corporate objectives is not adversely threatened.

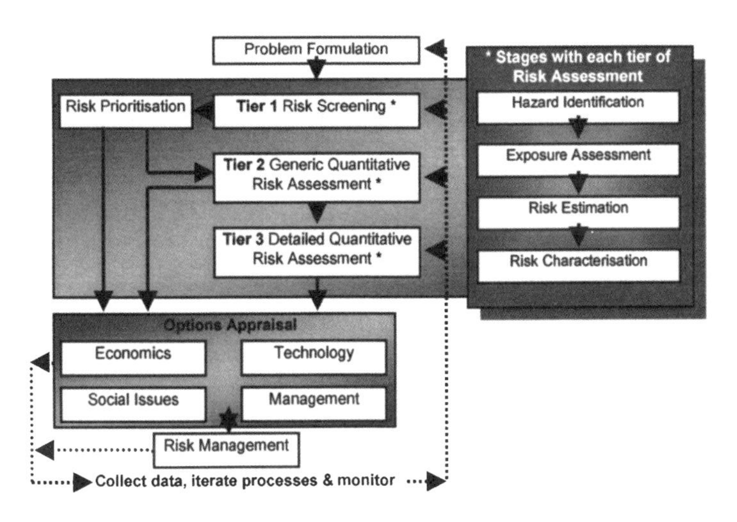

Figure 3 *Example risk management paradigm*[19]

Many water utilities seek to improve the processes involved with managing their risks and see this as an effective means of reducing exposure and improving risk-based decision making within their organisations. Understanding one's own risk management maturity has value in that it may (i) assist organisations in formalising their appetite for risk; (ii) help formalise and make more explicit the role of the group risk manager; and (iii) provide the opportunity for a 'climate' check on the implementation of risk management procedures on the ground within the organisation; thus acting as a check on corporate level statements on risk.

Our research at Cranfield University[20,21,22] has developed and piloted a risk management capability maturity model (RM-CMM) for the international water sector (Figure 4). There are 5 levels of capability that build on ideas from the theory of action and the concept of single and double loop learning[23]. Single loop learning occurs when risks are detected and the product or service is amended, thus permitting the organisation to carry on its present policies or achieve its present objectives. Double-loop learning occurs when risks are detected and managed in ways that involve the modification of an organisation's underlying norms, policies and objectives. Being able to manage the risks to your organisation extends beyond the ability to perform the risk analyses and options appraisal set out in Figure 3. There are key processes such as the ability to establish the organisational appetite for risk through setting risk acceptance criteria, and the ability to integrate risk management across business functions that are also important and reflect the wisdom (maturity) of approach (Figure 3).

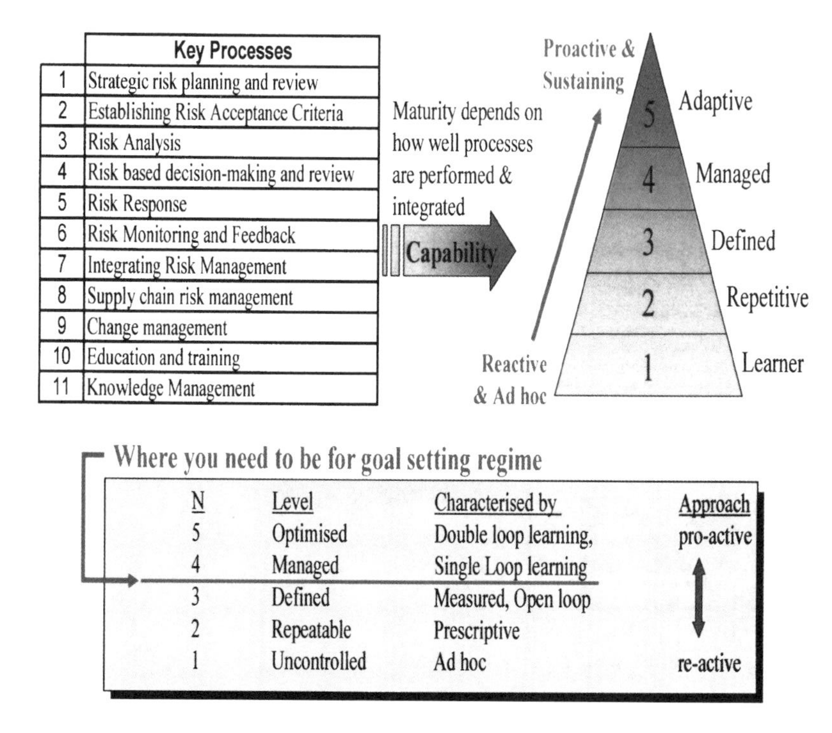

Figure 4 *Overview of the risk management CMM*[22]

The practical value of codifying this into an assessment model for water utility risk managers is in establishing the basis for risk improvement plans – structured and targeted action plans for better risk management within companies. The RM-CMM approach seeks to elicit where, on the ladder of improvement a utility wishes to be by reference to the importance of the risks it manages, and then to identify through critical analysis where the utility is on the ladder. An organisation's maturity in risk management can be schematically represented as snapshots of their current status[22]. The requirements to move between 'rungs of the ladder' (*i.e.* the levels of Table 1) provide the basis for risk management improvement plans within individual companies.

2.2 Applying the tool in practice

A preliminary RM-CMM framework has been developed and piloted within a number of water utility companies collaborating in the AwwaRF project for self assessment and review. The framework remains to be refined on the basis of initial responses from users. However, the initial responses offer insights into the risk management practices and cultures within the water utility sector. In summary, there is a growing capability, generally characterised by Level 3 of the RM-CMM. The pilot capability profiles returned by six organisations are summarised in Figure 5.

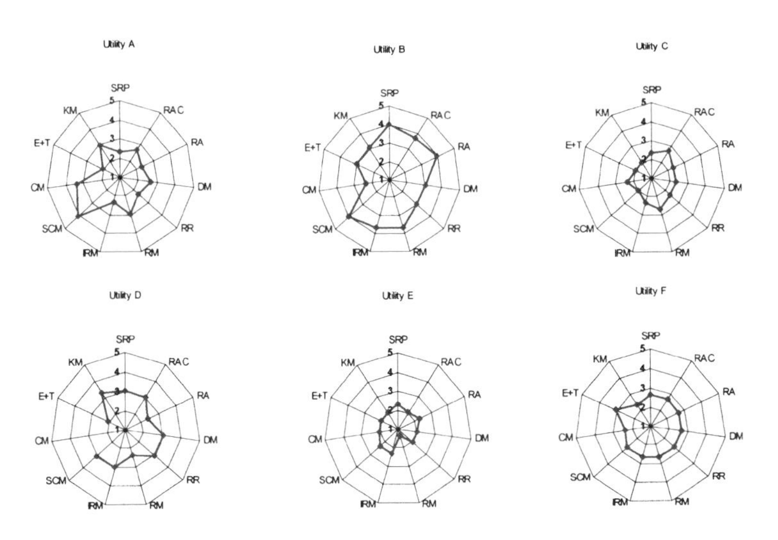

Figure 5. Capability profiles for 6 water utility companies[22]

Key for risk management processes:

SRP	Strategic risk planning and review of business goals
RAC	Establishing risk acceptance criteria
RA	Risk analysis
DM	Risk based decision-making and review
RR	Risk response
RM	Risk monitoring and feedback
IRM	Integrating risk management
SCM	Supply chain risk management
CM	Change management
E&T	Education and training
KM	Knowledge Management

Our initial observations indicate:

1. That most of the risk management processes reported are at a Level 3 or below, with most organisations at or becoming defined (Level 3) in terms of their risk management processes.
2. The transition to Level 4 capability is difficult because it implies that key management practices must influence decision making. This in turn implies that analysis is carried out early enough for it to influence decision making and that companies invest effort in effective response to risk. There appears to be widespread difficulty in transforming risk analysis and monitoring outputs into formats that inform decision makers.
3. The range of different methodologies used for risk analysis by water companies is limited, suggesting that there is a need for training in assessment methods and competency improvements.
4. Verification of risk management processes (the monitoring of residual risk) is often weak and validation activities often absent.
5. Risk monitoring is a core process but its practice appears surprisingly weak. One reason may be that organisations do not consider monitoring as a process in its own right but rather as validation of prior decisions (see 4 above, however).
6. The true integration of risk management with other core business processes appears some way off. It would appear that the interfaces are not fully understood, perhaps because of disciplinary barriers among technology, health and financial managers.
7. There appears to be only limited engagement of stakeholders in the risk management process, both internal (*e.g.* cross functional groups) and external (*e.g.* regulators).
8. An inability to measure and communicate the value (both tangible and intangible values) of risk management processes appears to have restricted corporate buy-in.
9. In many cases, organisational appetites for risk, in terms of tolerances, are weakly defined, often cited as a 'Board level issue'. Lack of risk acceptance criteria can adversely impact on risk response decision making.
10. An emphasis on individual business units 'owning' the processes has in some cases led to a proliferation of what are often disparate risk management initiatives, creating a barrier to integrated risk management.
11. Although senior and executive management have generally 'bought in' to the central role of risk management in utility operations, communicating its status throughout the company is proving somewhat difficult. 'Making it [risk management] part of the mindset' is often cited as the core function of the risk management team.

These are valuable findings. They suggest that most organisations have moved beyond the reactive state of Level 2 maturity for the range of risk management processes. However, the three core constraints they face in attaining full Level 3 status and beyond are:

- reaching the appropriate level of definition of their processes, *i.e.* identifying the tasks, activities, inputs and outputs of which they are comprised;
- the ability to enable these processes, *i.e.* establishing procedures for their initiation and providing the necessary resources;
- absence of a 'learning culture', which often proves to be the key constraint in proceeding beyond Level 4. On current performance this appears to be largely beyond the immediate control of corporate risk management teams.

2 DEVELOPING A RISK MANAGEMENT CULTURE

3.1 Getting the culture right – towards 'mindfulness' in the water utility sector

Much has been written on safety culture in the light of organisation accidents since the 1970s, much of it coincident with good risk management practice. Developing a risk management culture that is sustaining and continues to learn and improve in face of the inevitable peaks and troughs of organisation performance requires[24]: leadership, procedures, an appetite for conservative decision-making where safety it put first even under pressure; a culture of sharing reported close calls [*near misses have been described as inaccurate, more like "near hits", we use close calls later*]; good communication at the appropriate level, an open, learning organisational culture able to benchmark itself against the best-in-class, systematic competency checking, effective management of organisational change and the ability to prioritise.

Disasters and incidents have deeply-rooted causes[25, 9] that are often a combination of technical failures, an incapacity to manage change and of the underlying values within, or market forces acting on an organisation; as catalogued for waterborne outbreaks by Hrudey and Hrudey[8]. They are often a failure to convert hindsight into foresight and typically occur[24] when there is a loss of institutional foresight and corporate memory, in the face of strong market pressures for efficiency gains, when there are considerable elements of outsourcing, where organisations fail to maintain their status as an 'intelligent customer', with loss of internal technical expertise and particularly during, or following periods of business re-engineering. Cost pressures, priority-based working and changes that are rushed are all circumstances that can generate accidents. For organisations to become resilient and mindful, they must be able to anticipate and circumvent threats to corporate objectives and manage severe pressures and conflicts between performance and the risks that threaten it. Modern management culture sets a strong impetus on doing more for less ('lean') and on maintaining business continuity, and middle managers may find it difficult to challenge this philosophy in 'managing up' risk issues to the executive management or Board. When they do, risks that are not easily quantified in monetary terms may receive restricted air time at Board meetings and lie dormant within the organisation as latent causal factors[25]. It is clear from the prior art that leadership and management are key to establishing the right culture in terms of the expectations and example that are set, or not.

But how do organisations to develop a risk management culture without having first to suffer a major accident? How do we force ourselves to ensure risk issues are treated seriously? And how can we usefully process the volumes of risk information gathered by risk managers so as to make sense of it for accident/incident prevention? When should executive managers listen to the challenge from below? Above what threshold should they act? And are risk managers arguably an additional source of risk because, in taking institutional responsibility for coordinating risk assessment and management, they absolve others of their individual responsibilities for risk management? These are critical organisational questions germane to the organisation practice of risk management and requiring additional research. They remind us that managing risk requires wisdom and reflection, and that preventive approaches are creative and forward-looking. Best in class organisations are mindful about risks to their operations.

Weick and Sutcliffe[26] characterise 'mindfulness' in organisations that (i) are preoccupied with failure and the root causes of it; (ii) are reluctant to (over)simplify; (iii) are sensitive to operations; (iv) committed to resilience; and (v) are deferential to expertise. We propose that for water utilities seeking to develop mindfulness[27]:

- informed vigilance is actively promoted and rewarded;
- there exists an understanding of the entire system, its challenges and limitations is promoted and actively maintained;
- effective, real-time treatment process control, based on understanding critical capabilities and limitations of the technology, is the basic operating approach;
- fail-safe multi-barriers are actively identified and maintained at a level appropriate to the challenges facing the system;
- close calls are documented and used to train staff about how the system responded under stress and to identify what measures are needed to make such close calls less likely in future;
- operators, supervisors, lab personnel and management all understand that they are entrusted with protecting the public's health and are committed to honouring that responsibility above all else;
- operational personnel are afforded the status, training and remuneration commensurate with their responsibilities as guardians of the public's health;
- response capability and communication are improved, particularly as post 9-11 bioterrorism concerns are being addressed; and
- an overall continuous improvement, total quality management (TQM) mentality pervades the organisation.

3 CONCLUSIONS

Our conclusions are drawn from a series of extended interviews with risk managers within the utility sector and our initial piloting of the RM-CMM.

(1) Risk analysis is widely applied within asset management for assessing the likely condition, lifetime and projected management costs of asset maintenance and replacement. At the strategic level, however, risk and values are intimately linked and there appears to be little explicit expression of the risks utilities are prepared to accept and statements on those consequences that will not be allowed to occur. This is complex territory for many organisations and for water utilities depends, in part, on their private, public or corporatised legal status and corporate objectives.
(2) The promotion of water safety plans under revised World Health Organisation guidelines[6] is driving a more integrated approach to risk identification and analysis from catchment to tap[2]. Implementation will require sound and effective knowledge management.
(3) Along with many sectors, their remains a tendency to view risk assessment as an end in itself, rather than as the evidentiary analysis and input to a management tool for identifying company exposure and opportunities for innovation – the value of risk management requires greater advocacy in organisations.
(4) The provision of safe drinking water has not been historically viewed as a high reliability sector (as has aerospace and the nuclear sectors). This said, failure to adequately manage risk in a climate of 'efficiency gains' and 'optimisation', with risk analysis often being used to justify such actions, may leave the sector exposed to the types of organisational disasters we have witnessed with the fatal outbreak in

Walkerton, for example – attention to the *implementation* of risk management culture is required.

Acknowledgements - This work has been part-funded through an American Water Works Association Research Foundation (AwwaRF) award (RFP2939). BHM is co-funded through an EPSRC Doctoral Training Account award. The paper summarises themes addressed at the AwwaRF International Workshop "*Risk analysis strategies for better and more credible decision-making*", 6-8th April, 2005, Banff, Canada. The views expressed are those of the authors alone.

References

1 Prime Minister's Strategy Unit *Risk: improving Government's capability to handle risk and uncertainty*, The Strategy Unit, London, 2002, available at: http://www.number-10.gov.uk/SU/RISK/risk/home.html.
2 S.J.T. Pollard, S.E. Hrudey, L. Reekie and P.D. Hamilton (eds.) *Proc. AwwaRF International Workshop "Risk analysis strategies for better and more credible decision-making"*, Banff Centre, 6-8th April, 2005, Banff, Alberta, Canada, AwwaRF and Cranfield University, UK, 2005.
3 J. Colbourne, *The water safety plan approach – the Drinking Water Inspectorate viewpoint*. Presented at Risk assessment for drinking water safety, Chartered Institution for Water and Environmental Management Conference, 14th December 2004, Edinburgh, UK.
4 B.H. MacGillivray, P.D. Hamilton, J.E. Strutt and S.J.T. Pollard *Crit. Rev. Environ. Sci. Technol.*, 2005, *in press*
5 UK Offshore Oil Operators Association (UKOOA) *Industry guidelines on a framework for risk related decision support*, UKOOA, London, 1999.
6 A. Stirling (ed.) *On science and precaution in the management of technological risk*. European Commission Joint Research Centre publication 19056/EN/2, JRC Ispra, Italy, 2001.
7 WHO *Water Safety Plans, Chapter 4. WHO Guidelines for Drinking Water Quality* 3rd Edition. Geneva, World Health Organization, 2004, 54-88.
8 S.E. Hrudey and E.J. Hrudey, *Safe drinking water - lessons from recent outbreaks in affluent nations*, IWA Publishing, London, 2004
9 N.W. Hurst, *Risk assessment: the human dimension*, Royal Society of Chemistry, Cambridge, 1998.
10 S.J.T. Pollard, J.E. Strutt, B.H. MacGillivray, P.D. Hamilton and S.E. Hrudey *Trans. IChemE Part B: Process Saf. Environ. Protect.*, 2004, 82(B6): 453-462.
11 J.V. Sharp, J.E. Strutt, J. Busby and E. Terry *Proc. Intern. Conf. Offshore Mechanics and Arctic Engineering (OMAE)* 2, 2002, 383-390.
12 R.S. Kaplan and D.P. Norton, *The balanced scorecard: translating strategy into action*, Harvard Business School Press, Harvard, MA, 1996.
13 R. Phelps, *Smart business metrics*, FT Prentice Hall, London, 2004.
14 P.B. Crosby, *Quality is free*, McGraw-Hill, New York, 1979.
15 P.B. Crosby, *Quality is still free*, McGraw-Hill, New York, 1996
16 M.C. Paulk, M. Chrissis and C.V. Weber *IEEE Software*, 1993, 10(4), 18-27.
17 P. Fraser, J. Moultrie and M. Gregory, *The use of maturity models/grids as a tool in assessing product development capability*, IEEE International Engineering Management Conference, 2002.
18 ISO 9004 *Quality management systems - guidelines for performance improvement*, British Standards Institute, London, 2000.

19 DETR, Environment Agency and IEH, *Guidelines for environmental risk assessment and management, Revised Departmental guidance*, The Stationery Office, London, 2000
20 J.E. Strutt, J.V. Sharp, J. Busby, G. Yates, N. Tourle and G. Hughes *Proc. ERA Conference on Hazard Management of Offshore Installations,* London, 1998.
21 J.E. Strutt, J.V. Sharp, J. Busby, G. Yates, N. Tourle and G. Hughes *Proc. Offshore Europe Conference*, Aberdeen, 1999.
22 J.E. Strutt, B.H. MacGillivray, J.V. Sharp, S.J.T. Pollard and P.D. Hamilton, In: Pollard, S.J.T., Hrudey, S.E., Reekie, L. and Hamilton, P.D. (eds.) *Proc. AwwaRF International Workshop "Risk analysis strategies for better and more credible decision-making"*, Banff Centre, 6-8th April, 2005, Banff, Alberta, Canada, AwwaRF and Cranfield University, 2005.
23 C. Argyris and D. Schön, *Organizational learning: A theory of action perspective*, Reading, Addison Wesley, MA, 1978.
24 R. Taylor, Presented at *Achieving a good safety culture – the people dimension in health, safety and environmental performance*, Hazard Forum open meeting, 10th March 2005, Westminster, London, 2005.
25 J. Reason, *Managing the risks of organisational accidents*, Ashgate Publ., Brookield, VT, 1997.
26 K.E. Weick and K.M. Sutcliffe, *Managing the unexpected – assuring high performance in an age of complexity*, University of Michigan Business School, Josey-Bass Publ., San Francisco, CA, 2001
27 S.E. Hrudey, E.J. Hrudey and S.J.T. Pollard, *Environ. Int.*, 2005, *in press*

MASS SPECTROMETRY SCREENING TECHNIQUES

S.N. Cairns, D.P.A. Kilgour, J. Murrell, R. Parris, M. Rush, D.M Groves, and M.D. Brookes

Defence Science and Technology Laboratory, Building S18, Fort Halstead, Sevenoaks, Kent, TN14 7BP, UK

1 *INTRODUCTION*

Field detection and identification of chemicals at trace levels is a demanding application. In order to meet this requirement several types of laboratory instrumentation have been miniaturised and packaged for field use. This study examines some of the advances made in the use of portable mass spectrometers with novel field sampling techniques for the *in situ* screening of vapour and water samples.

1.1 Why Mass Spectrometry?

Mass spectrometry offers the advantages of sensitivity (detection of pg of analyte), specificity (a full fingerprint mass spectrum) and broadband capability (mass spectrometers can be configured to detect a vast array of compounds). However, mass spectrometers have traditionally been large laboratory-based instruments unsuitable for field portable applications.

Recent developments have generated several commercially available portable mass spectrometers that can be effectively used for field testing. Time-of-flight,[1] quadrupole[2] and ion trap instruments,[3] among others, have been successfully commercialised. This investigation explores the use of an ion trap instrument.

2 ION TRAP MASS SPECTROMETRY

In this type of mass spectrometer a sample is ionised and the ions captured by an alternating electric field in the 'ion trap'. The trap consists of three electrodes; two endcap electrodes and a ring electrode sandwiched between them. The original ion trap instruments had curved, hyperbolic electrodes in order to obtain the best possible trapping field. To aid miniaturisation this can be simplified to flat end caps and a cylindrical ring electrode. In this case the ion trap is known as a cylindrical ion trap (CIT), Figure 1.

To operate the ion trap a radio frequency (RF) field is applied to the ring electrode while the end caps are earthed. Ions inside the volume of the electrodes are trapped by being alternately attracted to then deflected from the electrodes as the voltage applied to the ring electrode alternates. The ions trapped in the volume can then be mass analysed by ramping up the applied RF amplitude to force the ions to become unstable in the axial direction one mass at a time. As they become unstable they are ejected through a hole in the end cap and impinge upon a detector. The time (related to the point on the RF ramp at which it was ejected) and amplitude of the detected ion signal is collected to form a mass spectrum.

Figure 1 *Image showing a cut away portion of the electrodes in a cylindrical ion trap analyser.*

The advantages of an ion trap mass spectrometer for producing field portable instruments are:

- It operates at relatively high pressures, which means pumping requirements are less demanding.
- It can be miniaturised. This gives a smaller volume and reduced power consumption.
- It has tandem mass spectrometry (MS/MS) capability.

In tandem mass spectrometry a full mass spectrum is recorded and then the ions producing a selected mass signal are isolated. These ions are then fragmented and the resulting product ions are analysed to produce a second mass spectrum. In this way two mass spectra are recorded, increasing the certainty of the identification. This is especially useful in a situation where a chemical background can interfere with identification, which is often the case with complex real samples. Normally these compounds would be separated using a technique such as gas chromatography (GC). Without this separation tandem mass spectrometry will often be needed in complex real samples.

It should be noted that using tandem mass spectrometry to screen for more than a few compounds at the same time could prove difficult, and a two tier screening process would be needed. In this scenario a full scan mass spectra could be taken and library searched. If a compound of interest is identified it could be confirmed by tandem mass spectrometry. If there is a very complex mixture then the instrument has a fast low thermal mass GC that can be attached. The GC can produce a full temperature ramped experiment in much reduced time. Controllable ramp rates of hundreds of degrees centigrade per minute are attainable.

3 EXPERIMENTAL

The instrument used for testing was the Griffin Analytical Technologies Minotaur 400 instrument, Figure 2. The instrument is cubic measuring approximately 46 cm on each side, weighs 27.2 kg and has a power consumption of 130 W, or 250 W with a miniature low thermal mass (LTM) GC attached. The instrument has unit resolution with a mass range of 400 Daltons. The mass analyser is a CIT (see Figure 1) and has a radius of approximately four millimetres.

Figure 2 ***Griffin Analytical Technologies Minotaur 400 mass spectrometer.***

3.1 Fibre Introduction Mass Spectrometry

For field deployment the sample must be collected and introduced into the mass spectrometer quickly and conveniently with the minimum of sample preparation. The method chosen for this study was fibre introduction mass spectrometry[4] (FIMS), alternatively named direct solid phase microextraction (SPME). Fibre introduction is essentially SPME sampling without any separation technique.

The SPME fibre holder (as purchased from Supleco®) is roughly pen shaped, approximately 20cm long. It contains, projecting from one end, a small polymer fibre, approximately 10 mm long and 0.1 mm diameter, in a metal tube. The holder can be used to expose the fibre in a headspace or placed directly into a solution. The polymer once exposed adsorbs the compounds in the medium it is in contact with. This is a selective process and depends on the composition of the polymer chosen for the fibre. The two fibres used in this work were coated in polydimethylsiloxane (PDMS) and divinyl benzene/ PDMS. Sampling the headspace above a liquid is shown in Figure 3.

Once the fibre has been loaded with sample it can be analysed immediately or, if a portable sampler is used, sealed with a GC septum and retained for subsequent analysis.

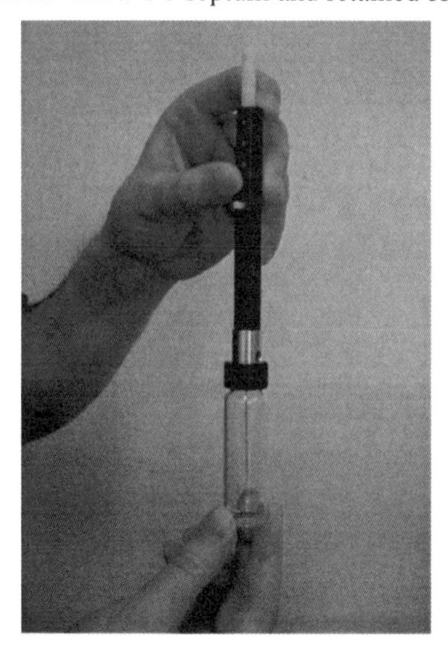

Figure 3 ***Picture showing sampling headspace with a field sampling SPME fibre.***

3.2 SPME Fibre Analysis

In conventional SPME analysis the fibre is placed into a gas or liquid chromatograph inlet and exposed to allow the compounds adsorbed on the fibre to be removed and placed on the column for separation. In FIMS the fibre is pushed through a GC septum directly into the vacuum manifold of the mass spectrometer. The drop in pressure from atmosphere outside the mass spectrometer to 10^{-5} Torr in the vacuum manifold leads to desorption of volatile compounds. For semi-volatile compounds the inlet can be heated to aid desorption. An ion trap mass analyser is well suited to this direct insertion method because it can accommodate the increase in pressure from bursting the GC septum and the sample volatilising into the chamber.

The analyte when desorbed fills the vacuum manifold and the ion trap volume. The sample molecules in the ion trap volume are ionised with an electron beam. Once ionised the sample is trapped and mass analysed.

The procedure for conducting a headspace sampling experiment was as follows:

1. The fibre was cleaned. This was conducted in advance by leaving the fibre exposed in a hot FIMS inlet or GC injector.
2. The SPME fibre was introduced into a vial containing the sample through a septum.
3. The fibre was exposed to the headspace for a set time depending of the volatility of the compound (approximately three seconds for volatile compounds).
4. The fibre was withdrawn and the SPME holder removed from the vial.
5. The SPME holder was taken to the mass spectrometer inlet and pushed through the GC septum immediately exposing it to the vacuum.

6. Mass spectra were recorded and the fibre removed from the inlet after approximately thirty seconds.

For the analysis of water samples the SPME fibre was immersed in the water for 20 minutes and continuously stirred before analysis. It should be noted that the fibres used are hydrophobic and so when the water solution is sampled a minimal amount of water is collected.

It is anticipated that FIMS could be employed for *in situ* screening tasks presently accomplished using SPME GC/MS such as post-blast screening of volatile explosives,[2] chemical warfare agent (CWA) decontamination verification[3] and water contamination assessment.

4 RESULTS

4.1 Headspace Analysis

An example of the results for the headspace analysis of 2-nitrotoluene sampled for two seconds is shown in Figure 4.

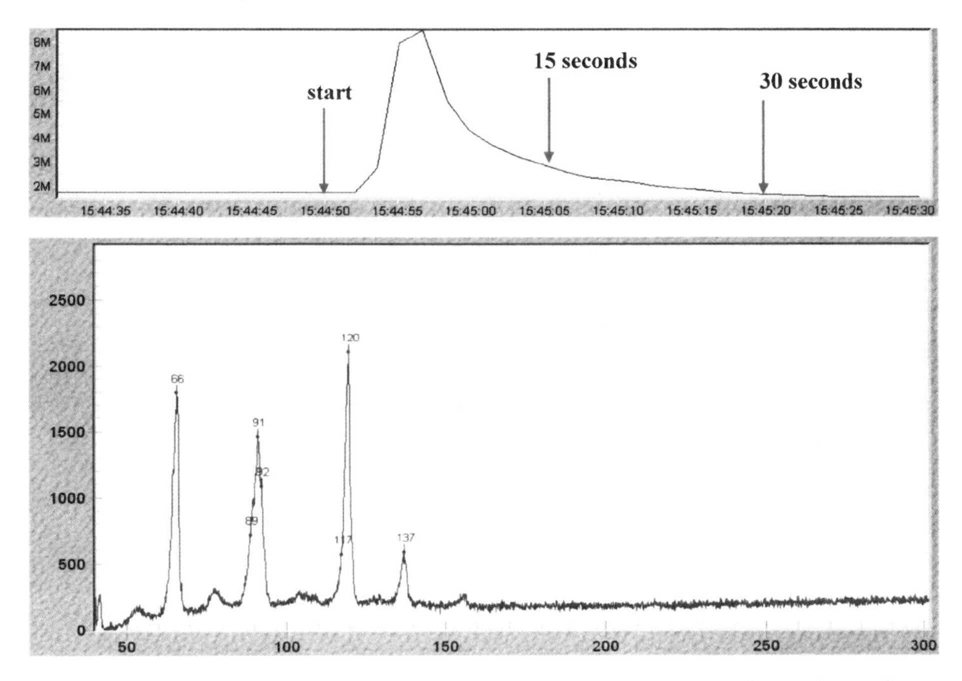

Figure 4 *Results of a 2-nitrotoluene FIMS experiment. The top trace shows the total ion current with time. The bottom trace is a mass spectrum recorded approximately 8 seconds after the SPME fibre was introduced into the vacuum. The resolution is rather poor because no helium bath gas was added during the experiment.*

It can be seen in Figure 4 that the sample analysis time was approximately 30 seconds. After this time the mass spectrometer total ion current is almost back to its original position. This could enable a sample to be analysed approximately every 30 to 40 seconds. This time scale for analysis of fibres, coupled with the use of multiple SPME fibres being used for sampling, would allow for a rapid high throughput screening process.

A wide range of compounds including CWA (chemical warfare agent) simulants, toxic industrial compounds (TIC) and explosives have been analysed. To find out how much of the analyte was being collected on the SPME fibre during an experiment a series of parallel SPME-GC/MS experiments were conducted using a calibrated GC/MS. In this way the amount of analyte on the SPME fibre in the FIMS experiment could be ascertained. The experiments were conducted using a small amount, approximately 0.5ml, of pure sample in a sealed glass vial. The results from these experiments are recorded in Table 1.

Compound	Mass (ng) on fibre after 3 second sample
Methyl Salicylate	3
Dimethyl methylphosphonate (DMMP)	11
Acetophenone	9
2-Nitrotoluene	3
Perflouro-1,3-dimethylcyclohexane	75
n-Butylbenzene	23
2,4-Dinitrotoluene (DNT)	3*

***10 second exposure**

All experiments conducted with a 100μm PDMS Supleco® SPME fibre.

Table 1 ***Results of FIMS and SPME-GC/MS comparison for a set of volatile and semi-volatile analyte for three second fibre exposure headspace samples.***

From these results it can calculated that in general approximately 1 ng of sample needs to be collected on the SPME fibre to obtain a clear mass spectrum. This result is not an absolute limit of detection for the experiment but a guide for field use.

4.2 Water Analysis

The detection of pesticides, CWA simulants and explosives in water has been accomplished. At this time definitive limits of detection for a series of analytes have not been ascertained. To this point solutions down to 0.001% $^{w}/_{v}$ (10 mg/l) have been tested. In should be noted that for some CWA simulants and pesticides the mass spectrum changes over time as the compounds hydrolyse in water. An example mass spectrum of the pesticide 1-naphyl-methyl carbamate (carbaryl) at a concentration of 0.02 % $^{w}/_{v}$ (200 mg/l) in water can be seen in Figure 5.

4.3 Tandem Mass Spectrometry

To demonstrate the use of the MS/MS experiment in conjunction with FIMS the ion trap was set up to collect a full mass spectrum and MS/MS pseudo-simultaneously. This was conducted by setting the mass spectrometer software to run a full mass spectrum and a tandem mass spectrometry experiment in one unit. This gives all the data as one mass spectrum with the full mass spectrum first and then the tandem data appended. For this method to be used conveniently in the field the software needs to be modified to correctly label the appended mass spectrum. An example of this experiment for headspace analysis of the methyl salicylate is given in Figure 6

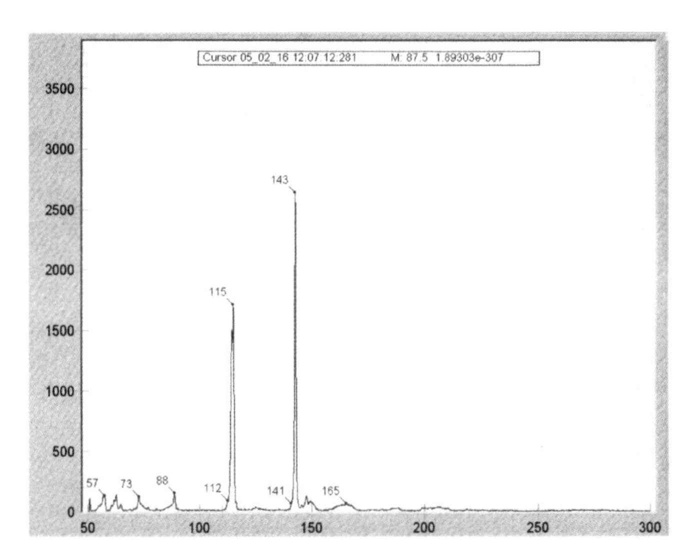

Figure 5 *The mass spectrum from a FIMS experiment of the pesticide 1-naphyl methyl carbamate (carbaryl) in water at a concentration of 0.02 %$^{w}/_{v}$ in water.*

5 SUMMARY AND CONCLUSIONS

The technique of FIMS has been demonstrated for use with a variety of compounds including CWA simulants, explosives, TICs and pesticides. The samples for the technique can be taken from the headspace or directly from a solution. The use of one mass spectrometer with a series of SPME fibres used for sampling would allow the rapid *in situ* screening in a variety of scenarios including post-blast screening of volatile explosives, CWA decontamination verification and water contamination.

Where the technique may not be able to give high enough certainty for identification of compounds in complex matrices a two stage strategy should be adopted of i) FIMS for initial screen and ii) confirmation with tandem mass spectrometry or GC/MS.

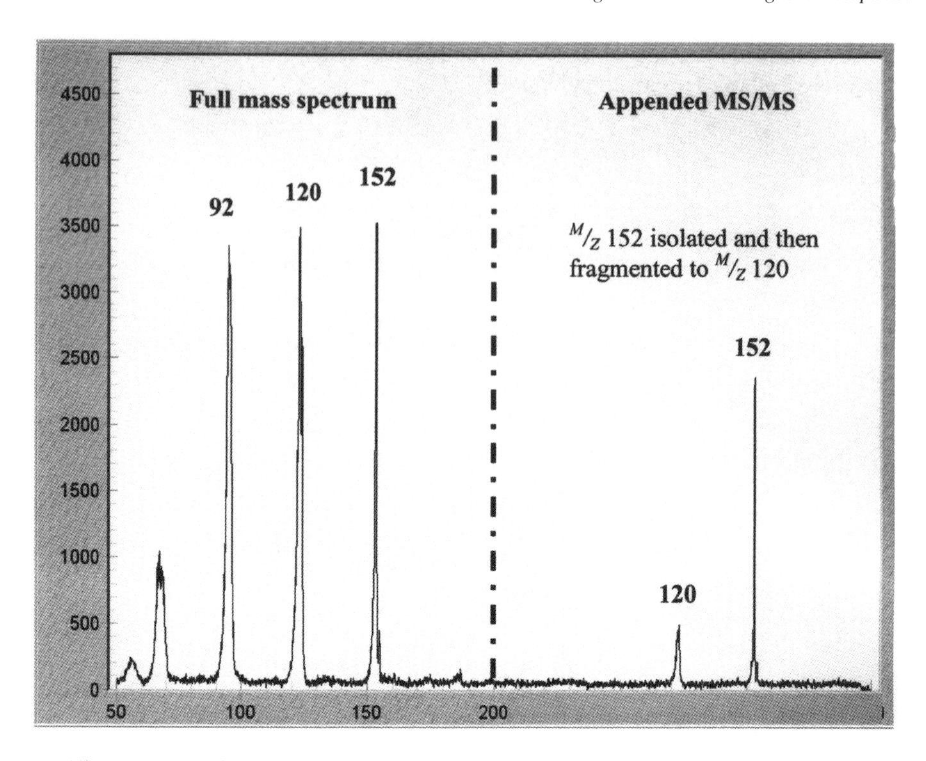

Figure 6 *Pseudo-simultaneous mass spectrometry and tandem mass spectrometry of methyl salicylate.*

References

1 www.Kore.co.uk
2 T.A. Spaeder and R.B. Walton, Abstracts of papers of the American Chemical Society 226:U128-U128 172-ANYL Part 1, Sep 2003.
3 G.E. Patterson, A.J. Guymon, L.S. Riter, M. Everly, J. Griep-Raming, B.C. Laughlin, Z. Ouyang, and R.G. Cooks, *Anal. Chem.* 2002, **74**, 6145.
4 E.C. Meurer, D.M. Tomazela; R.C. Augusto; and M.N. Eberlin, "Fiber Introduction Mass Spectrometry: Fully Direct Coupling of Solid-Phase Microextraction with Mass Spectrometry," *Anal. Chem.* 2002, **74**, 5688.
5 T. Tamiri, S. Abramovich-Bar, D. Sonenfeld, A. Levy, D. Muller, and S. Zitrin, "The Post Explosion Analysis of Triaetonetriperoxide". The 6th ISADE, 1998, Prague, Czech Republic.
6 P.A. Smith, C.J. Lepage, D. Kock, H. Wyatt, B.A. Eckenrode, G.L. Hook, and G. Betsinger, "Detection of Gas Phase Chemical Warfare Agents using Field Portable Gas Chromatography-Mass Spectrometry Systems: Instrument and Sampling Strategy Considerations". The 4th Harsh-Environment Mass Spectrometry Workshop, 2003, Florida, USA.

THE UTILIZATION ON-LINE OF COMMON PARAMATER MONITORING AS A SURVEILLANCE TOOL FOR ENHANCING WATER SECURITY.

D. Kroll and K. King

Hach Homeland Security Technologies, 5600 Lindbergh Drive, Loveland, Colorado USA 80539, phone 970-663-1377, email: DKROLL@hach.com

1 INTRODUCTION

Drinking water supplies are an integral component of society's infrastructure that has been recognized as being extremely vulnerable to deliberate attack. The recognition of this vulnerability is not new to the post 9/11 world. In 1941 FBI Director J. Edgar Hoover wrote, "Among public utilities, water supply facilities offer a particularly vulnerable point of attack to the foreign agent, due to the strategic position they occupy in keeping the wheels of industry turning and in preserving the health and morale of the American populace. Obviously, it is essential that our water supplies be afforded the utmost protection." [1] There is a long history of water being vulnerable to such attacks.

Terrorist groups in several countries have shown continued interest in using a Chemical, Biological, or Radiological (CBR) agent in their attacks and Islamic terrorist groups have also exhibited interest in water supply systems. While the threat from Islamic terrorists is dire, it is not the only threat. Domestic terrorists and disgruntled employees may represent a scenario just as serious and, possibly, more likely.

2 HOW COULD AN ATTACK OCCUR?

When observing a typical municipal water supply system (Figure 1) it is natural to assume that the main point of vulnerability to a CBR attack would be the introduction of an agent into the system at the source water (reservoir) or treatment plant. However; in order to create widespread casualties from an attack on the source water, the amount of contaminant required would, after taking dilution into account, be either too large to handle easily or be more expensive than other readily available terrorist weapons. Within the water industry, this concept is summarized by the phrase *dilution is the solution to pollution.*

After the attacks of 9/11, government experts publicly declared that, due to this dilution factor, our water supply systems were secure. The concept of dilution providing security for the system did not prevail. It wasn't long before government officials and industry experts realized that the crucial vulnerability to contamination was is the distribution system. By October 2003 a GAO report to the US Senate stated that the distribution system was the area most vulnerable to attack.[2] Conceding that an attack with CBR agents would most likely take place somewhere in the distribution system, several misconceptions about this type of attack still persist. Historic (and incorrect) dogma holds that such attacks require the assistance of several technicians, are expensive to carry out, and require complicated and expensive pumping equipment to inject contaminants into a pressurized system. More recent studies by the US Army Corps of Engineers and Hach Homeland Security Technologies personnel show that CBR attacks could in fact be carried out for 50 cents or less per lethal dose, that a single individual can obtain or produce effective contaminants in quantity, and that contaminants can be introduced into the distribution system with the aid of inexpensive and easy to obtain pumping equipment via a method called *backflow attack.*[3,4,5,6]

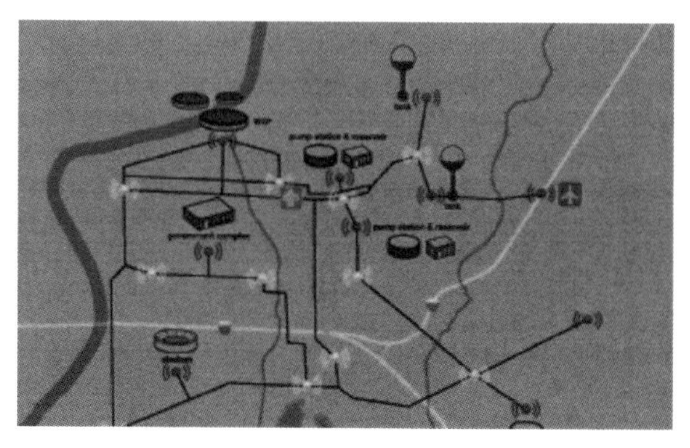

Figure 1 *A representative municipal water supply system.*

A backflow attack occurs when a pump is used to overcome the pressure gradient that is present in the distribution system's pipes. This is usually around 80 lbs/in^2 and can be easily achieved by using pumps available for rent or purchase at most home improvement stores. After the pressure has been overcome and a contaminant introduced, Bernoulli effects pull the contaminant into the flowing system and the normal movement of water in the system acts to disseminate the contaminant throughout the network effecting areas surrounding the introduction point. The introduction point can be anywhere in the system such as a fire hydrant, commercial building or residence. Studies conducted by the US Air Force and Colorado State University have shown this to be a very effective means of contaminating a system.[7] A few gallons of highly toxic material was enough, if injected at a strategic location via continuous feed, to contaminate an entire system supplying a population of 150,000 people in a matter of a few hours. A terrorist could launch such an attack and be on a plane out of the country before the first casualties begin to show up.

All distribution systems are vulnerable to backflow attacks. Currently, monitoring of drinking water supplies in the distribution system is limited. Previous to the terrorist threat, it was not a priority. The ability to detect an event in the distribution system and then identify it would be of incomparable value in responding to an incident in a timely and proper manner. Such an ability would also serve the purpose of mapping a system for clean up, and after words, it could be used as a forensic tool to identify the source of an event. Prior to this, there has not been a device capable of detecting such an event and alerting the system's managers so that the effects of an attack or accidental event can be contained. The general scientific consensus is that no practical, available, or cost-effective real-time technology exists to detect and mitigate intentional attacks or accidental incursions in drinking water distribution systems. The development of such a monitoring system was listed as by a panel of experts and industry leaders as a top priority in enhancing water security.[8]

2 WHAT SHOULD A MONITOIRING SYSTEM DETECT?

One of the problems when designing such a monitoring system for water is the vast number of chemical agents that could be utilized by a terrorist to compromise a water supply system tends to preclude monitoring on an individual chemical basis. Chemical warfare agents such as VX, Sarin, Soman, etc.; commercially available herbicides, pesticides and rodenticides; street drugs such as LSD and heroin; heavy metals; radionuclides; cyanide and a host of other industrial chemicals could be exploited as weapons. There are also a variety of biological agents and biotoxins that could be used. Which of the myriad possible agents would be the most likely to be deployed in a terrorist assault is sill a matter of conjecture.[9] To be truly effective a monitoring device needs to be able to detect any and all of the possible agents. A dedicated device capable of detecting anthrax for instance is interesting but not very practical, as it could be thwarted by the terrorist's use of another agent. This need to detect such a wide variety of diverse contaminants requires a realignment of thinking from the traditional development of a sensor for a given compound or agent.

With the goal in mind of creating a system that could be rapidly deployed in a cost-effective manner and was both robust in operation and diverse in its ability to detect contaminants, it was decided to investigate the possibility of using a variety of off-the-shelf sensors that were well characterized and proven to be robust for field deployment in a multi-parameter array. The sensors chosen for investigation were pH, Conductivity, Chlorine Residual, Turbidity, and Total Organic Carbon (TOC). Data was collected for Oxidation Reduction Potential (ORP), but it was not used in the final system because the probes are unstable and prone to fouling in long term installations.

3 THE PROBLEM WITH REAL WORLD DATA

Very little experience in the collection of real world data streams for multiple parameters existed at the onset of this project. To date over 64,000 hours of real time data from multiple sites across the United States has been collected during the course of this study. Real world baseline data is not always as neat and tidy as a laboratory system. Significant fluctuations can occur on a regular basis in real worlds systems. The problem then becomes, can we differentiate between the changes that are seen as a result of the introduction of a contaminant and those that are a result of everyday system perturbation? The secret to success, in a situation such as this, is to have a robust and workable baseline estimator. Extracting the deviation signals in the presence of noise is absolutely necessary for good sensitivity. Several methods of baseline estimation were investigated. Finally, a proprietary, patent pending, non-classical method was derived and found to be effective.

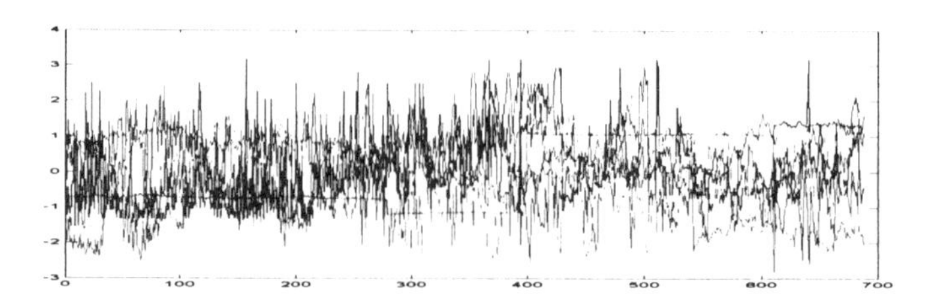

Figure 2 *Real World Auto-scaled Data.*

4 HOW THE SYSTEM WORKS TO TRIGGER ON AN EVENT.

In the system as it is designed, the signals from all of the instruments are processed from a 5-space vector to a scalar trigger signal in an event monitor (computer). The signal then goes through the proprietary baseline estimator. A deviation of the signal from the estimated baseline is then derived; a gain matrix is applied, and then compared to a threshold level. If the signal exceeds the threshold the trigger is activated.

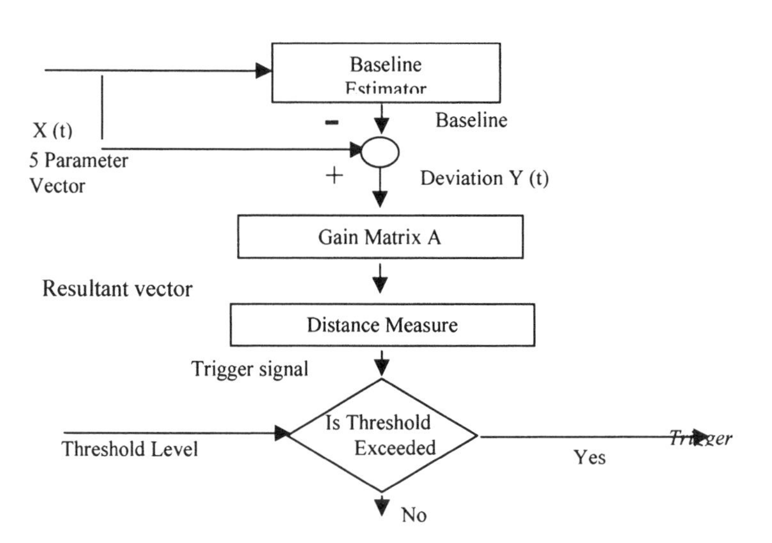

Figure 3 *Trigger Algorithm Flow Chart*

5 TESTING THE TRIGGER SYSTEM

Real world data was obtained from several sites and the most noisy data stream for each parameter was selected and used to test the system. Even with extremely noisy data the system does not trigger at a threshold level set at 1. Therefore; during normal operation, with no agent present, the process deviation should not be large enough to produce a Trigger Signal > 1. However when the data for a cyanide incursion at 1% of the LD-50 or approximately 2.8 mg/L is superimposed on the system the trigger level of 1 is easily exceeded. Other contaminants exhibit similar results.

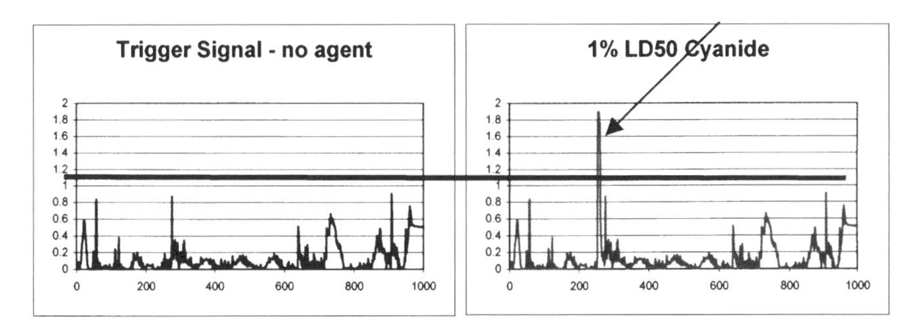

Figure 4 *A 1% LD-50 injection of Cyanide clearly exceeds the alarm limit.*

6 TESTING THE TRIGGER SYSTEM. FLOWING SYSTEM LABORATORY TESTING.

To determine if the system was capable of triggering on an actual event, a system was designed in the laboratory to allow the injection of an agent into the water stream as it flowed past the sensors. Do to safety concerns, the number and types of agents that could be evaluated on this system were limited. Cyanide, aflatoxin, sodium fluoroacetate, and nicotine were tested. The results for nicotine at 7.5% of the LD-50 are shown below. The system is actively triggering when the line is black or light gray. It is not triggering when gray.

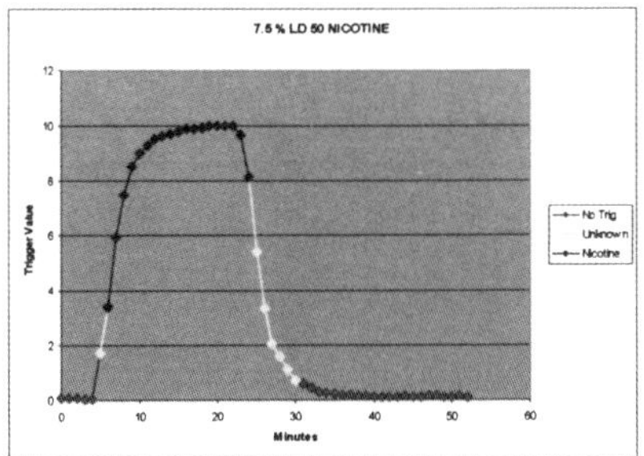

Figure 5 *Results for 7.5 % LD-50 Nicotine.*

The system is quite sensitive and has the ability to trigger at very low concentrations when exposed to the compounds of interest. The table below lists several compounds and the MDL that the system is capable of triggering on expressed as a percent of the LD-50. The LD-50 is defined as the amount of the compound that would kill 50% of the population of adult males with a weight of 70 kg after consuming 1 liter of water. The trigger level for these and most other compounds are at a level that there is not likely to be any acute toxicity exhibited.

AGENT	TRIGGER
Aflatoxin	0.37
Aldicarb	0.66
Cyanide	0.5
Nicotine	0.8
Oxamyl	2.5
Sodium Fluoroacetate	1
Strychnine	0.7

Table 1 **MDLs of Selected Agents as a % of LD-50**

The false alarm rate when the system is tracking real world data is also quite low. The system is also equipped with a learning algorithm, so that as unknown alarm events occur over time the system has the ability to store the signature that is generated during the event. The operator can then go into the program and identify that function and associate it with a known cause such as the turning on of a pump or the switching of water sources, etc. Then the next time that event occurs it will be recognized and identified appropriately. Over time as the system learns the probability of an unknown alarm that has not been previously encountered and identified will continue to decrease and will eventually approach zero. One point however that should be noted, is that as soon as the system is turned on, it will be actively working and will have the ability to trigger immediately if the signature of a known threat agent is encountered.

The probability of an unknown alarm due to a given event depends upon the frequency of the occurrence of such an event and the time that the algorithm has had to learn that event. Events that occur frequently will be quickly learned while rare or singular events will take longer to be learned and stored. This should result in a fairly rapid drop off in the number of false alarms as common events are quickly learned as is demonstrated by the real world scenario represented in the graph below.

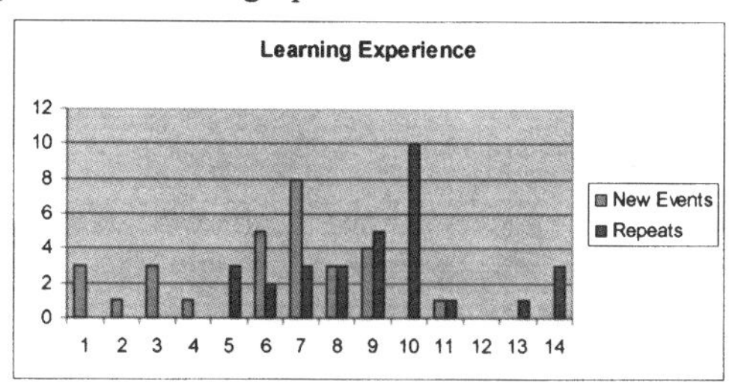

Fig. 6 *There were 26 unique events over 11 days of operation. All were learned, 16 of them were repeated*

Will the system recognize a new agent from the plant library once it has been included? In another experiment a new simulant was run past the sensors. In the initial case, the software does not recognize the agent, but records its signature in the Plant Library. After the initial signal has been recorded the simulant is run past the sensors again. In this case the simulant was a mixture of Potassium Acid Phthalate (KHP) and Formazin (turbidity standard). After the initial run, the system remembered the signature of the unknown and it was stored in the plant library and identified appropriately. At this point the agent was fed into the system again at different time durations ranging from 30 seconds up to 12 minutes and also, at various concentrations from a 5:1 ratio. The agent was identified in each case. This demonstrates not only the abilities of the system to identify but also to learn and remember.

7 DEVIATION/CLASSIFICATION ALGORITHM.

The deviation vector that is derived from the trigger algorithm contains significantly more data than what is needed to simply trigger the system. The deviation Vector's Magnitude relates to concentration and Trigger Signal, while the Deviation Vector Direction relates to the agent characteristics. Seeing that this is the case, Laboratory Agent Data can be used to build a Threat Agent Library of Deviation Vectors. A Deviation Vector from the water monitor can be compared to Agent Vectors in the Threat Agent Library to see if there is a match within a tolerance. This system can be used to identify what agent is present. Each vector results in a vector angle in n-space that, from the research conducted so far, appears to be unique. The graph below is a radar plot representation of some agent data that visually illustrates this point.

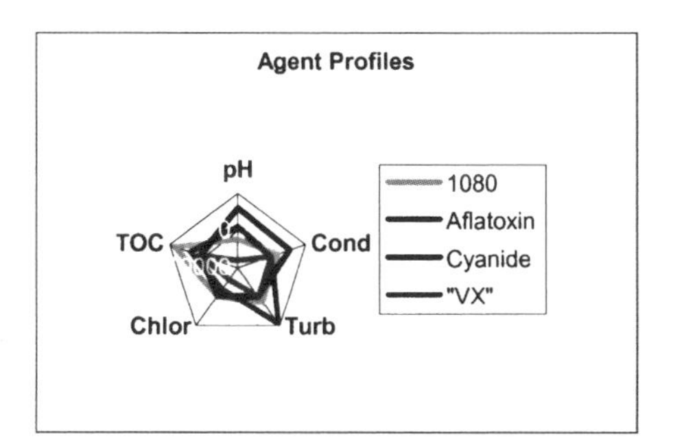

Figure 7 *Radar Plot of Agent Deviations*

The fact that the direction of the vector is unique for a given agent allows the use of an algorithm to identify the cause of a trigger being set off.

When the event trigger is set off the library search begins. The agent library is given priority and is searched first. If a match is made the agent is identified. If no match is found the plant library is then searched and the event is identified if it matches one of the vectors in the plant library. If no match is found the data is saved and the operator can enter an ID when one is determined. The agent library is provided with the system, and the plant library is learned onsite. If an attack occurs somewhere else in the country, the vector is saved and can be down loaded to any other plant that has the system. The agent library is also updateable as new profiles are generated. With over 80 agents currently in the library and over 64, 000 hours of real time data compiled and analyzed it has been found the plant events form vectors that are significantly different from those formed by the presence of agents.

Building the threat agent library is not trivial and requires extensive lab work. First, it is necessary to run laboratory experiments in several different matrices to derive instrument response to the agents. Then, it is necessary to mathematically examine the magnitude of response and unit vector structure to see if it will work in the method. The next step is to simulate addition of the agent to real water data at different dose levels. This helps to determine the ability to trigger on, and detect/classify the agent. At this point it is possible to add the agent to the library.

8 TESTING THE COMBINED DETECTION/IDENTIFICATION SYSTEM.

Multiple methods were used for testing the combined system. Spreadsheet models for agent addition were developed and tested. Monte Carlo simulation was done for noise analysis. Finally the system was verified with lab tests using Cyanide, Sodium Fluoroacetate (1080), Nicotine, and Aflatoxin. The experiments showed that noise has some effect on the ability to identify an agent, but identification still occurred well below the LD-50 level. The acceptance angle placed on the unit vector as to whether or not it is a match also plays a role in the ability to identify a compound. The higher the concentration of a compound the lower the acceptance angle can be. A reasonable acceptance angle appears to be around 10 degrees. The rate of identification for a given compound is therefore a function of the dose of that compound found in the water. At an acceptance angle of 10 degrees, there is a good ability to identify at reasonable dosage levels and there seems to be little or no overlap of the vectors for the agents that have been evaluated to date. The closest angles found in the data evaluated up until this point has been about 16 degrees. Therefore, an acceptance angle of 10 degrees should allow for adequate detection at the levels of interest and still prevent the occurrence of matching a single event to multiple vectors. In all cases, the system was able to correctly identify the compound in question at very low concentrations.

The system has the ability to both trigger and identify at very low levels. The table below lists the % of LD-50 for various compounds at which the system triggers and at which it is capable of making an identification.

AGENT	TRIGGER	IDENTIFY
Aflatoxin	0.37	1
Aldicarb	0.66	0.7
Cyanide	0.5	0.15
Nicotine	0.8	3.3
Oxamyl	2.5	2.6
Sodium Fluoroacetate	1	4.8
Strychnine	0.7	1.5

Table 2. *MDLs for Selected Agents as a Percentage of LD-50*

After the direction of a deviation vector has been used to determine the identity of an agent there is still more useful information that can be derived from the vector. After identification, which is done with vector direction only and is not related to magnitude, the vectors magnitude can be reinstated. Deviation vector magnitude relates to concentration. When an agent has been detected, the approximate concentration can be calculated from the deviations of TOC, Conductivity and Turbidity, which relate to chemical concentration. The determination of concentration of bio-agents is not possible with this system because it does not do cell counts, or differentiate between viable and non-viable microorganisms

Quantification can be very important when determining response steps. The concentration of a given agent may determine treatment of casualties by first responders and hospitals. A quick estimate for these purposes can make the difference between effective and ineffective treatment. Also, the cleaning of an area and remediation of the system can be streamlined when identity and concentration of the contaminant are known

9 CONCLUSIONS:

The system described has made use of robust off-the-shelf sensor technologies by placing them together in an array and using intelligent algorithms in a new and powerful manner to extract data that is of interest in devising an early warning system for water security. The system has been shown to be effective versus a wide variety of threat agents in a laboratory setting. The use of a unique system for estimating the baseline in real world systems allows for the identification of small deviations from normal readings in water analysis parameters. This in turn leads to a system capable of triggering on these deviations.

Once the system has been triggered, the algorithms have been shown to be capable of utilizing the unique profile represented by a threat agent's deviations to identify that threat agent. Laboratory procedures on over 80 agents to date have shown no significant overlap of profiles. As the database grows, there may be some overlap in the future, but it is likely that day-to-day plant events will not intrude on the vector space occupied by agents of concern.

The system also has the capability to learn day-to-day deviations that are unique to a given system. Events that occur commonly will be rapidly learned, and the rate of false positives to the trigger mode will rapidly decrease. The systems ability to identify threat agents is not affected by learning and if a system is compromised by a threat agent in the library the system should alarm and identify that threat from the first day of deployment. Over all, the system is an invaluable security tool for recognizing system incursions, but it has the ability to become much more than that. Hopefully most systems will never be in need of the security aspect of this system, but there are other dimensions to the system that should find use in any location. To date, most of the library work has been done on threat agents, but as time allows, the libraries will be expanded to incorporate common distribution system problems that may arise. Also, as an individual plant library recognizes deviations and the operators are able to identify them, the system will become a useful tool in evaluating the day-to-day system health and operational parameters.

10 FUTURE WORK:

The work of adding vectors to the library continues. There are some agents, which cannot be evaluated at the Hach laboratories due to hazards or regulations regarding their use. There is currently a program being undertaken at the Government's DOD Aberdeen/Edgewood facility in Maryland to generate profiles of agents that cannot be done in house at Hach including agents such as VX, GD, GB, Ricin and Anthrax. Loop testing of these agents will also be done at the facility. Plans are also underway to evaluate the systems ability to detect and identify agents on a large size test loop (4000 gallons) at the Aberdeen facility in conjunction with the Corp of Engineers Research Laboratory. These tests will most likely be comprised of surrogates rather than actual agents due to the generation of such large quantities of waste. At the time of the preparation of this manuscript (March 2005) an Environmental Technology Verification study (ETV) is being conducted by Battelle under guidance of the EPA at the EPA's loop test facility in Cincinnati, Ohio. This study will determine efficacy of long-term deployment of such a system and will evaluate the systems ability to trigger on and classify a variety of contaminants. This study should be complete in late March of 2005.

Beta site testing is currently underway at a variety of sites to verify operation, reliability and learning algorithms at real world locations. Over 64,000 hours of beta site data has been collected to date. The system is upgradeable as new sensors become available. There is active investigation of new online sensors that could be added to the system to update its capabilities as well as adding more specific sensors for bio-agent detection.

REFERENCES:

1 J.E. Hoover, Water Supply Facilities and National Defense. 1941. *Jour. AWWA,* 33:11:1861

2 *GAO-04-29* "Drinking Water Security: Experts' views on how future federal funding can best be spent to improve security"; October 2003.

3 D. Kroll, confidential paper, "Mass Casualties on a Budget", 2003, Hach HST

4 Army Corps of Engineers, Calculations on threat agents and requirements and logistics for mounting a successful backflow attack.

5 ASHRAE Satellite Broadcast: Homeland Security for Buildings; 14 April 2004.

6 V. F. Hock, S. Cooper, V. Van Blaricum, J. Kleinschmidt, M.D. Ginsberg, and E. Lory., Waterborne CBR Agent Building Protection, Proceedings of the National Association of Corrosion Engineers Exposition, 2003.

7 T.P. Allman, Drinking Water Distribution System Modeling for Predicting the Impact and Detection of Intentional Contamination. Masters Thesis. Summer 2003. Dept. of Civil Engineering. Colorado State University. Fort Collins, Colorado.

8 Office of Science and Technology Policy, The White House, "The National Strategy for the Physical Protection of Critical Infrastructures and Assets," February 2003.

9 D. Kroll, Utilization of a New Toxicity Testing System as a Drinking Water Surveillance tool, *Water Quality in the Distribution System,* Edited by W.C. Lauer, AWWA Press, 2004.

RISK ASSESSMENT METHODOLOGY FOR WATER UTILITIES (RAM-W™) – THE FOUNDATION FOR EMERGENCY RESPONSE PLANNING

J. J. Danneels

Materials Transportation Security/Risk Assessment, Sandia National Laboratories
P.O. Box 5800, Albuquerque, New Mexico 87185-0719, USA

1 INTRODUCTION

Concerns about acts of terrorism against critical infrastructures have been on the rise for several years. Critical infrastructures are those physical structures and information systems (including cyber) essential to the minimum operations of the economy and government. In the USA, The President's Commission on Critical Infrastructure Protection (PCCIP), established in 1996 by Executive Order 13010, probed the security of the nation's critical infrastructures. The PCCIP determined the water infrastructure is highly vulnerable to a range of potential attacks.

In October 1997, the PCCIP proposed a public/private partnership between the federal government and private industry to improve the protection of the nation's critical infrastructures. The water supply system was designated a critical infrastructure under the May 1998 Presidential Decision Directive 63, a National Security Council directive. The responsibility for the security of the water infrastructure was assigned to the Environmental Protection Agency (EPA).

In early 2000, the EPA partnered with the Awwa Research Foundation (AwwaRF) and Sandia National Laboratories to create the Risk Assessment Methodology for Water Utilities[1] (RAM-W™)[2]. Soon thereafter, they initiated an effort to create a template and minimum requirements for water utility Emergency Response Plans (ERP). All public water utilities in the U.S. serving populations greater than 3,300 are required to undertake both a vulnerability assessment and the development of an emergency response plan by Title IV of the Public Health Security and Bioterrorism Preparedness and Response Act, Public Law 107-188.

This paper explains the initial steps of RAM-W™ and then demonstrates how the security risk assessment is fundamental to the ERP. During the development of RAM-W™, Sandia performed several security risk assessments at large metropolitan water utilities. As part of the scope of that effort, ERPs at each utility were reviewed to determine how well they addressed significant vulnerabilities uncovered during the risk assessment. The ERP will contain responses to other events as well (e.g., natural disasters) but should address all major findings in the security risk assessment.

2 SECURITY RISK ASSESSMENT METHODOLOGY – RAM-W™

The security risk assessment methodology for assessing the vulnerability of water utilities has seven basic steps:[3]

1. Background and Planning
2. Threat Assessment
3. Site Characterization
4. System Effectiveness
5. Risk Analysis
6. Determination of Acceptable Risk
7. Upgrades and Impacts

For those utilities with large numbers of critical assets, a pairwise comparison technique was designed to facilitate prioritization of facilities/components. The next several sections will discuss the first four steps in the methodology, but will not cover all seven sections.

The final 3 steps of RAM-W™ complete the risk analysis and help determine what risks are acceptable to the utility, any PPS and/or system upgrades needed, and highlight new resources or processes required. Eventually this information may impact the ERP, but until upgrades are in place, the current system configuration should be used.

2.1 Background and Planning

The first step in the security RAM-W™ is to determine the prioritized mission objectives of the water utility. For most water utilities, the prioritization generally results in something similar to the following:

1. Provide sufficient pressure for fighting fires.
2. Provide potable water.
3. Reach the greatest geographical extent.

If there are multiple mission objectives, they are placed into a pairwise comparison to determine clear priorities. This analysis helps determine where to apply resources first to lower risk. Next, the goals of the risk management program are clearly articulated. The goals may contain all or parts of the following:

- Defeat a certain level of adversary
- Detect adversarial actions
- Reduce high consequences to an acceptable level
- Deter an adversary

Next, the utility will prioritize the facility and/or system components that are most at risk. If the water utility has a large number of assets, a pairwise comparison is performed to prioritize the facilities/components for risk reduction, including major treatment plants, major pump stations, major storage facilities, critical wells, critical pipelines, critical reservoirs, etc. The facilities/components are compared to each mission objective determined in the first step above, resulting in a weighted, prioritized list. For example, each of the facilities/components are pairwise compared for their ability to "provide sufficient pressure for fighting fires." Obviously, if the item being compared in the list does not play a major role in supporting this objective, it will be ranked lower than items that do play a major role. This approach helps determine the absolute minimum number of facilities that must remain operational for the majority of the system to continue to support the mission objective. Once the pairwise comparison step is completed for all mission objectives, this prioritized list of facilities/components is utilized in the remainder of the methodology.

2.2 Threat Assessment

Before a security risk assessment can be completed, a description of the threat is required. A threat assessment is "...a judgment, based on available intelligence, law enforcement, and open source information, of the actual or potential threat to one or more facilities or programs."[4] This description includes the type of adversary, tactics, and capabilities such as the number in the group, weapons, equipment, and transportation mode. Information is also needed about the threat to estimate the likelihood that the adversary might attempt to accomplish the undesired events. For water contamination, this process requires a review of potential contaminants and the determination of which ones pose a significant threat to the water utility. There is an endless list of chemicals and biological agents that could be used, but the amount of dilution, chlorine resistance, difficulty to manufacture, efficacy in water, and several other variables have to be evaluated to determine just how much of a threat they really are. The final list needs to contain specific contaminants, amounts, and methods of introduction into the system.

The specific type(s) of threat to a facility is referred to as the Design Basis Threat (DBT). The DBT is often reduced to several paragraphs that describe the number of adversaries, their *modus operandi*, the types of tools and weapons they would use, and the types of events or acts they are willing to commit. Ultimately, the DBT chosen is a management decision.

2.3 Site Characterization

The next step in RAM-W™ involves characterization of the facility operating states and conditions. This step requires developing a thorough description of the utility, including the location of the site boundaries, building locations, floor plans, and access points. A description of the processes within each facility is also required, as well as identification of any existing physical protection system (PPS) features. This information can be obtained from several sources, including facility design blueprints, process descriptions, safety analysis reports, environmental impact statements, and sanitary sewer surveys. An important part of the Site Characterization includes understanding how the utility has implemented security policies and procedures; security training; employee background checks; access control for vendors, suppliers, and contractors; and other items that help build the foundation of a solid security program. Questionnaires have been developed as part of RAM-W™ to assist in gathering Site Characterization data.

A structured approach is needed to identify critical components for prevention of the undesired events. These facilities/components and their locations become the critical assets to protect. The analysis methodology uses a fault tree as a primary tool for analyzing and describing the site vulnerabilities. A fault tree is a logic tool. It itemizes critical assets, those that must function to prevent an undesired event, and shows scenarios that could produce such an event. The tree guides analysts in estimating the degree of risk associated with threats.

The next step is to categorize undesired events or loss of critical assets. The categories of consequences generally will be defined by descriptors such as number of deaths, number of hours without sufficient pressure for fighting fires, number of hours without potable water, and value of economic losses.

Low, medium, or high levels of the consequences are determined for each undesired event. The levels are unique to each water utility. Values can then be assigned to these levels to provide a relative ranking of assets whose loss hold the potential to result in unacceptable consequences.

2.4 System Effectiveness

The current PPS features must be described in detail before the existing physical protection system effectiveness (See RAM-W™ manual for complete discussion of system effectiveness – available through the American Water Works Association) can be evaluated. An effective security system must be able to detect the adversary early, delay the adversary long enough for the response to arrive, and stop the adversary before their mission is accomplished. In particular, an effective security system provides balanced detection, delay, and response. In RAM-W™, questionnaires have been developed to facilitate the evaluation of the existing PPS. The performance of the PPS is dependent on the DBT chosen in the Threat Definition portion of the assessment. In other words, a door alarm might be very effective against an outsider without knowledge of the PPS, but would be ineffective against an outsider with knowledge, who would simply choose another path of attack. The performance of the PPS is also very dependent on the implementation policies and procedures followed by the water utility. Once this analysis is completed, the probability of effectiveness for the existing PPS can be determined for use in the risk equation.

3 EMERGENCY RESPONSE PLANS

Now that we've described the initial steps of RAM-W™, their importance to the emergency response effort will be demonstrated. Emergency response planning is undertaken at many levels, from small local disasters to Department of Homeland Security National Recovery Plans. Many elements are common to all the ERPs: chain of command, emergency responders, communications, evacuation procedures, training, etc. Generally the goals of these plans are to protect lives and recover as quickly and safely as possible. The majority of the ERPs are written to deal with various types of natural disasters.

Historically, malevolent events have not been included in the ERPs for most water utilities, with naturally occurring water contamination the one exception. With the rise of the threat of terrorism, ERPs need to address malevolent acts that could have devastating impacts. Because analysts can envision countless attacks on the water utility's assets, some bounding assumptions need to be made. When considering which scenarios to include in the ERP, there should be a direct correlation to the scenarios developed in the security risk assessment, especially those involving high consequences that don't require sophisticated attacks or large numbers of attackers. From the EPA Water security website: "Emergency response plans describe the actions that a drinking water or wastewater utility would take in response to a major event, such as natural disasters or man-made emergencies. They should address the issues raised by the utility's *vulnerability assessment*.[5]"

Fortunately the security risk assessment can make for a streamlined development of the ERP and provide the necessary bounds. The Background and Planning step not only identifies all the critical facilities/equipment, but also prioritizes them. Site characterization collects all the necessary information about the operation of critical facilities and associated protective measures. The Threat Definition step bounds the types of adversaries and attacks that the water utility will consider. The contaminants considered in the risk assessment should be the ones dealt with in the ERP. System Effectiveness determines what types of attack the facilities/assets are likely to withstand and which ones are likely to be successful. When the information is put together in a risk assessment, the water utility now has a good understanding of its weaknesses.

This information should become the foundation for the malevolent portion of the ERP. If it has been determined that intentional chemical releases (water utility owned chemicals) cause significant health impacts, and the chemicals will continue to exist at the utility, then how this would be dealt with in an emergency should be clearly stated, affected parties trained, and large-scale, multiple agency practice sessions undertaken. For many water utilities, large pump stations contain very unique equipment (often no longer manufactured) that would be difficult to replace. If the adversarial capabilities could cause loss of the pump station, the ERP should spell out how the water utility would respond. All of the weaknesses and/or high consequence events identified in the security risk assessment need to be dealt with through operational planning or in the ERP.

4 CASE STUDIES

Many of the water utilities are obviously concerned with terrorist activity. Terrorism is considered a high level threat and the terrorist's goals (developed in the DBT) generally include causing major disruptions to one or more of the mission objectives of the water utility as well as many deaths and injuries. For many water utilities, chemical handling facilities become one of the critical assets. Such facilities are an attractive target due to their level of potential consequence if disrupted. The water utility then has to make a determination to either increase the effectiveness of the PPS around these facilities or to reduce the consequences of their disruption. Due to the life-cycle costs and the operational impacts to significantly improve the PPS, most utilities are deciding to use alternate methods of water treatment, thereby reducing consequences. When reviewing existing ERPs at major water utilities, the only chemical event that was dealt with on a regular basis was a chlorine leak. None of the ERPs dealt with the response to a malevolent, catastrophic failure of the chlorine tanks(s). No drills had been conducted on large-scale chlorine events, especially those involving other emergency responders. The response to a catastrophic leak is entirely different and critically time dependent when compared to a small leak.

Due to their primary mission of providing adequate fire-fighting capability, the water utilities often find that pumping stations are high on the list of critical assets. Here a combination of efforts are employed to improve the effectiveness of the PPS and reduce the consequences through off-site spares, alternate energy sources, and other forms of consequence reduction. The ERPs reviewed did not address an intentional attack on the pumping station and a complete loss of pumps. Most utilities estimated that it would take a minimum of six months and possibly as long as eighteen months to recover from such an event, but had not developed a plan on how they would manage that event.

These are two of many possible scenarios considered during a security risk assessment. If the utility decides they are significant enough to include in the security risk assessment, then inclusion in the ERP should be mandatory.

5 CONCLUSIONS

An analysis methodology for assessing the vulnerability of the water infrastructure has been described and used at hundreds of water utilities. The assessment provides a wealth of information for developing an ERP. High consequence events, especially those requiring low capabilities or low numbers of adversaries to effect, must be addressed. Most of the information needed to address the malevolent portion of the ERP is found in the security risk assessment and the two documents should be in agreement.

References

1 Security Systems and Technology Center (2002). *Risk Assessment Methodology for Water Utilities (RAM-W^{SM}) Second Edition*, Awwa Research Foundation and U.S. Environmental Protection Agency, Awwa Research Foundation, Denver, CO.

2 J.J. Danneels,. "Terrorism: Are America's Water Resources and Environment at Risk?" Statement before the United States House of Representatives Committee on Transportation and Infrastructure, Subcommittee on Water Resources and the Environment, October 10, 2001.

3 Y.Y. Haimes, D.A. Moser, E.Z. and Stakhiv, (eds.) *Risk-Based Decision making in Water Resources IX,* Proceedings of the Ninth Conference, ASCE (American Society of Civil Engineers), Reston, VA 2000.

4 M.L. Garcia. *The Design and Evaluation of Physical Protection Systems*, Elsevier Butterworth-Heinemann, Burlington, MA. 2001

5 EPA Website: http://cfpub.epa.gov/safewater/watersecurity/home.cfm?program_id=8

FASTER, SMALLER CHEAPER: TECHNICAL INNOVATIONS FOR NEXT-GENERATION WATER MONITORING

W. Einfeld

Sandia National Laboratories, Chemical and Biological Technologies Department, MS-0734
Albuquerque, New Mexico, USA 87185-0734

ABSTRACT

Following a discussion of the water security context into which advanced monitoring technologies fit, an overview of four specific areas of ongoing micro-analytical sensor research at Sandia National Laboratories is presented in this paper. Micro-analytical sensor system developments are part of a laboratory supported effort to address a variety of monitoring challenges related to distribution system security in the drinking water industry. Included in this paper is a brief review of the status of a liquid microChemLab capillary electrophoresis system for onsite biotoxin and biological monitoring and a gas microChemLab chromatographic system for the analysis of volatiles in water. An innovative sample conditioning method under development known as insulator dielectrophoresis for selective pre-concentration of biological species in water is also discussed. Finally, an overview of a nano-electrode sensor system for the analysis of electro-active species in water is presented. These and other technologies are discussed in the context of how they might be deployed either prior to or following a water contamination event.

1 INTRODUCTION

Response and countermeasure options associated with a water contamination event may be exercised both prior to and following the event. Administrative pre-event activities may include conducting vulnerability assessments, emergency response planning and the use of mock events and table top exercises to assess the readiness of the utility staff to respond. Other pre-event activities may include the improvement of field screening and laboratory analytical capabilities or the design and installation of detect-to-warn systems. This paper will concentrate in particular on evolving technologies that may form a complement of an early warning system (EWS) or a rapid screening capability for a water contamination event. In the longer term, major infrastructure changes may be implemented out to reduce system vulnerability. Such actions might include point-of-use water treatment devices, back-flow prevention devices and major layout changes to pipeline networks that would maximize water routing options and adaptable valving configurations for the rapid isolation of contaminated sections of pipe.

Post-event responses may include attempts by a utility to more rapidly integrate the various data streams that might be suggestive of a contamination event, (e.g. customer complaints, hospital admissions, on-line detection systems). The execution of pre-planned grab sampling missions that are accompanied by a relevant laboratory analysis or on-site screening capability is another possible response. Consideration is also being given to the development of hydraulic flow models that can run in real-time during a contamination event in such at way that they can provide decision support to water system operators in carrying out such tasks as identifying the source of contamination, deciding how best to isolate the contamination as well as planning and executing a decontamination process.[1] Risk-based system design approaches are attractive and can help allocate limited utility capital resources in such a way that they effectively target areas of higher risk.[2]

Water utility managers are quick to point out that any monitoring solutions proposed for reducing the risk from a malevolent attack should also serve in a dual-use capacity, such that they can also be used for routine assessment of distribution water quality and improved treatment process control. The capital, maintenance and operating expense associated with the design and installation of an early warning system is likely to be significant and, as a result, will necessitate multiple-use benefits in order to be considered a cost-effective investment by the utility industry.

2 EARLY WARNING SYSTEM DESIGN

A tiered instrument approach is one of the EWS design architectures being considered by the water utility industry and the supporting research community.[3,4] In this approach, as illustrated in Figure 1, the first tier of sensors are designed to measure various physical parameters (e.g. pH, conductivity, temperature, residual chlorine, redox potential) using sensor technology that is already commercially available. Advancements in this area include the research and development of chip-based sensors that incorporate many of these analytical functions on a single chip, thereby reducing the cost and complexity of the overall system. It is also generally understood that some sort of statistical interpretation of these multiple-parameter data streams will be required to reliably signal a contamination event and efforts are underway by various researchers to develop robust algorithms that will enable the detection of a significant baseline shift that is outside the normal operating parameters of the water system that is being monitored. The ability of a given suite of physical sensors to detect a wide range of contamination events is largely unknown at present and rigorous testing of such systems will be required with either actual agents or reliable stimulants prior to water utility managers making significant investments in such sensors for a distribution network installation. Ideally, these first-tier sensor systems would provide an indication of a significant baseline shift in general water quality parameters that could be taken as a reliable indicator of a compromised system; however it is unlikely that significant information about the specific nature of the contaminant will be ascertained from first-tier systems.

2.1 Higher-tier Sensors

A second tier of sensors could be deployed in an on-line, rapid-wake-up capacity or as a part of a rapidly deployable manned mobile laboratory used to provide a more detailed analysis of the signaled event. They might be associated with a sample pre-concentrator or a remotely activated sample collection device since normal water flow through distribution system pipes and unavoidable delays in activation of second-tier of sensing capabilities will result in the contaminated stream moving on through the distribution system to downstream points. In this situation, hydraulic flow models continuously running as a component of the utility data management system could provide timely information for the identification of the best location for sampling. Second-tier instruments, examples of which are to follow in this paper, could provide an on-site triage capability, allowing system operators to narrow the list of suspected contaminants by definitively ruling out other possible candidates through reliable on-site screening analysis.

Figure 1 *The tiered instrument approach*

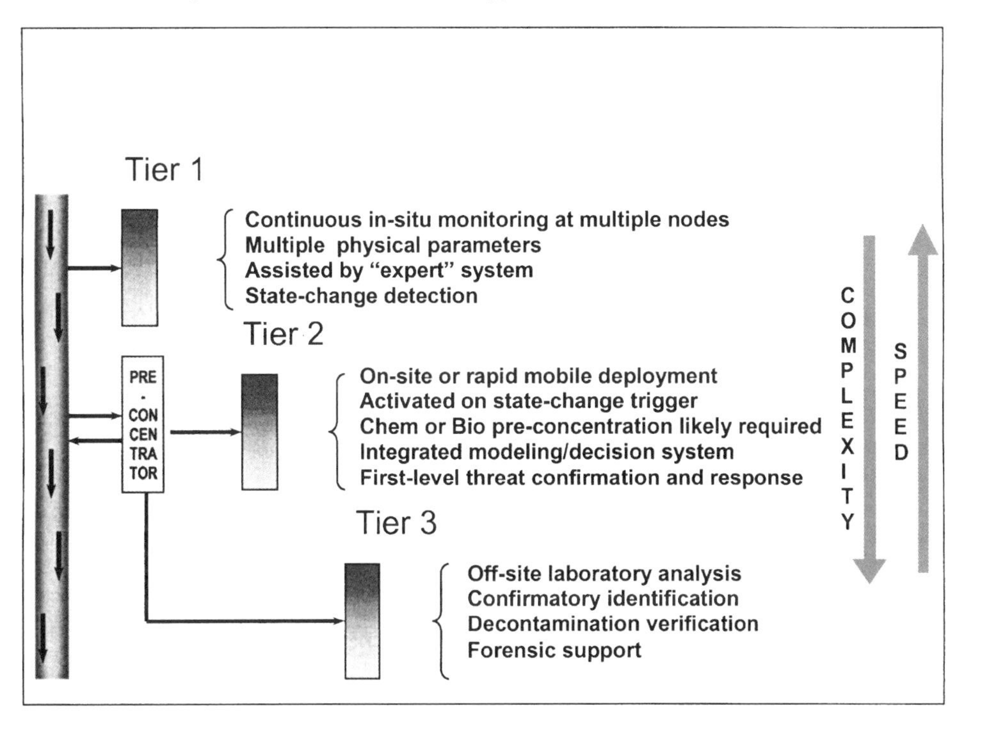

The final tier of analytical capabilities in this design architecture is understood to be the conventional laboratory. Although conventional laboratory capabilities are generally not considered to offer a fast response, it is likely that any positive confirmation of a contamination event with significant public health consequences will come from this conventional analysis pathway. Reliable indications of contamination from a second-tier instrument may provide the utility operator with enough valid information to make early decisions about utility response option to the contamination event.

2.2 Critical EWS Performance Issues

Important performance issues for a water system EWS are shown in Table 1. Among the system attributes listed, the analyzer false positive rate is clearly one of the most important since a high false alarm rate will stress utilities' response capabilities and further result in a lack of confidence in the system over the longer term. The individual false alarm rates in multi-node sensor systems throughout a distribution system can have important implications for overall false alarms rates, further emphasizing the necessity of low false positive rates.[5] A second important attribute is the maintenance and operational costs of a system, given that fact that most utilities are constrained in terms of the availability of trained field personnel and available financial resources for an expanded labour force to effectively operate a large network of sensors.

Contaminant (or surrogate indicator) detection levels
False positive and negative rates
Instrument time response
Degree of broad-spectrum contaminant coverage
Extent of spatial coverage
Degree of system autonomy
Maintenance and operational constraints
System cost
Degree of self-diagnostics
Ability of provide a multiple-use benefit

Table 1 *Performance Attributes for Candidate EWS Sensors*

Risk-based approaches to EWS design and implementation are warranted in light of the typically limited financial resources that utilities possess, and the fact that any EWS design is unlikely to provide complete spatial coverage of a distribution system. Pre-design simulation and modeling can provide useful insight as to where best to locate a limited number of sensors.

Other more global issues surrounding the implementation of EWS systems include the need for design guidelines and standards with the caveat that at some fundamental level each water distribution system will need to be considered as a unique entity with specific design requirements and adaptations from a standardized approach. The overall readiness of the market to adopt and deploy next-generation sensors for water monitoring has not been determined. In light of the fact that few sensors are available for long-term reliable autonomous operation, particularly for pathogenic species, it comes as no surprise that many utilities are taking a guarded, "wait and see" approach before making significant investments in this area.

Finally, at the present time, only limited research and development funding is available for the further development of next-generation analytical devices. There has been a strong emphasis from various US agencies for rapid deployment of new technologies that can address the water problem; however, few viable commercial options exist at the present time, and although many micro-analytical approaches hold promise, they will only come to fruition through government-supported development efforts.

3 MICRO-ANALYTICAL SYSTEM DEVELOPMENTS

The remainder of this paper will focus on several micro-analytical systems which have directly application to water analyses, which are under various stages of development at Sandia National Laboratories. Sandia has strong competencies in device design, microfabrication processes and the supporting analytical chemistry. Four technology areas in particular will be discussed: a liquid-phase microChemLab, a related gas-phase microChemLab, a dielectrophoretic technique for the pre-concentration and separation of biologicals and a microfabricated electrode system for the measurement of electro-active species in water.

3.1 Liquid-Phase MicroChemLab

Researchers at Sandia have developed a hand-held analyzer that utilizes various chip-based technological innovations such as microfluidics, capillary gel and zone electrophoresis columns combined with small laser induced fluorescence detectors for the analysis of biotoxins and other proteinaceous materials in water. Features of the analyzer, shown in Figure 2, include a dual-channel analysis pathway and integrated data processing capabilities for increased detection confidence.

The analytical chip is used to separate protein mixtures that are labeled with a fluorescent dye prior to injection onto the chip. Electrokinetic pumping—a method that does not require valves and mechanical pumps—is used to move buffer solutions that contain the sample on the chip. The chip is presently fabricated in glass; however, future plans call for plastic as the substrate in order to reduce production costs. Protein separation methods can include both capillary gel and capillary zone electrophoresis. The detector consists of a commercially available UV diode laser and accompanying photodetector. The system design includes modular features that allow simple and rapid changeout of an entire analysis channel. The current device also includes side-by-side analytical channels to facilitate improved analyte detection confidence.

The system is presently configured for the analysis of biotoxins and can provide nanomolar sensitivity with an analysis time of under 5 minutes. Additional studies have shown that the system can also be used to detect various marker proteins associated with bacteria and viruses. The challenge in this particular application is to develop a suitable biological pre-concentrator and cell lysis device as a front-end to the analyzer. Ongoing research is underway in both areas. Another project is underway to convert the present system configuration into an autonomous analyzer for periodic on-line assessment for biotoxin content in water. A sampling interface to a distribution system water pipe is being fabricated, as is the incorporation of automated addition of fluorescent tags and sample injection onto the analytical column.

3.2 Gas-Phase MicroChemLab

Using a similar micro-analytical approach, Sandia had originally developed a hand-held gas chromatograph on a silicon chip that was optimized for monitoring airborne chemical agents. The analyzer consists of a micro-fabricated vapor pre-concentrator, a separations column etched into silicon and a micro-fabricated, multi-channel surface acoustic wave detector, as shown in Figure 3. This system has been successfully demonstrated in various long-term air sampling applications. A fast analysis time is an important feature of the device, with typical analysis times for volatile airborne species on the order of 2-3 minutes.

Currently under development is a "front-end" water sampling module that will further expand the use of this instrument for the analysis of volatile and semi-volatile organic species in water. The system incorporates an automated purge and trap design and is initially focused on the on-site, fast analysis of disinfection byproducts such as trihalomethanes. Further applications will include the analysis of petroleum hydrocarbons, chemical warfare agents and their hydrolysis products.

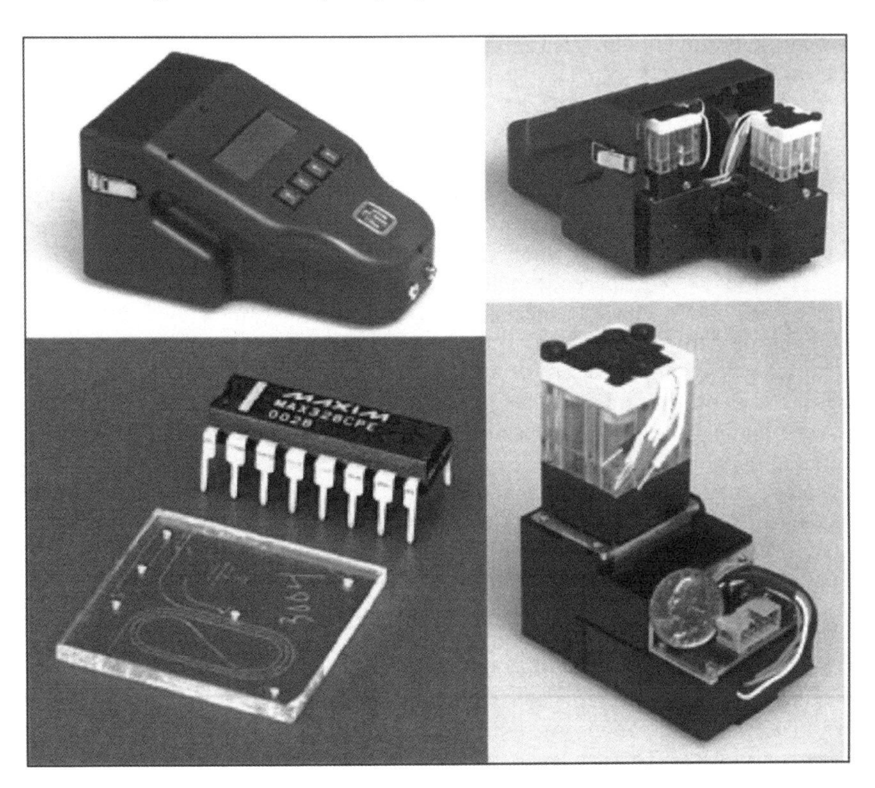

Figure 2 *The Sandia-developed liquid microChemLab*

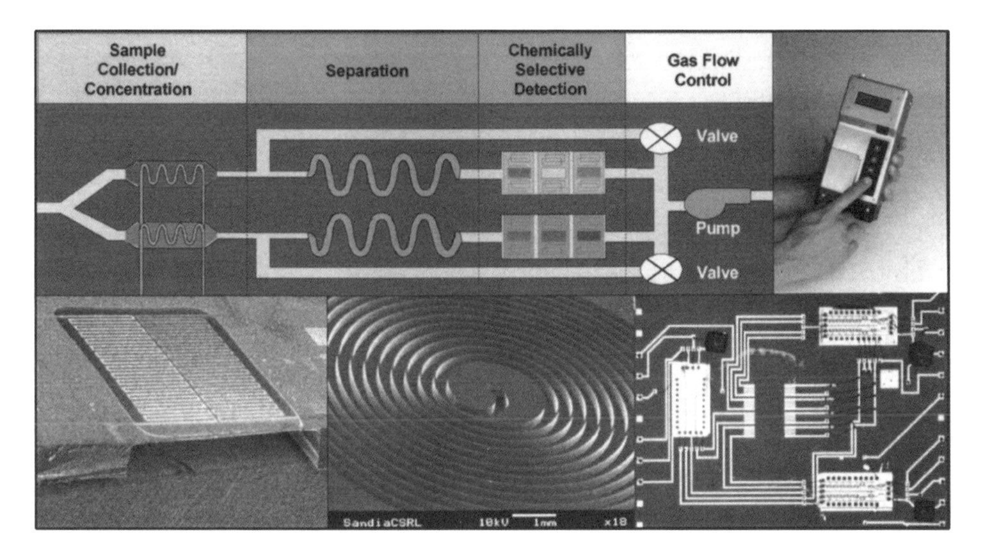

Figure 3 ***The Sandia-developed gas-phase microChemLab***

3.3 Novel Bacterial Pre-Concentration Techniques

Researchers at Sandia are also exploring the use of dielectrophoretic (DEP) trapping in micro-fabricated silicon arrays as a way to pre-concentrate and sort various water-borne biological species prior to subsequent analysis on micro-analytical platforms. Researchers have demonstrated that the principle of dielectrophoresis—the motion of a polarizable particle relative to the fluid medium in an electric field—can be used to differentiate between live and dead bacteria as well as between bacterial species.[6] As shown in Figure 4, electric field strength can be used to selectively trap different bacterial species.[7] Recent experiments have also demonstrated that bacterial concentration and trapping can be achieved and that trapped species can then subsequently be selectively released for downstream analysis of bacterial content. Maturation of this technology as a sample-preparation front-end for micro-analytical platforms will require that mass fabrication techniques for low-cost plastic electrode surfaces be developed. Furthermore, significant scale-up of these individual micro-fabricated dielectrophoresis chips will be required in order that litre quantities of water can be efficiently processed.

3.4 Micro-fabricated Nano-electrodes

Other Sandia investigators are exploring the use of micro-fabricated electrodes for the measurement of electro-active species in water. As shown in Figure 5, nano-electrode surfaces with high number density (~ 10^7electrodes/cm^2) have been micro-fabricated and used to measure lead in water using differential anodic stripping voltammetry methods. The use of high number density nano-electrodes makes direct measurement of electro-active species in water possible without the need for added buffers or electrolytes. The goal of this effort is to develop reagent-free analysis of metallic species such as chromium and arsenic in water using low-cost micro-fabricated systems.[8]

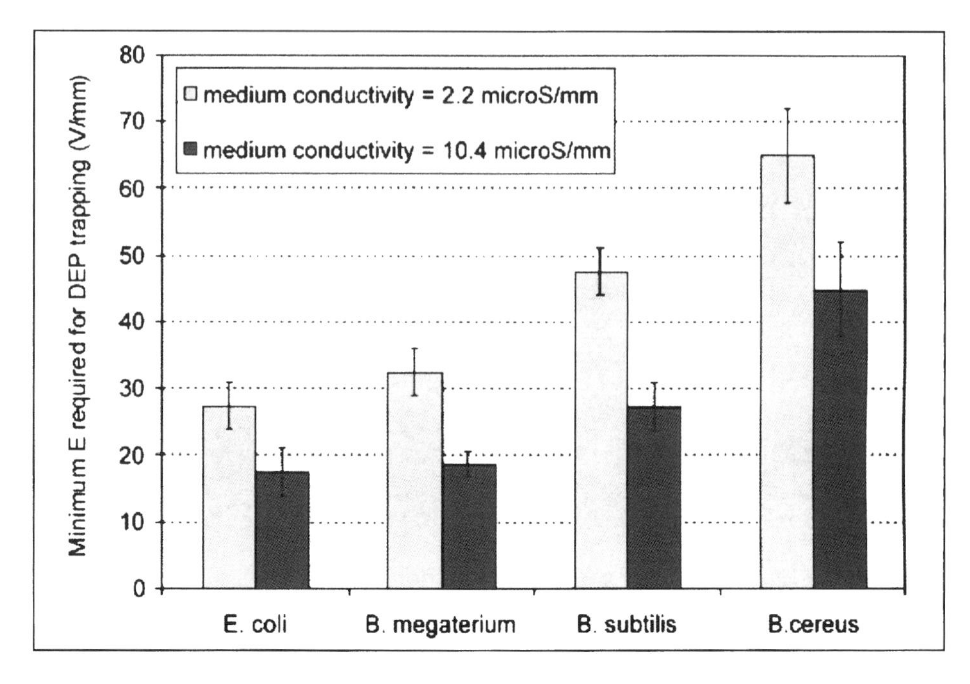

Figure 4. *Experimental data showing that electric field strength can be used to selectively trap different bacteriological species. Conductivity can be roughly correlated to the ionic strength of the fluid medium containing the bacteria.*

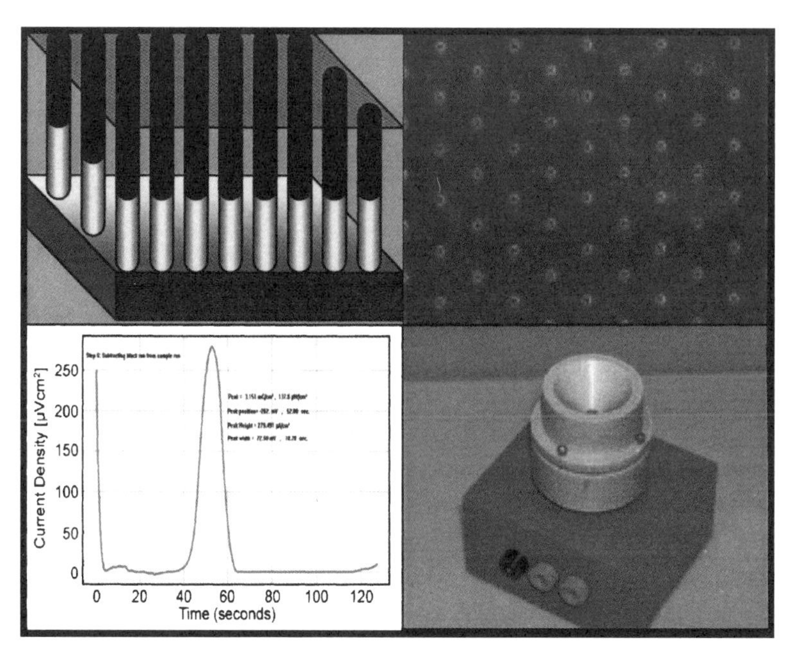

Figure 5. *Micro-fabricated electrodes for the measurement of electro-active species in water. A 3-D representation of the two-electrode cell template (upper left). A photomicrograph of a fabricated nanoelectrode with 0.2 micron diameter electrodes and 1.4 micron spacing (upper right). A differential anodic stripping voltammetry trace for lead in water (lower left). The prototype electrode and holder (lower right).*

4 SUMMARY AND CONCLUSIONS

Advanced monitoring technologies constitute an important option among the selection of countermeasures that can be applied to an unexpected water contamination event. In some cases an early warning system might be effectively applied to provide a detect-to-warn capability for water system managers; however, important challenges must be faced in terms of the design and desired functionality of such a system. Risk-based hydraulic modeling of water distribution systems will play an important role in the design and deployment of monitoring networks intended to increase systems security. It is unlikely that fully autonomous agent or organism-specific detectors will be implemented in an early warning system in light of the maintenance and operational constraints faced by water utilities and the breadth of the potential threats. Autonomous monitoring may in fact be limited to the measurement of physical parameters such as pH, conductivity, turbidity and residual chlorine. Statistical interpretation of these multivariate data streams may give an indication of a contamination event, however additional measurements will be necessary to further ascertain the nature and extent of the contaminant. A new class of instruments based on micro-analytical principles could effectively fit into a second-tier analysis capability. Hand-portable instruments such as briefly described in this paper could be quickly deployed to the site of interest to provide additional information as to the nature of a contamination event, thereby saving valuable time for a utility operator faced with the need for a rapid response.

Several classes of micro-analytical systems currently under development at Sandia National Laboratories have been briefly reviewed in this paper. The liquid microChemLab for biotoxin and bacterial analysis and the gas microChemLab for the measurement is disinfection byproducts are examples of next-generation systems that could be used for more extensive water quality measurements throughout a water distribution system. The dielectrophoretic trapping technology shows particular promise for sample pre-treatment and concentration, particularly for biological species. A hand-held electrochemical analyzer for electro-active species such as lead and arsenic could find important uses for rapid onsite screening of water quality throughout a water treatment system.

ACKNOWLEDGEMENT

Sandia is a multi-program laboratory operated by Sandia Corporation, a Lockheed Martin Company, for the United States Department of Energy under contract DE-AC04-94AL85000.

References

1 R. Murray, R. Janke, and J. Uber, "The Threat Ensemble Vulnerability Assessment Program for Drinking Water Distribution System Security," Proceedings of the American Society of Civil Engineers EWRI Congress, 2004.

2 W. M. Grayman, R. A. Deininger, R. M. Males. Design of Early Warning and Predictive Source- Water Monitoring Systems, 2001. AWWA Research Foundation. Denver, Co. 297 pp.

3 S. J. States, *et al*, "Rapid Analytical Techniques for Drinking Water Security Investigations," Journal AWWA, 96:1:52, 2004.

4 Interim Voluntary Guidelines for Designing an On-line Contaminant Monitoring System, American Society of Civil Engineers, Washington DC. Available online at: http://www.asce.org/static/1/wise.cfm#WISE

5 S. E. Hrudey. and S. Rizak, "Discussion of: Rapid Analytical Techniques for Drinking Water Security Applications," Journal AWWA, 96:9, 2004.

6 B. H. Lapizco-Encinas, B.A.Simmons, E. B. Cummings and Y. Fintschenko, Dielectrophoretic Concentration and Separation of Live and Dead Bacteria in an Array of Insulators. *Anal. Chem.* 76, (2004)1571-1579.

7 B. H. Lapizco-Encinas, B.A., Simmons, E. B. Cummings, and Y. Fintschenko, Insulator-Based Dielectrophoresis for the Selective Concentration and Separation of Live Bacteria in Water. *Electrophoresis* 25 (2004) 1695-1704.

8 W. G. Yelton, et al; "Nano Electrode Arrays for In-situ Identification and Quantification of Chemicals in Water"; Sandia National Laboratories Technical Report, SAND2004-6229, Sandia National Laboratories, Albuquerque, NM, 87185 Available online at http://infoserve.sandia.gov/sand_doc/2004/046229.pdf

A DUTCH VIEW OF EMERGENCY PLANNING AND CONTROL

O.J. Epema, J.M. van Steenwijk and W.G. van Gogh

Institute for Inland Water Management and Waste Water Treatment RIZA, Rijkswaterstaat (RWS), Department of Transport and Water Management, Ministry of Transport, Public Works and Water Management (V&W), Zuiderwagenplein 2, NL 8224 AD Lelystad, The Netherlands.

1 INTRODUCTION

The Dutch Institute for Inland Water Management and Waste Water Treatment RIZA is part of Rijkswaterstaat, a state body that governs the infrastructure of the Netherlands. RIZA plays an important role in the Netherlands by providing vital information and knowledge for both general surface water management as well as for incident management. In this paper the resources and position of RIZA will be described in relation to the management of contamination incidents on inland waters in the Netherlands.

In most cases RIZA is not the party that is directly responsible for actually managing the incident. RIZA's role is limited to an advisory one, giving expert judgement and providing reliable measurements when asked for. RIZA has broad resources to play this role properly.

2 THE DUTCH INLAND WATERS

The Netherlands are made up of four international river basins, the Rhine, Meuse, Scheldt and Ems (Figure 1). Both the quantity of water as well as the quality of the water that is entering and leaving the country is the concern of Rijkswaterstaat, as this body is expected to prevent flooding, to provide sufficient and sufficiently clean water for all users, and to maintain waterways in order to ensure safe and timely passage.

Due to the fact that 60% of the area of the Netherlands lies below sea level a fine meshed water system is required to enable steady discharge of surplus water from deposition or seepage (Figure 2). The dense infrastructure is inherently vulnerable to local or widespread contamination incidents. Rijkswaterstaat collaborates with foreign as well as local water authorities as regional and state waters are closely intertwined. The management of the surface waters involves 12 provinces, 55 regional water boards and more than 500 municipalities.

The importance of water management can be illustrated in many ways. *e.g.* more than half of the volume of drinking water is produced using river water, most of the areas of high economic value are situated in the lower parts of the Netherlands and are densely populated, etcetera.

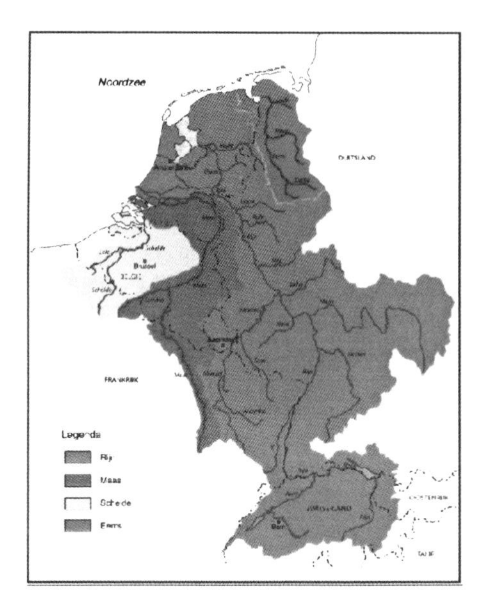

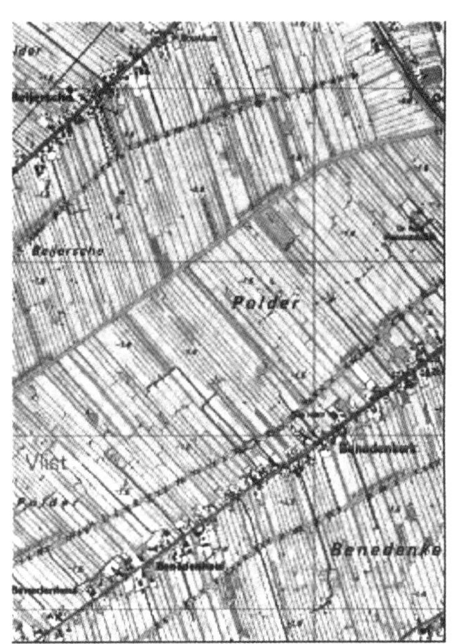

Figure 1 *The 4 river basins*

Figure 2 *Fine meshed water systems*

RIZA supports the regional services of Rijkswaterstaat by provision of information and knowledge. RIZA governs long term monitoring programs and research projects, but it also is able to give an up-to-date picture of the current state of key water systems in various aspects like naval traffic, flux and presence of contaminants. This thorough understanding of the water systems (Integral Water Management) is a major factor for successfully controlling contamination incidents.

3 RIZA FRONT OFFICE

3.1 General

The specific task of incident control is supported by RIZA. Some bodies within the RIZA organisation are dedicated to incident control and or permanent monitoring, for most departments however giving support in the case of an incident, whether it has a nuclear, biological or chemical nature, is just part of the job. In the following paragraphs the bodies of RIZA that are involved and the tools and applications that are employed in case of an emergency are described.

3.2 Information Centre for Inland Waters (Info Centre)

At the Head Office location of RIZA in Lelystad, in a central part of the Netherlands, the Information Centre for Inland Waters (Info Centre) is based. It provides current data on levels, temperatures and cooling capacity, and distribution of water in the Netherlands to all sorts of users. Moreover it publishes shipping and traffic reports on a daily basis (or hourly if necessary), issues weather reports and storm warnings (dike patrol), reports on the quality of swimming waters and supports incident management by collecting, processing and relaying of information.

The Info Centre is very well suited for the latter tasks as it fitted with backed up state of the art communication equipment and employs extensive electronic sources of available information. All data and reports can be communicated by phone, internet (Website: www.infocentrum-binnenwateren.nl) or teletext broadcasts.

3.3 Alarm group

RIZA has established an expert advisory team of experts, the so-called *Alarm Group,* to deal with urgent queries on nuclear, biological and chemical incidents on or in the vicinity of the Dutch inland surface waters, which is on duty 24 hours a day and 7 days a week. It is supported by various tools and experts on stand-by and closely allied to the Info Centre. Its task is to assess the incident in order to minimize the short term risks to the environment, nature and people. The assessment should therefore result within a short period in tailor-made advice to the authority that is handling the incident.
In order to be able to perform the assessment properly and extensively but also timely the alarm group uses several tools and sources to gather and process information and to relay it to people and bodies involved. The alarm group has access to several databases containing chemical, toxicological and physical data. In addition quick modelling can be performed in order to determine distribution characteristics of specific compounds, all in relation to background levels derived from long term monitoring programs.

3.4 Infra Web

Infra Web is an internet application solely designed and developed to track and trace (water) incidents and to facilitate (inter)national collaboration. Infra Web is a non-public bulletin board on which all measures that are taken to control an incident are published (Figure 3).

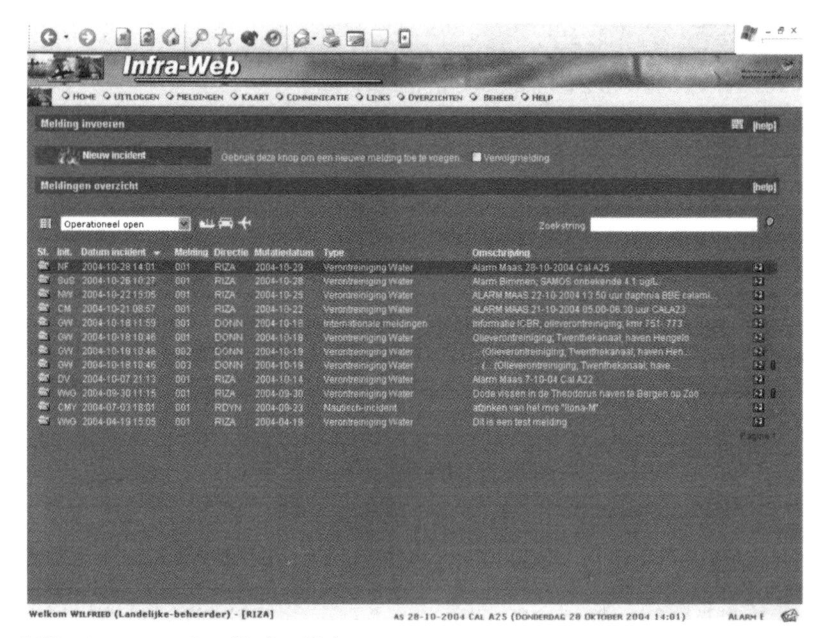

Figure 3 *Viewing example of Infra-Web*

Due to the nature of the medium, all people involved in the incident are readily informed about recent measures being taken. This greatly speeds up communication and inhibits the distortion of messages. The alarm group uses Infra Web as a primary means of communication.

Infra-Web contains a history of former incidents. Due to the availability of geographical information an assessment of the location is easily made as it is readily seen which areas and authorities will (have to) be implicated in the event(s).

Registration of a new incident is strongly directed. A number of standard incident questionnaires have to be used. Depending on the scale of the event all the people and bodies involved are notified that a new entry in Infra-Web has been entered and they are informed via standard reports to allow quick assessment of new information. All entries are logged allowing a precise chronological reconstruction afterwards to be used in evaluations.

Users can access various databases directly connected to Infra-Web in order to obtain information about *e.g.* hazardous chemicals, distribution characteristics, local shipping, responsible local authorities, etc.

3.5 Monitoring stations

At key points the quality of the water that is entering the country is monitored on a permanent basis. On the Rhine a joint Dutch-German station has been established right at the border. On the Meuse the primary station is also situated at the border (location Eijsden, Figure 4). Further down stream minor stations have been established.

Figure 4 *Permanent monitoring stations along the Rhine and Meuse*

The monitoring stations are permanently guarding water quality using biological, chemical and physical techniques. On- and off-line measurements are employed to determine parameters like salinity, temperature, acidity, turbidity, and concentration of oxygen, fluoride and ammonia, of volatile and semi volatile and polar organic compounds. In addition the general quality is monitored employing organisms (algae and water fleas), which may respond to the presence of contaminants outside the analytical window. Data are made available on the internet using the application *Aqualarm* (Figure 5) and when pre-defined levels are exceeded the alarm group and drinking water companies that use river water downstream to produce drinking water are among the first that are automatically alerted. In the near future the station at Eijsden will be fitted with an on-line radioactivity monitor. Other in-situ or on line equipment is still being assessed for implementation on board.

Bio monitoring is useful in addition to chemical monitoring because it detects a biological effect that is a response to the coincidence of substances. The combination of chemical and biological methods can be very powerful as in general only 10 % of the total toxicity of river water as seen by bioassays can be explained by the presence of known substances. A bio alert is therefore treated as a standard chemical alert. In a particular case the level of the pesticide dimethoate did not exceed the alert level of 1 μg/l whilst the *Daphnia* monitor did give a strong response. A standard warning was issued and drinking water inlets were shut down for the duration of the alert.

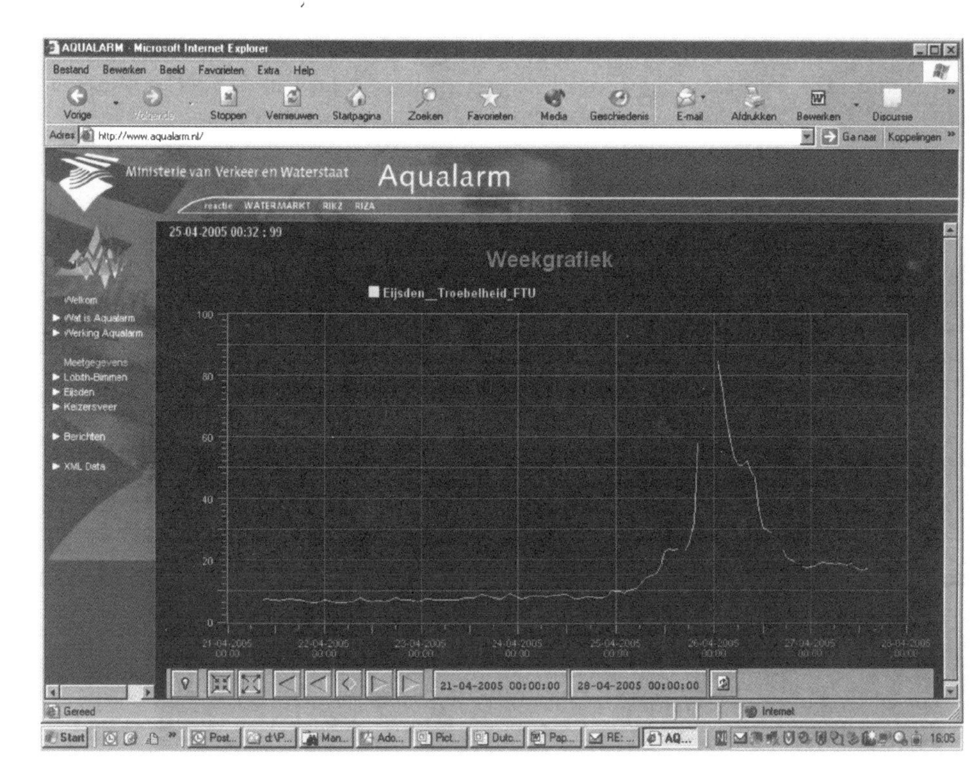

Figure 5 Turbidity of the Meuse at Eijsden as published on Aqualarm (www.aqualarm.nl)

4 RIZA BACK OFFICE

RIZA is involved in various monitoring and research programs for which it requires modern laboratory facilities. In case of an incident these facilities can act as a back office and provide reliable measuring data, tailor-made method development, screening and identification of unknown substances, and toxicological and environmental assessments. In addition a large network of experts can be accessed if specific questions, *e.g.* on human toxicology or food topics, have to be addressed.

4.1 Expertise at RIZA

The analytical and research laboratory departments of RIZA house about 80 people working on a broad range of subjects connected to environmental chemistry and ecology of water systems. In case of an emergency the RIZA-labs are able to do immediate testing, perform screening and target analyses (toxicity, organic, inorganic, radioactivity, nutrients, etc.), attempt to identify unknown substances, do impact assessments, perform ad hoc method development, give an independent second opinion and supply background data (of the uncontaminated water before the incident situation). In this section branches and abilities will be briefly discussed.

4.1.1 Organic analysis Surface and waste water and suspended matter samples can be analysed quantitatively for presence of semi volatile as well as polar organic compounds by GC and LC separation techniques and mass spectrometric detection. In addition oil identification and oil comparison is performed in order to track illegal discharges mainly. from ships Hyphenated screening techniques are employed in conjunction with the RIZA-GCMS database (*vide infra*) to track and identify unknown contaminants.

4.1.2 General inorganic analysis In this department a small group of people are performing a broad range of analytical techniques and analyses like the determination of numerous nutrients with both photometric as well as ion chromatographic techniques or the characterisation of samples on Kjeldahl nitrogen, chemical oxygen demand, total and dissolved organic carbon, phosphate and total phosphorus, organic halogen compounds etc.

4.1.3 Elemental analysis Trace elements are detected by optical and highly sensitive (high resolution) mass spectrometric techniques in conjunction with inductively coupled plasma (ICP) and also by atomic fluorescence spectrometry (AFS). Elemental screening is easily performed by ICP-MS allowing fast characterization of unknown samples.

4.1.4 Radioactivity measurements This department has a capability unique in the Netherlands to measure very low background levels of all kinds of radiation. Total parameters and specific nuclides can be determined allowing preliminary identification of sources.

4.1.5 Hydrobiology and ecology Next to the chemical state of a water body the ecological or biological state can be established by determining the variety of macro fauna and zooplankton present. These kinds of analyses are very time consuming and require highly skilled personnel and in relation to contamination incidents are usually employed in order to establish after the clean up whether the original uncontaminated state has been restored or not.

4.1.6 Toxicology In many cases the presence of toxic contaminants cannot be detected by chemical techniques due to inadequate sensitivity of the equipment. The damaging effect of the water may however be established by measuring the effect it has on bioassays or populations of sensitive water organisms like water fleas. RIZA has much experience with tests for toxicity determinations.

4.1.7 Environmental chemistry As measurements just deliver data the generation of information and knowledge from these data requires comprehensive scientific expertise in environmental chemistry (and biology) that is readily available within the RIZA organisation.

4.1.8 The RIZA GCMS database of known and unknown substances The database and expert system (Access 2000) was developed between 1995 and 2002 as a tool to facilitate GCMS screening. Data from six GCMS systems located at RIZA and other institutes have been processed by the NIST deconvolution program AMDIS to obtain condensed data of identified and unidentified compounds detected in surface and waste water samples from specific locations at specific times and dates. Compounds are characterized by their specific column retention time index (Kovats Index) and reduced mass spectrum.

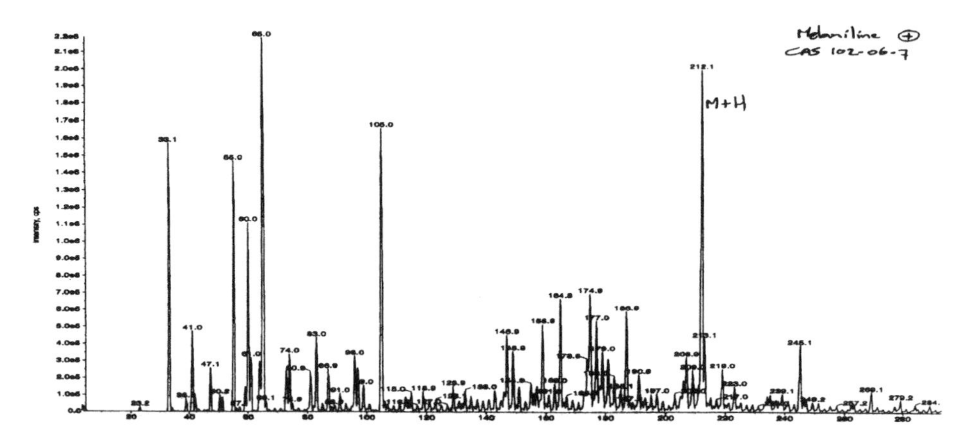

Figure 6 *Mass spectrum example*

GCMS screening at RIZA is aimed at industrial contaminants, pesticides, PCBs, PAHs, pharmaceuticals and in fact at all contaminants that may be found in the environment whether they are listed in monitoring programs or not. The database is a very useful tool in determining which compounds are frequently detected in the environment at alarming levels and should therefore be properly investigated. Moreover the database can be used in many other instances where efficient data handling and interpretation of GCMS data is needed. The database can be accessed by external parties via the internet site: www.riza.nl/gcms

4.1.9 Alliances with other institutes As no institute will be able to be well informed on all subjects, RIZA relies on its relationship with a number of research institutes within and outside the country to get the answers on specific questions.

5 HANDLING AN ALERT

When an incident occurs the RIZA alarm group will be informed and start to work along a pre-defined sequence of steps. In the next paragraphs several steps are briefly commented upon. An extensive description is useless as in the case of an emergency various steps will be executed in a different order and pace.

5.1 Communication

Incidents that are reported to the alarm group or to the Info Centre (office hours) and are registered in Infra-Web. An alert is issued to a fixed list of involved parties that may or may not spring into action depending upon the assessment of the situation.

5.2 Assessment

From the initial information and questions the alarm group member on duty has to determine what action is appropriate. A basic checklist is used to ensure all aspects are addressed.

Identification	- which components?
Characterisation	- environmental behaviour?
Toxicity	- effects on organisms (inc humans)
Quality standards	- are risk levels exceeded?
Emission	- dispersion to other environments?
Amount/distribution	- how much / where?
Criteria match	- when are functions restored?

Depending on the severity appropriate advice is essential to protect as many involved as possible with respect to both the short as well as the long term. As the identification or at least indication of contaminants is vital to the assessment process it may be necessary in the first stages of the incident to have analytical laboratory backup available. As RIZA is well equipped with a broad range of laboratory departments this can usually easily be organized.

Issues on the list are not necessarily checked one after the other. Advice on how to contain contamination (issue emission) can often be given irrespectively of what has been spilled or involuntarily discharged.

5.3 Determining the scale

If the extent of an incident is becoming clear it can be decided to scale things up or down. The task of the alarm group ends here. In case of escalation it may be necessary to involve other departments nationwide. The first body to be involved by Rijkswaterstaat is the Departmental Coordination Centre (DCC) of the Ministry of Transport. From DCC it may be taken up to the national organisation for large scale incident control that is operational (Figure 7).

This Policy Support and Advisory Team for Environmental Incidents (BOT-mi) is alerted e.g. in the case of a nuclear incident as these kinds of incidents tend to be have implications far beyond the scope of RIZA and Rijkswaterstaat.

BOT-mi acts within a network in which many ministries and institutes cooperate. It is important to appreciate that BOT-mi has the leading role and is related to a different ministry, the Ministry of Transport, of which RIZA is no longer part .

5.3.1 Ministries involved in national incidents

Ministry of Housing, Spatial Planning and the Environment (VROM)
Ministry of Transport, Public Works & Water Management (V&W)
Ministry of the Interior (BZ)
Ministry of Defence

5.3.2 Related institutes

National Institute for Public Health and the Environment (RIVM)
Institute for Inland Water Management & Waste Water Treatment (RIZA)
Institute of Food Safety (RIKILT)
Royal Dutch Meteorological Institute (KNMI)
Food and Consumer Product Safety Authority
Centre for External safety of the RIVM (CEV)
National Poisoning Information Centre of the RIVM (NVIC)
National Environmental Incident Service (MOD)

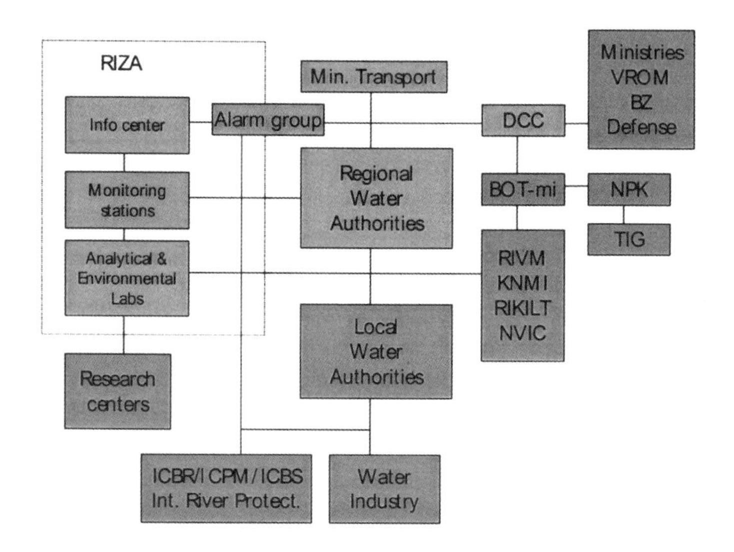

Figure 7 *Position of RIZA and DCC in national network for control of environmental incidents*

5.4 Aftercare

In particular cases a successfully contained incident may have caused a widespread contamination. Imminent danger has been reduced but long term damage to the environment may still be possible. In most cases a clean up is organised if possible in conjunction with a monitoring programme. Both chemical analyses as well as ecological assessments will have to be employed to study the behaviour of the contamination or establish whether the original functions have been properly restored and to determine whether additional measures are still necessary.

6 CONCLUDING REMARKS

RIZA has plenty of resources to support water authorities in their struggle to control incidents and contain its impact on the environment. Different departments of RIZA may be collaborating in all stages of the incident due to the presence of a large body of experts on a wide range of subjects.

WATER DISTRIBUTION SYSTEM MODELING: AN ESSENTIAL COMPONENT OF TOTAL SYSTEM SECURITY

R.E. Finley

Department of Geohydrology, Sandia National Laboratories, Albuquerque, NM 87185-0735

ABSTRACT

One of the major concerns for water utilities in the United States is how to protect the consumers from chemical/ biological contamination resulting from surreptitious malevolent attacks within the water distribution system or from accidental contamination from any means. This threat is especially problematic because the ability to detect and then meaningfully describe the movement (migration) and dilution of contaminants is one part of an overall systematic management tool for sensing, predicting, controlling, and treating contaminants within the distribution system. Over the course of the last several years, the need to resolve these important issues has taken on greater urgency. Following September 11, 2001, the United States government directed that vulnerability assessments be completed for the medium and large utilities in the country. While these assessments were an important step forward, the outcomes indicated that distribution systems weren't adequately understood from a security standpoint, nor were the actual threats. Furthermore, given the potential diversity of contaminants, sensor technologies were (and generally still are!) by and large unavailable to detect and identify even a small percentage of the possible contaminants. To make matters even worse, water utilities generally don't have the research budgets or available financial resources to create an early warning system for the sole purpose of security; early warning systems will need to also serve the dual purpose of assessing general water quality.

This paper will discuss early warning systems in the context of the need for water distribution system modeling as the tool that can be used to *a priori* define contamination risks and optimize sensor and response locations, but also will serve to transform data from early warning systems into actionable knowledge. Ongoing activities at Sandia National Laboratories (SNL) and collaborations with the United States Environmental Protection Agency/National Homeland Security Research Center (EPA/NHSRC) have resulted in development of a number of numerical models and schemes for determining the consequences from a contamination event (in terms of human health effects), tools to efficiently optimize sensor and sampling locations for large data sets, methods to identify the contamination source location in near real time, the ability to include uncertainty in the analyses, and a model to evaluate the likelihood of attacks to water systems. This paper provides an overview of these activities.

1 INTRODUCTION

In the mid-1990's the United States recognized the need to evaluate and improve security for its diverse infrastructures. President Clinton issued Presidential Decision Directive Number 63 in May 1998. This document identified the various critical infrastructures and the Federal agencies whose responsibility it is to understand their vulnerabilities and to protect them. On June 12, 2002, President Bush signed the Public Health Security and Bioterrorism Preparedness and Response Act of 2002 (Bioterrorism Act) into law (PL 107-188). This act required every community water system in the United States serving a population of 3,300 persons or more to (1) conduct a vulnerability assessment (VA); (2) certify and submit a copy of the assessment to the Administrator of the US EPA; and (3) prepare or revise an emergency response plan that takes into consideration the results of the VA and to certify to the EPA Administrator that the system has completed such a plan within 6 months of completing the VA. Sandia National Laboratories (SNL) had, prior to the September 11th attacks, begun the development of a vulnerability assessment tool for water utilities based on a similar one developed for large Federal dams. This tool, RAM-W or Risk Assessment Methodology for Water was developed in conjunction with the US EPA and the American Water Works Association Research Foundation (AWWARF) and was used on the vast majority of the large water systems in the United States. From the development of RAM-W and through conducting numerous VA's and training sessions, it became clear that although RAM-W effectively identified vulnerabilities to physical attacks it left largely unanswered the question of contamination in the water distribution system. From this question, SNL began development of an internally funded research program whose goal was to evaluate the fate and transport of contaminants in drinking water distribution systems and the vulnerability of such systems to contamination attacks. This research program formed the basis for a second generation RAM-W which focused on understanding the risks of intentional contamination by coupling the EPA's system hydraulics code EPANET with SNL's optimization and uncertainty quantification code DAKOTA. In addition, evaluation of the variability in standard water quality is being evaluated because of its importance in understanding the baseline conditions in a water distribution system prior to a contamination event. Finally, tools for evaluating and quantifying the threats to drinking water distribution systems has been evaluated and an approach using a Markov Latent Effects(MLE) approach that essentially aggregates the "possibilities" for likely threats and threat subsets. This tool essentially allows for the first time a systematic approach to quantify the likelihood of attack for water systems resulting in the possible selection of a non-unity term in the simplified risk equation in RAM-W. For additional information on the ongoing research at SNL and the collaboration with the EPA/NHSRC, the reader is directed to the recent publications of the research team[1-15].

2 EARLY WARNING SYSTEMS AND VULNERABILITY ASSESSMENTS

Early warning systems for water have been discussed extensively within the water sector, especially after September 11th. Extensive efforts at developing and implementing sensing technologies to detect the range of threats imagined for the water infrastructure have been undertaken throughout the United States and the world. While sensors capable of detecting all possible contaminants may still be the ultimate goal, the likelihood of achieving this goal for even the most important contaminants is still in question. Because of the vast array of potential contaminants, sensors specific to one or a small number of contaminants present a challenging problem to the water industry. It is highly unlikely that large arrays of multiple contaminant-specific sensors can economically be installed and operated in most systems. Complicating this even more is the recognition that sensors themselves only provide data at a location for a given time. An early warning system must be capable of integrating data from multiple sensors at multiple locations temporally as well as spatially, essentially transforming data into actionable knowledge. Figure 1 illustrates the concept described above.

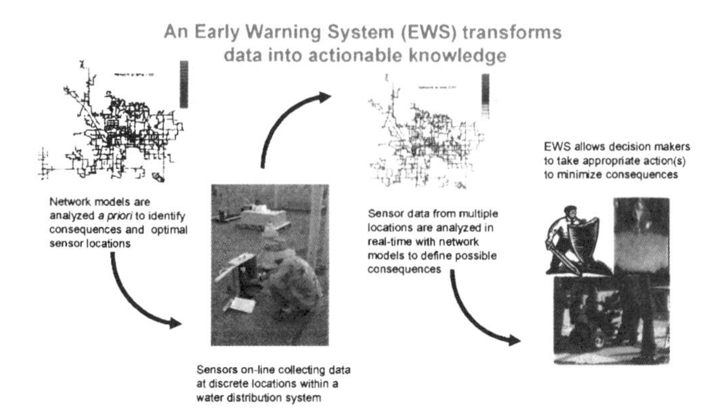

Figure 1 Conceptual illustration of an early warning system for water distribution systems.

As shown in Figure 1, an early warning system is comprised not only of sensors, but also models that can adequately and accurately describe the fate and transport of contaminants and can thus analyze the sensor data in real time so that the water system managers can make decisions and take actions to minimize the consequences of the contamination event. For an early warning system to be effective, the number, location, scan rates, and sensitivity of sensors must be incorporated. To accomplish this, models of the system hydraulics must be developed that can optimize sensor locations, identify the source of the contamination in real time, and include the uncertainties of the system. As will be described in the following section, the available models were not sufficiently robust for such complicated analyses particularly in real time. The early warning system (sensors/models) fits into the RAM-W vulnerability assessment process as shown in figure 2. This figure shows the method that was used to evaluate the risk to terrorist threats to water systems for the majority of the US drinking water infrastructure. The early warning system described above is really represented in the "Proposed Upgrades" compartment as they were generally not in place for the types of terrorist threats envisioned in the Bioterrorism Act.

The original VA's conducted in the United States tended to concentrate on the physical vulnerabilities and policy/procedures in place at the time. The dark blue text boxes (Background and Planning, Threat Assessment, Site Evaluation and Consequence, System Effectiveness, and Risk Analysis) essentially completed the assessment of vulnerabilities from physical attacks to critical assets. The results were described in terms of relative risk (between assets) in a semi-quantitative sense and Proposed Upgrades were limited to improved procedures/policies and/or physical protection around those assets with the highest relative risk.

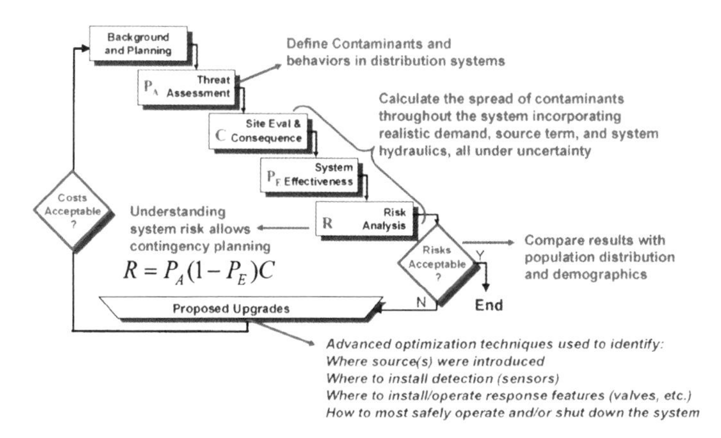

Figure 2 *How early warning systems integrate with the vulnerability assessments conducted in the United States.*

The distribution system, although identified as a high risk was generally not analyzed further. The red notation identifies how the early warning system water distribution system network modeling can be used within the RAM-W vulnerability assessment framework to vastly improve understanding of the risks to contamination within a distribution system. The coupled EPANET/DAKOTA model is capable of determining the spread of contamination within a network while taking into consideration the key uncertainties in the system hydraulics and contaminant. These results are compared with estimates of population demographics to obtain a number for the potential population affected above a given dose rate. This then represents the potential consequence from that specific contamination scenario. Naturally, multiple realizations (e.g. Monte Carlo) are needed to identify in a quantitative sense which parts of the system represent the greatest risk from a contamination event. These types of analyses would normally be performed *a priori* to give the water manager the understanding necessary to minimize the consequences from such an attack. Once this consequence assessment is completed for a given water distribution system, then upgrades can be recommended to further minimize these consequences (control risks). Some of these options are shown in light blue text accompanying the "Proposed Upgrades" box in Figure 2. These upgrades can include fully implementing an early warning system taking into consideration the optimum locations of sensors and development of response plans and protocols. Finally, when these analyses are completed and the proposed upgrades evaluated, the cost-benefit of any upgrades can be assessed and the upgrade implemented if necessary.

3 EARLY WARNING SYSTEM MODELLING: HYDRAULICS, OPTIMIZATION AND SOURCE INVERSION

As outlined in the previous section, an effective early warning system is comprised not only of robust hardware, but also robust models capable of transforming the data from the sensors to actionable knowledge. This section presents a brief overview of the modeling component of an effective EWS describing the hydraulic model, optimization techniques, and real-time source inversion. We have employed the USEPA hydraulics analysis package, EPANET for evaluation and design of water distribution systems although this commonly used tool does NOT incorporate risk, uncertainty, or optimization. How a contaminant will disperse in a distribution system will depend on the time of day it is injected, the location of injection(s), type of contaminant, the concentration, and injection rate. As there is no way to define each of these variables deterministically, the solution must be approached probabilistically. Within a water distribution system, pipe hydraulics is driven by consumer demand. However, there has been almost no effort to develop a realistic picture of water demands that is suitable for use in current network models. Hydraulic conditions change continuously over time and space in response to random demands imposed by many consumers dispersed throughout the service area. Current water quality models rely on arbitrary fixed demand patterns that effectively force quasi-steady plug flow conditions everywhere in the distribution systems. They also assume turbulent flow everywhere in the system (e.g. perfect mixing of solutes), do not include dispersion, and are limited to relatively simple first-order reactions. While these expedient assumptions are acceptable for general system design needs, they are insufficient (even inappropriate) for designing and evaluating water systems with contamination security in mind.

We have teamed with the EPA/NHSRC to develop a more robust modeling capability by coupling EPANET with DAKOTA, a SNL developed advanced non-linear optimization code capable of optimizing large datasets (e.g. water distribution networks) and accounting for uncertainties. These tools are being developed and tested on real water networks in the US on an experimental basis to evaluate the potential spread of contaminants. Figure 3 is a graphic illustration of the results of a series of Monte Carlo analyses of a water distribution system showing the location of the node representing the "worst-case" injection location.

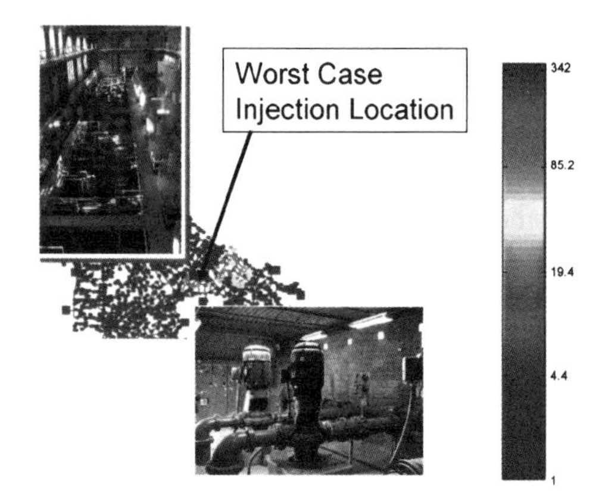

Figure 3 *Graphic illustration of the results of Monte Carlo analyses of contamination scenarios.*

Figure 3 highlights the potential number of nodes where the contamination rate exceeds a certain threshold and illustrates how important it might be to know before an attack where the most critical nodes are in a system. These analyses of potential consequence must then be evaluated for ways in which to upgrade the system. For a contamination event, these upgrades take the form of sensors (the hardware part of the EWS) or response features that allow changes to the system hydraulics to control the flow of the contaminant. For sensor placement, the optimum location is important because of the likelihood that only a limited suite of sensors can be placed within the system. The reasons to optimize are many: rapid detection, limit the spread of contaminants, provide for rapid in situ treatment, and to assist in developing realistic emergency response plans are but a few. From the perspective of an EWS, optimizing provides the very attractive opportunity to minimize the number of sensor locations, thus minimizing the cost. An example of a simple small-network optimization is shown in figure 4. This figure shows that for this small network of 470 nodes, that a large number of sensors doesn't decrease the risk for the assumed population of 7600 people significantly. In fact an increase from 10 sensors to 50 sensors decreases the number of affected nodes to 1/3 of the original number of nodes. Naturally, the water system manager must determine whether this level of risk (and cost-benefit) is of value for their particular situation. The figure does show that these analyses can assist in planning and risk minimization. It is also important to note that these analyses can also inform the management of a water utility as to the level of risk that is likely to be necessary for a given budget scenario and constraints.

In addition to optimizing the sensor locations for an effective EWS, the ability to identify where the source was introduced is critical to effective management of an emergency response. This is accomplished through multiple realizations of inverse simulations essentially resolving the system hydraulics backward from the point(s) of detection until a unique solution is found. Because of the complexity associated with inverse solutions and large data sets, we have developed sub-domain inversion techniques to expedite the analyses and reduce the time required to complete the analyses. We have developed both optimization and inversion algorithms capable of solving for skeletonized networks exceeding 5000 nodes with the inversion completing in near real time (<30 minutes). We continue to work on improving the algorithms to improve efficiency and scale.

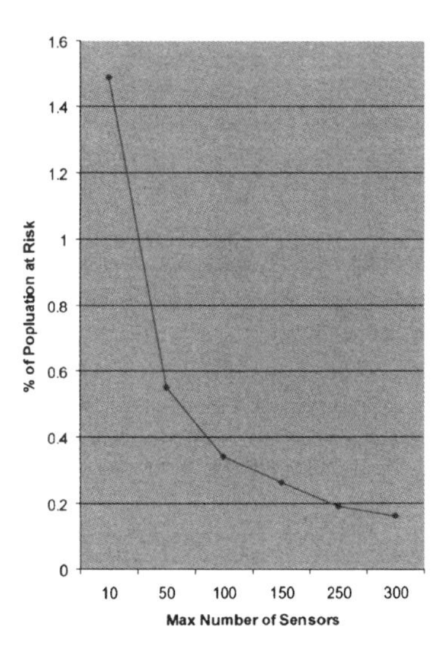

Figure 4 *Example optimization of sensors for a small water distribution system network*

4 ASSESSING THE THREAT

The threats to water distribution systems from potential contamination are numerous and inadequately understood. The VA's conducted for water systems in the US concentrated on the physical attacks and only recognized that contamination of the distribution system represented a critical threat. Also, the simple risk equation in RAM-W was only able to acknowledge threats in broad categories and was unable to provide additional fidelity to the risk estimation. To assist in wise expenditure of scant resources, the likelihood of attack should be a discriminator in estimating the vulnerability of a drinking water system. This led to the realization and development of a new approach to estimating the threats to water systems utilizing a Markov Latent Effects (MLE) approach. This technique was originally developed to evaluate the effects of aging aircraft components on safety. The technique essentially decomposes the problem (in this case the range of threats – physical, SCADA, and contamination) and decomposes each category and specific threat into decision elements. The process is accomplished through elicitation with the water utility experts using best available information and the interviews are driven by internal logic already incorporated in the software. Each decision element is assigned through this process with values and weights and the data are aggregated to quantify the "possibility" of events. Figure 5 shows a simplified analyses for a contaminant (undefined) and illustrates how the latent effects are aggregated with direct decision elements.

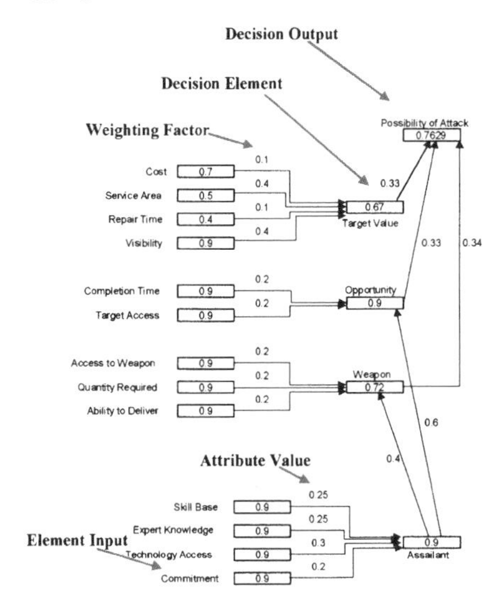

Figure 5 *Example Markov Latent Effect evaluation for an undefined contaminant*

Note that the technique results in a possibility of attack for this example of about 0.76. This number can then be incorporated into the simplified risk equation to yield greater fidelity than was heretofore used in US VA's. As a reminder, the simplified risk equation is:

$$R = P_a(1 - P_e)C$$

Where:

R = Relative Risk from a malevolent attack
Pa = Likelihood of attack (from the MLE analysis)

Pe = The effectiveness of the existing system to withstand the attack
C = Consequence from the attack (measured in terms of human health effects or $)

In Figure 5, the decision elements include attributes of the system, attributes of the weapon (contaminant, explosive, etc.), and attributes of the assailant(s). Each decision element is given an assumed likelihood, and a weighting factor. The weighting factors must sum to unity for each grouping of decision elements, including possible latent effects from other groupings of decision elements.

The overall outcome from application of the MLE threat tool will be to allow the water system managers to apply reason and logic to result in greater fidelity in their analysis of risks for critical assets within their system. In essence, it allows for the recognition that not all threats are equally probable and if it is applied uniformly, can result in a more realistic evaluation of the overall risks for their water system.

5 SUMMARY

This paper described the broad research program conducted by Sandia National Laboratories and in collaboration with the United States Environmental Protection Agency/National Homeland Security Research Center in the area of development of computational tools to support water system security. Specifically, these tools support the application of Early Warning Systems by acknowledging that an effective EWS incorporates effective hardware and effective models that transform the hardware output data into actionable knowledge. The linking of hardware with software, although incomplete and not fully ready for universal application shows promise for greatly improving the protection of water distribution systems as does the threat analysis tool for improved estimation of overall relative risk. Future efforts will concentrate on improving the computational capabilities by concentrating on the following areas of research:

- Effects of variable demand
- Effects of dispersion
- Role of time-steps in analyses
- Incorporation of uncertainties
- Continue to increase the size of the networks
- Improved chemical reactions
- Validating the threat analysis tool at real water systems

Acknowledgement

Sandia National Laboratories is a multi-program laboratory operated by Sandia Corporation, a Lockheed Martin Company, for the United States Department of Energy under contract DE-AC04-94AL85000.

References

1 J. Berry, L. Fleischer, W. Hart and C. Phillips, "Sensor placement in municipal water networks," World Water & Environmental Resources Congress, 2003.
2 J.W. Berry, W.E. Hart, and C.A. Phillips, "Scalability of Integer Programming Computations for Sensor Placement in Water Networks," World Water & Environmental Resources Congress, 2005a.

3 J. Berry, W.E. Hart, C.A. Phillips and J. Uber, "A general integer-programming-based framework for sensor placement in municipal water networks", World Water & Environmental Resources Congress, 2004.
4 J. Berry, W.E. Hart, C.A. Phillips, J.G. Uber and T.M. Walski, "Water Quality Sensor Placement in Water Networks with Budget Constraints," World Water & Environmental Resources Congress, 2005b.
5 J.W. Berry, W.E. Hart, C.A. Phillips, J.G. Uber and J. P. Watson, "Validation and assessment of integer programming sensor placement models," World Water & Environmental Resources Congress, 2005c.
6 R.D. Carr, H.J. Greenberg, W.E. Hart and C.A. Phillips, "Addressing modeling uncertainties in sensor placement for community water systems," World Water & Environmental Resources Congress, 2004.
7 C.D. Laird, L.T. and Biegler, B.G. van Bloemen Waanders, "A mixed integer approach for obtaining unique solutions in source inversion of drinking water networks," World Water & Environmental Resources Congress, 2005.
8 C.D. Laird, L.T. Biegler, B.G. van Bloemen Waanders and R.A. Bartlett, "Time dependent contamination source determination: A network sub-domain approach for very large networks," World Water & Environmental Resources Congress, 2004.
9 S.A. McKenna, B. van Bloemen Waanders, C.D. Laird, S.G. Buchberger, and Z. Li, R. Janke, "Source location inversion and the effect of stochastically varying demand," World Water & Environmental Resources Congress, 2005.
10 V.C. Tidwell, J.A. Cooper, C.J. Silva and S. Jurnado, "Threat assessment of water supply systems using Markov Latent Effects modeling," World Water & Environmental Resources Congress, 2004.
11 B.G. van Bloemen Waanders , "Application of optimization methods to calibration of water distribution system," World Water & Environmental Resources Congress, 2004.
12 B.G. van Bloemen Waanders, R.A. Bartlett, L.T. Biegler and C.D. Laird, "Nonlinear programming strategies for source detection of municipal water networks," World Water & Environmental Resources Congress, 2003.
13 B. van Bloemen Waanders, G. Hammond, J. Shadid, S. Collis and R. Murray, "A comparison of Navier Stokes and network models to predict chemical transport in municipal water distribution systems," World Water & Environmental Resources Congress, 2005.
14 J.P. Watson, H.J. Greenberg, and W.E. Hart, "A multiple-objective analysis of sensor placement optimization in water networks," World Water & Environmental Resources Congress, 2004.
15 J.P. Watson, W.E. Hart and J.G. Berry, "Scalable high-performance heuristics for sensor placement in water distribution networks", World Water & Environmental Resources Congress, 2005.

STRENGTHENING COLLABORATIONS FOR WATER-RELATED RISK COMMUNICATIONS

R. Parkin[1], Lisa Ragain[1], Heidi Urquhart[2] and Paula Wilborne-Davis,[3]

[1] The George Washington University Medical Center, Washington, DC, USA;
[2] National Association of County and City Health Officials, Washington, DC, USA
[3] Association of Occupational and Environmental Clinics, Washington, DC, USA

ABSTRACT

Collaborations between water utilities, public health and clinical personnel are essential during water contamination events. Few formal mechanisms are in place, however, to ensure that effective risk communications will occur either on an emergency or routine basis. The focus of this AwwaRF-sponsored project is on advancing three-way collaborations by building on lessons learned from recent initiatives, developing a Framework for Action to facilitate collaborations, and evaluating it under different scenarios. The foundation for the Framework includes literature reviews, surveys, interviews, workshops, and tabletop exercises. More that 100 utilities, 160 health departments and 40 clinicians have contributed to this project. Most of the utilities were publicly owned and served up to 100,000 customers. Ninety percent had worked with local public health agencies (LPHAs); one-third with clinicians. The participating clinicians were mostly environmental health specialists. Although they were willing to work with utilities, very few clinicians had ever done so. Utility, LPHA and clinical personnel felt that the public's concerns about drinking water issues are growing. Key themes that emerged across all project activities relate to establishing and sustaining long-term, trust-based relationships; identifying appropriate organizations and individuals with whom to interact; developing mechanisms to institutionalize relationships; and building knowledge about each others' interests, priorities, roles, and responsibilities. Utilities and clinicians prefer health departments to inform clinicians about water issues, but LPHAs do not have the resources for this role. Our Framework for Action assists potential partners with: ascertaining their current resources, roles, and responsibilities; creating routine and emergency response plans; and evaluating collaborations. The Framework was evaluated in tabletop exercises. The project results related to emergency conditions are the focus of this paper.

1 INTRODUCTION

Following several recent drinking water-related health emergencies in the United States, water utility professionals raised concerns that clinicians were not available or sufficiently informed to collaborate effectively with utilities. Additionally, medical expertise had not been well integrated into planning, drilling or implementing risk management protocols for emergencies. Although some local public health departments coordinated with utilities and clinicians, there were no clear or consistent mechanisms linking clinicians and utilities. Some utility professionals reached out to clinicians, but the response was usually minimal. Through a Request for Proposals, the Awwa Research Foundation (AwwaRF) sought to sponsor research that would result in new knowledge for improving collaborations among utilities, LPHAs, and clinicians.

The project results that serve as the primary basis for this paper relied on peer-reviewed methods and literature from several fields – especially psychology, risk communication, statistical modeling, and decision analysis. The project (AwwaRF Project #2851) included a series of scientifically based activities designed to provide water utilities, the public health community, and clinicians with new knowledge, strategies and tools for collaborating on water quality issues.

From our earlier research in Iowa (AwwaRF #2776)[1], it was apparent that even well informed consumers had difficulty with the term "emerging contaminants." The phrase was unclear to many and suggested differing things to different people. Some thought it referred to emerging conditions, rather than contaminants; others thought it was an inevitable problem with an uncontrollable timeline. Many non-experts – including people who could accurately define "emerging" – rapidly linked the term to other concepts, most notably "emergency" conditions. This cognitive connection is likely to block people's ability to hear or understand non-emergency messages about "emerging contaminants." One of the impacts of this particular conceptual linkage was that people seemed confused when water utilities did not respond to "emerging" contaminants in an "emergency" mode. Consequently, some people may think that utilities are non-responsive or uninterested in public concerns when they treat "emerging" issues as if they do not need active responses. Although experts appear largely unaware of this non-expert view of "emerging," it is important to acknowledge that public risk perceptions may involve more than what experts would consider real emergencies.

On the organizational level, the participating water utility (Des Moines Water Works, DMWW) had an Emergency Operations Plan designed to address hazardous materials in water. Measures of success under this plan included timely, clear, and complete information provided to the public by appropriate officials. The effectiveness of the information would in part rely on the knowledge available to develop public information, understanding of risk communication methods and tools, and DMWW's sense of obligation to inform the public. Additionally, having positive relationships with the mass media was and remains one of DMWW's strategic priorities because their customers expect to hear about emergencies first through the media.

The earlier emergencies – real or perceived – can be recognized, the more lead time organizations have to design and implement responses to them. Effective risk management strategies and approaches need to be in place before public responses become necessary, especially under emergency conditions. The management of environmental health risks is a multi-step process that relies on sound science,[2] as well as effective risk communication and collaboration strategies and tools.[3] Though the value of scientifically based information for designing effective risk communication approaches has grown over the past few decades, it has been valued more slowly than the science for other parts of the risk management framework. As a result, scientific findings have often not been considered or used in the development of risk communication or collaboration strategies. Our research was designed to expand the knowledge base that would inform the development of these strategies and tools for both routine and emergency issues. The purpose of this paper is to highlight the project results related to emergency conditions.

2 METHODS

The project described in this paper involved gathering and assessing peer-reviewed literature and field data about multi-organization communication and collaboration experiences; and the subsequent development, testing, and distribution of an evidence-based framework. The specific activities in the project were: literature reviews; interviews of water utility, public health, and clinical personnel; a two-day interdisciplinary workshop involving nine focus groups; and one-day exercises, including two scenarios each, in each of the five participating regions (Glendive, Montana; Lansing, Michigan; New York City, New York; Redmond, Washington; and Tucson, Arizona). All research materials and procedures were reviewed and approved by The George Washington University Institutional Review Board (IRB) (IRB #U070222).

We analyzed peer-reviewed and field-based literature using keywords to search a variety of electronic sources of peer-reviewed public health and science literature. We used non-academic (often governmental) resources to identify field evidence. Local public health agencies included in the field evidence search were identified through peer-reviewers, national associations, and referrals obtained from public health professionals. The search for field evidence focused on expanding our understanding of existing collaboration structures.

1. To obtain information about recent and existing communication collaborations, we conducted telephone surveys of water utility and public health agency personnel who were responsible for communication and/or water quality issues in the United States. Membership lists from the American Water Works Association (AWWA) and the National Association of County and City Health Officials (NACCHO) were sampled. The final telephone survey samples included 98 water utilities and 160 health agencies. Thirty practicing clinicians, who were members of the Association of Occupational and Environmental Clinics (AOEC), were interviewed by telephone to determine their views about water-related health risk issues and communications.
2. Nine focus groups were hosted during a two-day workshop convened in March 2004. Facilitators guided the groups' discussions using a set of open-ended questions. The workshop participants included water utility personnel, public health agency staff, physicians, elected officials, and academicians. Four discipline-specific focus groups were convened: water utilities, public health agencies, clinicians, and other stakeholders. A focus group was then convened for each of the five regions.
3. From November 2004 through January 2005, a one-day group discussion – including two different scenarios – was held in each of the participating regions. The prior workshop participants and anyone else the utility wanted to include were invited; about 45 people participated in these on-site activities. While "tabletop exercises" are typically designed to test a specific framework or protocol in order to achieve a particular goal, the March 2004 workshop results indicated that no single framework could be developed that would be

suitable for all five regions. Additionally, follow up phone calls to each utility indicated that limited progress toward collaboration had been made, suggesting that a needs assessment and planning tool would be more appropriate and useful for the participants. We therefore designed on-site exercises to evaluate a draft assessment and planning Framework, and to have each site work through two different hypothetical drinking water-related health scenarios. F ive scenarios were based on concerns stated in the follow up telephone calls. Each scenario was different – one was an acute microbial pathogen contamination and one was a sudden chemical hazard. Each of the five scenarios was tested by two utilities and their partners; no combination of scenarios was repeated.

4 RESULTS

4.1 Literature and Field Evidence

Despite our extensive literature searches, no past or current drinking water-related collaborations between water utilities, public health, and clinical medicine were found. However, the field evidence indicated that LPHAs have good working relationships with water utilities. Whether the communication is mandated, voluntary, formal, or informal, the overall finding from the literature is that communication *is* taking place at the local level.

Nine of 11 jurisdictions evaluated had evidence of water-health collaborations, but the collaborations varied in nature and subject matter. Formal agreements with water utilities tended to be written at the state, not local, health agency level. New York City and Boston have communication agreements as a result of United States Environmental Protection Agency (EPA)-initiated lawsuits concerning unfiltered water systems; these agreements focus on enhanced surveillance systems and data sharing, but also include general emergency plans, monthly conference calls, and mutually developed contact lists. Some large utilities (e.g., San Francisco and New York) fund full-time positions in the LPHA to conduct disease surveillance and provide other support related to waterborne contaminants. If a violation or other emergency situation occurs in Portland (Oregon) and poses an acute health risk, a public notice must be provided to all persons who reside and are served by the water system in that jurisdiction. The city has a general plan for boil water alerts (BWAs) but not for other emergencies. The comprehensive emergency management program utilizes the incident command system, so the Incident Commander is ultimately responsible for making the decision to issue a BWA. Prior to public delivery of any water emergency notification, the public information officer, water quality and treatment manager, members of the state's Health Division, County Health Officer, and other officials must be notified.

The Sydney (Australia) health department and water utility have a formal Memorandum of Understanding that describes their mutual responsibilities to protect the public's health; calls for sharing of strategies, long-term and annual plans, monitoring data and information; and guides access, consultation and advice under routine and emergency conditions.[4]

4.2 Individual Interviews

Most of the water utilities surveyed were publicly owned (92%) and served up to 100,000 customers (66%). Participants indicated that the utility worked with local and state health agencies (89% and 88%, respectively) more than federal agencies (64%); most communicated on a monthly basis. Seventy-nine percent of the utilities had designated a specific individual to communicate with health agencies; about half said this was the utility director. Only 30% had a formal agreement with a PHA. Respondents characterized their relationships with PHAs as excellent (53%) or good (34%). In the five years before the survey, 62% of the utilities had worked with a Public Health Agency (PHA, at an unspecified level of government); 87% had collaborated to develop emergency response plans related to water security. Very few utilities (32%) had worked with Health Care Providers (HCPs) on drinking water issues; among these, 74% contacted them once a year or less. In the five years before the survey, less than 28% of the utilities had communicated with HCPs about general *or* emergency drinking water issues. Only 17% said that their organization worked with HCPs to develop water security plans, but 68% included communications with HCPs in the plans.

Most of the LPHAs in the survey were county-level organizations (67%) that served up to 100,000 people (56%). Agencies in the Midwest (42%) were over-represented compared to the rest of the nation. Thirty-one percent of the LPHAs communicated with utilities weekly or more often. Two-thirds said their LPHA had designated a specific person, usually a manager (28%), to coordinate with water utilities. Eighty percent of the LPHAs did not have a formal agreement with a utility; 18% did. Fifty-seven percent had worked with a water utility in the past five years; 69% had worked on an emergency response plan. Overall, LPHAs rated their relationships with local water utilities as excellent (29%) or good (53%). Fifty-four percent stated that they had ever worked with health care providers; among these, half communicated with HCPs about drinking water once a year or less. In the five years before the survey, LPHAs had communicated with clinicians about general (41%) and emergency (34%) drinking water issues.

The clinicians in the telephone interviews tended to be male (63%) physicians (87%), who had been in practice for over 10 years (83%) in small specialty groups (60%) serving areas of over 500,000 people across the U.S. (50%). Ninety percent said patients asked questions about drinking water issues. When they sought information about waterborne diseases, most turned to online tools – such as Medline (83%), and the Centers for Disease Control and Prevention (CDC) (77%) – rather than water utilities (27%). Two-thirds of the participants characterized their relationships with PHAs as excellent or good. Also, 66% ever contacted a PHA about a drinking water related issue; among these, 50% contacted a specific program and 60% contacted a specific person, preferably the Health Official or a clinician. Thirty percent had been contacted by a PHA for a referral or expertise. Clinicians viewed their PHA partnerships as clinically relevant, and successful or very successful (both at 71%); the least successful activities were responses to emergencies. Thirty percent of all respondents had worked with a PHA on emergency response planning. Most (90%) clinicians had never been contacted by a water utility; only two participants had worked on emergency response plans. Nearly all respondents were willing to collaborate on drinking water issues; half would do so monthly or quarterly.

4.3 Group Interviews

In the workshop, all four discipline-specific focus groups saw positive relationships as crucial to success in handling water-related health issues. Many people stated that these relationships need to be built slowly and consistently over time; they should not be expected to just happen during a crisis. Participants stressed that problems need to be addressed upfront with a maximum degree of trust and credibility. All groups indicated that they felt the public has growing concerns about water issues; it was thought that some of these concerns result from the public seeing confusion rather than clear leadership during emergency water events.

Utility participants expressed their frustration with PHA staff not reading water quality reports and/or not being able to interpret the reports. The utilities had very limited relationships with clinicians, and did not see the need for more active relationships. They expected to interact with clinicians through PHAs. While they wanted customers to take their health questions to clinicians, they said clinicians have little knowledge of water, water treatment or water systems, and therefore may not be able to address patients' concerns.

The LPHA representatives stated that personal relationships with utility staff were crucial; knowing the specific utility person with whom they must work was very important. LPHA personnel acknowledged that there are many utilities for them to know about, so they tended to focus on the biggest utility's Public Information Officer (PIO) because they knew PIOs were expected to return calls the same day. LPHA personnel want to know more about specific contaminants, crises, water supply systems, and what communities need to know. They said they have strong relationships with clinicians, but communications may not be frequent except in emergencies. LPHAs would like to involve more clinicians in emergency drills.

The clinical participants felt that food-borne hazards, especially pathogens, were more important than waterborne ones. They discussed how environmental health concerns, including water, are not within most clinicians' mindsets. Most indicated that their medical training did not emphasize or even include environmental health training or relationships with the PHAs. Half indicated that they had easy, direct and positive relationships with the LPHA, but in crises the nature of these relationships changes. One participant said, "Identify a person to communicate with other populations... [Find] a trusted person in the community. Find them *before* the crisis."

In the scenarios conducted on-site in the five participating regions, most participants were unfamiliar with each other's roles and responsibilities, did not have clear means to involve Public Information Officers, and did not have an accurate understanding of what "risk communication" is. Risk communication was understood by most as crisis communication, but it actually involves much more.[1] Questions asked and assumptions expressed indicated that confusion would occur, especially under emergency conditions. One participant said "governing myths" could cause confusion that would impede responses and frustrate the public during fast-breaking events. One area of confusion was who should contact whom. For example, clinicians expected utilities to contact them, but utility personnel did not know when to contact clinicians or PHAs. Physicians assumed PHAs would review and synthesize key information, and send it to them in relevant and timely summaries. However, PHA personnel said they did not know what physicians want or need, and did not have contact lists to reach all clinicians. Further, PHAs did not have the resources to expand or keep these lists updated. One person suggested establishing a protected website where clinicians statewide could go in and change their own contact data whenever necessary.

There was a lack of consensus about how to foster collaborations. Some participants liked formal Memorandums of Understanding or other specific instruments, but others raised concerns about them. One person said, "They're not worth the paper they're written on. Good relationships are more important."

Several regions discussed interagency emergency response debriefing sessions and their failure to reconvene, avoid blame, and transform important observations into operational improvements. Participants felt reforming the debriefing process may be important for more effectively meeting their public service obligations. Additionally, effective emergency response plans (ERPs) and water security plans should involve public health officials, clinicians, and various medical and nursing facilities.

Both acute and chronic risk scenarios were used in the regional exercises; most people found the acute (emergency) ones easier. There were two emergency scenarios that were used in three regional workshops; one was a microbial issue and the other was a chemical contamination. Both were drilled in the smallest participating town, and one each in a suburban area and a mid-sized city. Personnel in all areas were fairly clear about what they or their agency would do in the initial phases of the scenarios, but most became less clear about legal authorities, roles, and obligations as the scenarios evolved. People in the suburban area were the least familiar with each other and the town's ERP or communication systems, but they readily used their ERP with its chain of communications. Although the lead agency for each scenario in each location was clear, the smallest town's utility personnel said they likely would have handled the scenario without involving the PHA or clinicians. The two non-suburban towns raised liability and debriefing concerns as the scenarios progressed. Needs for better training, planning, and improving emergency risk communications and collaborations were identified in each region, but not all participants valued the same approaches. For example, one person did not see the need for formal agreements, while another person in a different area felt strongly that a formal agreement was essential before another emergency arose.

The Framework for Action and a tool to facilitate cost-benefit assessments of collaboration options are currently being revised based on the regional workshop results. These will be published in the final report for the project.[5]

4.4 Synthesis of Results

From these findings, we noted that individual professionals had some correct and some incorrect knowledge about each other's perspectives, interests, authorities, roles, and responsibilities. Further, many participants revealed gaps in their knowledge or important assumptions that would affect their performance and collaborations, especially under crisis conditions. Utility personnel were somewhat surprised when health and clinical professionals did not know the meaning of CCR (Consumer Confidence Reports) and other common terms used by utilities.

Additionally, participants had important conceptual constraints that limited the effectiveness of their mutual problem-solving. Although they mentioned other clinicians (e.g., nurses, pharmacists, dieticians, dentists, specialty physicians, etc.) during the scenarios, none of the participants used "clinician" to mean people other than physicians and none seemed to realize the key roles these other clinicians could play. Participants also consistently revealed limited abilities to identify a broad range of potentially important community "partners" (e.g., charitable organizations for indigent or vulnerable populations, neighborhood leaders, faith-based leaders, etc.). This raises the concern that even if utilities, PHAs and physicians begin to collaborate they may not develop other partnerships that could be crucial especially in crisis conditions.

An organizational issue that emerged during nearly all regional exercises was the confusion utilities, PHAs and clinicians felt about the many entities with which they could work. Only New York City had PHA and water service areas that were the same; all other regions had several utilities and/or several PHAs serving different areas. Consequently, utilities were often unclear about which PHA they should work with and PHAs were unclear about which utility to work with. In a few cases, personnel had already worked with multiple entities, but questions lingered about whether it was politically wise to work with no, one, some or all of the possible entities. Typically, utilities communicated with neighboring utilities and LPHAs shared information with adjacent LPHAs. What was surprising was how unaware organizations were of each other's confusion regarding organizational relationships.

While the participants readily recognized some problems with their communication and collaboration, the groups did not seem to realize other important problems. The participants knew they needed to do more comprehensive planning, receive more training, conduct better drills and debriefings, and update their contact lists and procedures. Some said that addressing public confusion during emergencies was essential, but did not mention the potential value of involving community partners to help reduce confusion. Participants also knew that, before they have a crisis to deal with, they must build stronger, trusting personal relationships with each other; however, they were uncomfortable with the time and effort that process would require. Problems identified by the research team but not noted by the participants included: their limited framing of "clinicians" and "community partners," their often limited approach to acknowledging and addressing public fears, their lack of pre-testing risk communication "messages" and materials, and lack of knowledge about systems-oriented and strategic approaches to risk communication and collaboration.

5 DISCUSSION

The project described here had many scientifically based data gathering steps, each designed to complement the information gained by the other methods. When the insights derived from all components were combined, they provided a richer understanding – than any one method could have provided – of the issues regions face when they address water-related health risk concerns.

Another strength of the project was the participation of utility, health agency and clinical personnel from all areas of the United States and all types of jurisdictions. Further, participants in both the individual interviews and group processes were professionals who typically had ten or more years of work experience to draw on.

The participation rate for the individual interviews was less than we had expected. Even though the participants were similar to the non-participants, the survey results may not be reliable enough to generalize to a national scale. Additionally, specialty clinicians were selected based on their expertise and self-identified interest in environmental health issues. However, these two "predisposing factors" to the subject of our work still was not sufficient to elicit the number of physicians, nurses, and physician assistants that we had anticipated.

Despite these limitations, the results are consistent across the individual and group interviews. Collaboration is commonly reported as difficult and time consuming, due to the complex challenges and sustained resources required for making it successful. Physicians find it difficult to justify the time away from serving patients and other obligations to work on building emergency response capabilities that they may never have to use. Building trust among potential partners is widely recognized as important, but people often are not fully aware of the impacts their behaviors and choices have on the trust that others place in them.

6 RECOMMENDATIONS

Considering all project results, recommendations were developed to help organizations work toward effective collaborations. For example, very few of the participants seemed to be aware of risk management frameworks and how they could be used to conceptualize potential partnerships for each step in the risk management process. Whether a risk is routine or an emergency, the partners needed may vary across the process. Risk management frameworks could be used as tools to facilitate dialogues for building common understandings and strategic approaches to communication and collaboration.

Being strategic also requires greater knowledge of what "the risk problem" is, not just from professional viewpoints but from community perspectives as well. The more heterogeneous the population served and the more complex the risk issues, the more important science-based and community-involved approaches will be for developing in-depth knowledge about stakeholders' risk perceptions, beliefs, and expectations. The advanced insights gained –from rigorous, sustained approaches for building this knowledge base – will be important throughout the risk management response, whether it is an emergency or long-term response. Assuming what people think and feel about environmental health risks has repeatedly produce poor results, especially in crisis conditions.

Organizations need to create incentives for their staff to commit to collaboration efforts, find resources (e.g., grants) to support staff for their participation, and encourage them to build relationships with personnel in other entities that will be present in crisis and routine risk management environments. Organizations also need to broaden their concepts of partnerships both within and outside of their domain, allow sufficient time for collaborations to mature, and recognize assumptions and conceptual constraints that may hold staff and the organization back from new approaches. It will also be important for managers to foster a work culture in which "governing myths" and incorrect knowledge can be modified constructively, thereby opening up possibilities for innovative problem-solving and partnering. The process of change is typically challenging for most people and very stressful for some, but it can be enjoyable and rewarding, especially when organizational leaders support shared learning. One option would be for utilities to contract specialty clinician time a few days a year to ensure clinician involvement in meeting routine and emergency needs. As one utility manager said, "If we don't have more understanding up front, our bottom line will be personalities. That's a bad place to be when supplying public services."

7 ACKNOWLEDGEMENTS

The George Washington University (GWU) and the authors gratefully acknowledge that the Awwa Research Foundation is the joint owner of the technical information upon which this paper is based. The George Washington University thanks the foundation for its financial, technical, and administrative assistance in funding and managing the project through which this information was discovered. The authors also thank the USEPA for its support of AwwaRF Project #2851, and our subcontractors – Decision Partners, LLC, (AwwaRF Project #2776) and OpinionSearch (AwwaRF Project #2851) – who assisted with the research presented in this paper. The reviews conducted by members of NACCHO and AOEC, and by the Peer Advisory Committees and Project Officers (India Williams and Linda Reekie) for our AwwaRF projects strengthened our research activities, books and other products. The authors would also like to acknowledge the contributions of GWU, NACCHO, and AOEC administrative and technical staff members who assisted us with these projects. Further, the results presented in this paper would not have been possible without the generous participation of many water utility, public health agency, and clinical personnel.

References

1 R. Parkin, L. Ragain, M. Embrey, C. Peters, G. Butte and S. Thorne, "Risk Communication for Emerging Contaminants," Awwa Research Foundation, Denver, CO, 2004.
2 P/CCRAM (Presidential/Congressional Commission on Risk Assessment and Management). "Risk Assessment and Risk Management in Regulatory Decision-Making; Final Report, Volume 2." Washington, D.C., Presidential/Congressional Commission on Risk Assessment and Management, 1997.
3 CSA (Canadian Standards Association). "Q850 Risk Management: Guideline for Decision-Makers." Toronto, Ontario, Canadian Standards Association, 1997.
4 Sydney Water Corporation. "Memorandum of Understanding between NSW Health and Sydney Water Corporation." Sydney, New South Wales, Australia, Sydney Water Corporation, 2001. (Last accessed at http://www.sydneywater.com.au/Publications/#MoU on May 4, 2005.)
5 R. Parkin, L. Ragain, R. Bruhl, H. Urquhart and P. Wilborne-Davis. "Advancing Collaborations for Water-Related Health Risk Communication," Awwa Research Foundation, Denver, CO, to be published in 2006.

RISK ASSESSMENT, PERCEPTION AND COMMUNICATION - WHY DIALOGUE IS POLITIC

H. Mallett.

Consulting Group Director, Enviros.20-23 Greville St, London EC1N 8SS (0207 421 6388). Hugh.mallett@enviros.com

1 BACKGROUND

Like many other countries in the industrialised world, the UK is now dealing with many sites of contaminated land arising from our industrial past. Policies and legislation are now being implemented to deal with that legacy which has been brought into focus by Government's targets for the re-development of such sites. The management of this contaminated land legacy is based upon a risk based approach. Such an approach should be systematic and objective and it should provide a scientifically sound, robust and defensible basis on which the options for mitigating such risks can be considered.

The scientific and technical aspects concerned with the assessment of potential risks associated with land contamination are generally well known and understood by many of the involved parties. However, perhaps what up until now has been less well recognised particularly by problem holders and their consultants, is the need for and mechanisms of, communication with all stakeholders in this process.

2 PRINCIPLES OF STAKEHOLDER DIALOGUE

It is important to recognise that when contaminated land projects are initiated, an appropriate degree of attention must be given to dialogue with stakeholders. Such a dialogue must be an inclusive process involving all groups who may have an interest in the outcome of the project. This may run counter to the instincts of some problem holders, but experience has shown that a properly conducted process, involving all stakeholders (including those who are often deliberately excluded) will often maximise the "buy-in" of stakeholders and thus ensure that the solutions arrived at will be supported and will survive over the long term.

The principles of effective stakeholder dialogue can be summarised as;

- Identifying who all the stakeholders are;
- Recognising why you are engaged in the process;
- Clearly defining what you are trying to achieve, and
- Determining how you will attain those objectives.

3 THE INVOLVEMENT OF STAKEHOLDERS

There are several ways in which information between stakeholders can be obtained and exchanged. Each of these have their own particular characteristics and thus influence the potential outcome. It is therefore important that before engaging with stakeholders, consideration must be given to what all parties are likely to want to come away with from any such involvement. So for example, if the problem holder determines that they want only to provide information to stakeholders they may choose to provide limited information by means of a simple announcement. They may not wish to gather or listen to any views of the other stakeholders. The results of such "involvement" are likely to depend upon degree of authority of the problem holder and the perceptions of that authority by stakeholders. However, there is a real possibility that stakeholders may react adversely to such a one-way communication and positions will become polarised as a result.

Therefore at the commencement of the process it is important to be clear about what objectives there are for both the problem holder and the various stakeholders. This will then enable determination of what type of engagement is most appropriate. So for example, awareness raising or information giving will tend to be one way communication with specific objectives (e.g. of simply providing the results of a decision or explaining some issue to change prevailing attitudes). Two-way communication becomes an essential element where engagement involves consultation, involvement or partnership. These processes involve an increasing level of engagement with stakeholders.

There are occasions however when serious consideration must be given to not entering into a dialogue with stakeholders. For example, if a decision that cannot be changed has been made, a 'consultation' process will merely raise expectations and will inevitably disappoint. Similarly if there is no time available, no commitment from senior management, or if key stakeholders will not attend, there is no real prospect of success for a process of 'dialogue'.

In 2003 the government (HM Treasury and Cabinet Office) listed five principles for good risk management, namely;

- Openness and transparency – make public; risk assessment, the decision making process, admission of mistakes.
- Involvement – actively involve significant stakeholders, two-way communication at all stages, clarification through open discussion, balance conflicting views.
- Act proportionately and consistently – action to be proportionate to protection needed and targeted to risk. Consistent approach to risk. Precautionary principle. Revisit decisions as knowledge changes.
- Evidence-based decisions – consider all evidence and qualify before making decisions. Seek impartial/informed advice. Absence of evidence is not absence of threat.
- Responsibility – those that impose risk also take responsibility for control and consequences of inadequacy. Where feasible, give individuals choice (where others not exposed to disproportionate risk / cost). Identify where responsibility lies.

4 RISK ASSESSMENT

Risk assessment is the process of collating known information on a hazard or set of hazards in order to estimate actual or potential risks to receptors. The receptor may be human health, a water resource, a sensitive local ecosystem or even future construction materials. Receptors can be connected with the hazard under consideration via one or several exposure pathways (e.g. the pathway of direct contact).

Risks are generally managed by isolating the receptor, or by intercepting the exposure pathway. Without the three essential components of a source (hazard), pathway and receptor, there can be no risk. Thus, the mere presence of a hazard at a site does not mean that there will necessarily be attendant risks.

By considering the source, pathway and receptor, an assessment is made for each contaminant on a receptor by receptor basis with reference to the significance and degree of the risk. In assessing this information, a measure is made of whether the source contamination can reach a receptor, determining whether it is of a major or minor significance. The exposure risks are assessed against the present site conditions.

Such an assessment of risk is based upon the procedure outlined in DETR Circular 02/2000. In addition DEFRA (formerly DETR), with the Environment Agency and the Institute of Environment & Health, has published guidance on risk assessment (Guidelines for Environmental Risk Assessment and Management). This guidance states that the designation of risk is based upon a consideration of both:

- The likelihood of an event (probability); [takes into account both the presence of the hazard and receptor and the integrity of the pathway].
- The severity of the potential consequence [takes into account both the potential severity of the hazard and the sensitivity of the receptor].

Probability (likelihood)		Consequence			
		Severe	**Medium**	**Mild**	**Minor**
	High Likelihood	Very high risk	High risk	Moderate risk	Low risk
	Likely	High risk	Moderate risk	Moderate / low risk	Low risk
	Low Likelihood	Moderate risk	Moderate / low risk	Low risk	Very low risk
	Unlikely	Moderate / low risk	Low risk	Very low risk	Very low risk

Table 1 *Probability and consequence to determine risk*

Such consideration can then lead to a classification of risk, with each of the risk categories having a well defined meaning.

Term	Description
Very high risk	There is a high probability that severe harm could arise to a designated receptor from an identified hazard at the site without appropriate remedial action.
High risk	Harm is likely to arise to a designated receptor from an identified hazard at the site without appropriate remedial action.
Moderate risk	It is possible that without appropriate remedial action harm could arise to a designated receptor. It is relatively unlikely that any such harm would be severe, and if any harm were to occur it is more likely that such harm would be relatively mild.
Low risk	It is possible that harm could arise to a designated receptor from an identified hazard. It is likely that, at worst if any harm was realised any effects would be mild.
Very Low risk	The presence of an identified hazard does not give rise to the potential to cause harm to a designated receptor.

Table 2 *Classification of risk*

The process of risk assessment in contaminated land projects is therefore now generally well developed, understood and delivered by the many specialists involved in the area. The use of simple tools has improved consistency in qualitative assessment and the continuing development of numerical systems has lead to widespread acceptance (at least in the industry) of the value and validity of quantitative risk assessments.

However, there is a common failing within problem holders and their consultants in understanding that this stepwise process of risk assessment, which to us appears straightforward, and scientifically robust, is not so logical to many stakeholders. It does appear to be a commonly adopted position that because our 'expert' assessment shows the level of risk to be 'acceptably low' this should be automatically and universally accepted by other stakeholders (who are necessarily less 'expert' than ourselves). In adopting this stance, we are failing to recognise that factors other than the technical assessment can have a significant influence on people's perception of risk. In fact "Perception is Reality" [Sniffer 1999].

5 RISK PERCEPTION

People with different social, economic and cultural backgrounds living in different places will perceive risks in different ways which reflect their own particular knowledge and their environment. That is, people's response to a particular hazard depends upon their perception of the hazard and their knowledge/awareness of both themselves, and "society" to deal with that hazard.

The risk perceived by people may also reflect; the potential for the hazard to be controlled, the potential for catastrophe and their "dread" of that hazard. So for example if a site has radioactive contamination associated with it, the perceived level of risk is likely to be high, reflecting peoples dread of radioactivity. Such dread reflects for example the known effects of fallout from nuclear explosions and visions of Hiroshima. The less familiar people are with a hazard and the less control they have over the potential for exposure, the greater the perception of the risk.

Perception may also be affected by whether or not exposure to the risk is voluntary (when people are more prepared to accept risk, or their perception of such voluntary risk is less than when they have no choice about the matter). For example the risk of knowingly ingesting known carcinogens directly into your lungs several times a day is perceived to be lower by smokers because this activity is undertaken voluntarily. This must be contrasted for example with the circumstances where residents are told that they have been living on the site of historic contamination where say polyaromatic hydrocarbons (PAHs) may be present at concentrations above background levels. The combination of the dread associated with the term 'carcinogenic' and the involuntary nature of the exposure will serve to increase the perception of this risk well above the level of any technical assessment.

We must therefore recognise that risk assessment carried out by experts can not be "absolute" because;

- Experts themselves may be biased or motivated by their own values/self interests; and
- This expert assessment takes no account of the beliefs of 'the public'.

What often takes place in the contaminated land debate is that the expert defines the objective risk and then tries to align the perception of the public and the regulator with this version of reality (or "the truth"). As described above, often no recognition is made by the expert that factors other than the technical estimation of risk influences how people perceive and behave in the face of particular hazards.

Decisions taken with regard to a particular risk are not driven only by calculation of probability. For example, relative estimates of risk have not figured at all in the recent 'debate' about GMOs, and a major factor in the opposition of this technology is a lack of trust in those "in control" of the technology or regulating the risks. It has also been shown that an important element in the perception of risk is the particular personal disposition of an individual: – i.e. people perceive risk as more or less difficult depending on the way in which they see the world (e.g. whether they are fatalists, egalitarians etc).

So it is important that we recognise this complex series of issues which influence the perception of risk, because if we don't, we will not be able to develop an effective means of communicating that risk. The key is in the risk – benefit communication which enables people then to make an informed choice regarding exposure to a particular hazard. Risk communication should not be seen as top down (i.e. expert to public) but as a constructive dialogue between all parties.

We must also understand that the media have a key role in the public perception of risk. This is because the media are likely to influence judgements about risk much more than people's objective assessment of the 'facts'. The media may or may not directly influence what someone thinks, but the amount of coverage given to an issue can make issues appear significant / important. Social amplification of risks may also occur when an event associated with a hazard interacts with the psychological, social and cultural make up of people, raising (or reducing) the perception of risk and affecting how people then behave. Some researchers have noticed that in some cases, increased coverage actually leads to an increase in factual information (and thus people's ability to make "proper" risk judgements). Conversely, the use of headlines and photographs can disproportionately affect the perception of the hazard and the emotional tone of the article.

It is clear that success or otherwise of communications regarding risk involves the consideration of both;;

- the message about the hazard itself (its likelihood, potential costs and benefits in mitigation)
- the trust people have in those giving the message.

Failure in risk communication is usually caused by public distrust in both the makers of policy, and officers of companies/regulatory bodies, due to problems of credibility. ["They would say that wouldn't they"]. Experts presenting technical numerical information about risk which discounts the public perception of that risk as 'irrational' become distrusted by the public who view them as arrogant and lackeys of vested interests. It is also important to understand that it is much easier to loose someone's trust than to build it. Once trust is lost, it is very difficult to regain. The key elements in trust are the competence/credibility of the people putting the argument. People's perception of this is often significantly influenced by the track record of the organisation/expert. It should be recognised that there has been a general decline in the trust of scientists since the 1950s [DDT, thalidomide, Three Mile Island, Chernobyl, BSE etc.].

Of course, we must also understand that the potential hazards associated with contaminated land and their associated risks can also be used by objectors to particular schemes/projects. Such groups of people also have a vested interest, often summarised as "not in my backyard". Increasingly land contamination and potential risks to people's health (often the new residents who will live in the proposed new housing development) is being used by objectors as a stick to beat the developer/local authority with in their opposition to a proposed development scheme.

Experience has shown that to those involved in the assessment/redevelopment of contaminated land, importance must be attached to both the technical assessment of risk and also to the concerns of all stakeholders and their perceptions. This is well illustrated in two case histories.

6 CASE HISTORIES

The first of these histories concerns a site in many ways typical of brownfield redevelopment. The site had a long manufacturing history but had been vacant for more has ten years when a developer began considering what they considered a major development opportunity. The subsequent events illustrate very well the consequences of a failure to;

- Recognise the importance of risk communication
- To develop at the outset, an appropriate communication strategy

The second case history concerns a site which initially appears to face some insurmountable obstacles in respect of perceived levels of risk. It was owned and operated by the Atomic Weapons Establishment, the contaminants present were difficult and emotive, the site was surrounded by housing and the local press was attracted to what it undoubtedly saw as a "good story". The results of the attention given by the MoD to risk perception and communication issues illustrate the benefits of positive engagement with stakeholders.

7 CONCLUSIONS

- Risk assessment is not just a scientific calculation. Perception is and must be, the reality.
- We must get it right at the beginning (ie both the assessment of risk and the communication strategy);
- We must recognise that stakeholder concerns are 'real' – we cannot just discount levels of perceived risk and steam roller through our scientific judgement
- We must do all we can to be (and appear to be) credible and trustworthy
- We must ensure that we are able to properly communicate the risks that face people and the environment to all of the stakeholders involved
- This need to communicate is new to many of us and is undoubtedly a challenge that we will face in our various fields over the coming years.
- It can be done and it is imperative that we do engage positively in communicating our risk assessments to all stakeholders to ensure the delivery of projects formulated on appropriate risk based decisions.

8 FURTHER READING

Dialogue for Sustainability [Training Manual]. Environment Council, 2002.
Contaminated land risk assessment – A guide to good practice. CIRIA 2001
Environmental Protection Act 1990: Part IIA Contaminated Land, Circular 02/2000. DETR 2000.
Guidelines for environmental risk assessment and management. DEFRA 2000
Communicating understanding of contaminated land. Sniffer (1999).
Risk communication, a guide to regulatory practice. Ilgra (1998)
Risk: Analysis, perception and management. Royal Society (1992).
Risk and modern society. Lofstedt and Frewer (1991).

BOUNCING BACK

E. Lewis Jones

Liquid Public Relations Ltd, 9, Tutnall Grange, Tutnall, Bromsgrove, Worcestershire B60 1NN

1 THE COMMUNICATORS CRISIS

Crisis management and damage limitation have become real buzz words over the last ten years. In our 24/7 media culture we all understand the importance of managing our reputation and portraying ourselves and our organisations in the best light possible.

The reality is that most companies pay lip service to crisis management, but ideas for updating crisis procedures or planning a table top exercise to cope with a new management structure or company expansion often ends up falling off the edge of the desk.

A crisis doesn't have to be the result of a major incident though – it can develop out of change. All companies and organisations have to experience change and it's our ability to manage this change that can help diffuse a crisis. The change can be internal, for example a new chief executive, or external, such as a fall in the stock market.

What makes a problem into a crisis is the likelihood of media attention. A problem which might be on the agenda for the future becomes a crisis when it is pushed, by the media into the present, ahead of its schedule. A potential problem becomes a crisis when the media announces its existence.

When a true disaster strikes, for example an industrial accident, it is the media's treatment of the event that determines to a great extent whether you have a problem or a full blown crisis.

2 CRISIS MANAGEMENT TODAY

Today over 80% of senior executives in Europe expect to have to deal with a major crisis during their career. With a 24/7 news operation, as well as the impact of the internet, information – both good and bad, can be distributed very quickly.

Consumer programmes such as Watchdog, have raised corporate profiles leaving the public demanding and expecting corporate responsibility and accountability. A lame excuse isn't good enough – today people demand an apology, they want to know how you are going to rectify the situation and make sure that it doesn't happen again.

As pressure groups through the media gain increasing awareness and in turn, support, it becomes even more important that companies behave in an appropriate way and develop a positive image and relationship with their stakeholders.

2.1 Influence of image

The influence of image in a modern crisis cannot be underestimated. As the world becomes smaller and as companies compete in similar markets, it is the image of an organisation that distinguishes it from a competitor. For example, Vodafone and Orange offer similar services – it is only how they are perceived in real terms that actually set them apart.

The image of a company jointly with the way that company handles a crisis is crucial. Perrier is a good example – they recalled their product, they acted responsibly and today we think of them as a strong, quality brand, reliable and trustworthy. They undertook the huge step of product recall –they weighed up the cost of image versus the very real cost of recall and realised that they had to be seen to be acting responsibly.

2.2 Corporate Culture

Crisis management is much more than a manual, a few holding statements and a lucky horseshoe, black cat or heather. Today, crisis management goes hand in hand with corporate culture – it is one of the main features determining whether or not organisations suffer a crisis.

When we think of a landmark crisis – the good – British Midland, the bad – Ratners and the Ugly – the Paddington Rail crash, it is widely believed that they were most powerfully defined by the culture of the organisations concerned and the attitude of senior managers within them.

3 CRISIS INCUBATION

Denis Smith, Professor of Management at the University of Sheffield has undertaken extensive research into 'crisis incubation – a product of corporate culture'. He argues that crisis incubation occurs when organisations do not fully identify the cause of a crisis and therefore allow the potential for failure to continue to incubate. This is often very much the result of management activity.

In public relations, effective management must deal with the process of crisis incubation as well as contingency planning. This is easier said than done – when involved in a company there can still be reluctance to identify a problem due to scape goating and the stigma and trauma attached to people involved in the event.

Changes can only happen if the management perceives that changes need to be made – especially if the problem you come across challenges an organisation's core beliefs or industry.

4 PUBLIC RELATIONS IN A CRISIS

For Public relations professionals, crisis management activity is all about three main areas:-

1. Reducing the loss of a client base or market share
2. How we can maintain or enforce a positive company image or profile
3. Look at ways in which the organisation can use the current crisis experience to improve the organisational situation in terms of market share, image and performance.

While you might have the goal of protecting your client base, a realistic situation is likely to be one in which customers may go elsewhere. Rail companies experience this –

for many companies a brief withdrawal from operations for some days may mean permanent loss of clients.

5 DEVELOPING A SURVIVAL STRATEGY

Crisis managers need to carefully plan how they make use of a crisis situation and look fo ways in which positive outcomes and visible improvements can be presented. As is don with conventional strategic management and marketing analysis – such as SWOT managers need to look at how to turn threats into opportunities and weaknesses int strengths.

A 'survival strategy' needs to be developed and needs to address four key areas of;-
1. Image – the public image of an organisation
2. Support – activities that support the achievements of goals and objectives but do no necessarily provide value to the organisation on their own, such as administration and customer service
3. Stakeholders – these are the people - the staff, customers, creditors, insurers etc
4. Value adding – the tasks that achieve the goals or objectives for an organisation, these are cash flow generators or value adding activities.

When a crisis happens one of more of theses four elements may be damaged. If imagining the four areas as four walls to a room - The organisation reacts to reduce the damage and recover the wall. The 'saggy floor' is made secure and alternate facilities are put into action and the appropriate contingency or recovery plan to return the wall to its pre crisis state is activated.

The 'floor' of the room illustrates performance – it's obviously going to dip in the area that has been damaged. Therefore in order to 'bounce back' the level of the floor needs to be raised i.e. performance needs to be raised. This doesn't just happen to the area that has been damaged – the other walls also need to improve to help support the damaged wall. This increased level of performance has huge benefits to companies.

The intense efforts made during the crisis are unlikely to be sustained, but any improved focus on stakeholders and organisational image, extra effort in delivering services and support and search for innovative solutions to prevent the loss of market share may leave the organisation with a better than pre-crisis performance level.

4. BOUNCING BACK

There are seven basic common sense ideas to help companies bounce back:-

a. Organisations should look at developing and maintaining a wish list – these are things members of an organisation would like to change should the opportunity arise. These wish lists may include new designs for facilities or products, different staffing procedures, changes to operational processes or new equipment and locations. These are small steps that will impact on the corporate culture but need to be clearly measurable and identifiable.

b. Increase the skills of organisational members. Few organisations appear to seriously improve the crisis management skills of their members even though many of these skills can be used in day to day operations – such as:

- time management

- communications
- information management
- stress management

c. Develop experts rather than jacks of all trades.

d. Look at communication channels.
There are two main channels of communication in a crisis – those connecting personnel dealing with a crisis and those connecting the organisation. At a time of crisis, when most needed they can become grid locked. There's also the increasing likelihood that major users of communication channels will not absorb messages as time goes by. Messages need to be accurate, readily available and frequently broadcast.

e. Develop an effective image and management programme
Organisations need to develop an effective image that suits their self perception and objectives, which match the perception of the outside world. The image needs to be maintained after the heat of the crisis otherwise it will be seen as inconsistent and therefore even fraudulent and manipulative.

f. Get closer to and involve stakeholders
A way of cementing relationships is actually to involve the stakeholder in helping you deal with the effects of a crisis. This includes morale and motivation of staff – who can create rumours, adverse speculation or critical comments in the media.

g. Focus on resolving the situation soon and positively
While being sensitive to the wishes of those affected by the crisis, use of ceremonial acts and celebrations can help signal an end to a particular situation or stage and help people to recover.

5 CONCLUSION

There is no set formula to deal with a crisis – if there was no company or organisation would have one! However, it is important to be up to date on crisis management procedures and to have regular table top exercises in order to plan, prepare and update.

Corporate culture is a key influence on how a crisis is managed and ideally should be addressed in the planning process – but it is also important to realise that a crisis can take place due to the corporate culture too. Finally, if an incident does happen it is often an opportunity to build a better, more robust organisation and to address key issues, such as corporate culture that may have been difficult to manage in the past.

Above all, in successful crisis management is common sense –a management and public relations tool that should never be underestimated.

POOR COMMUNICATION DURING A CONTAMINATION EVENT MAY CAUSE MORE HARM TO PUBLIC HEALTH THAN THE ACTUAL EVENT ITSELF

P.R. Hunter and M. Reid

School of Medicine, Health Policy and Practice, University of East Anglia, Norwich NR4 7TJ, UK

1 INTRODUCTION

Water contamination events pose significant challenges to those public health services that are called on to lead in their management. This is well illustrated by a review of such events that have happened in the UK and elsewhere.[1] The vast majority of such events to-date has either been accidental or due to natural events but it is almost certainly the case that the challenges would be even greater should a large scale contamination event be proven or suspected to be due to deliberate action on the part of individuals hostile to the UK.

The public health imperative in managing such events is to reduce the harm to people who may have been exposed to the contaminant. Most attention in incident management resolves around the assessment of exposure and determining whether such exposure at the levels predicted will cause adverse health effects in the population. Relatively little attention is directed at the psychological effects or the psychological mechanisms that may cause illness independently of any direct toxicological effect.

It is the basic proposition of this chapter that p sychological mechanisms may be as important as direct toxicological mechanisms. Indeed, in many incidents, such psychological mechanisms may even be more important. If this proposition is accepted then the implications for incident management are immense. In particular, the degree of harm to individuals' health following a contamination may be subject to amelioration by the way the event is presented to the public through the media.

In this chapter we shall use a few case studies that support the principal hypothesis, discuss some of the psychological mechanisms why this may be the case and then discuss the implications for public communication.

2 CASE STUDIES

On the 6th July 1988 a relief driver accidentally contaminated the water supply to the Cornish town of Camelford by pouring 20 tons of aluminium sulphate into the contact chlorine reservoir.[2] An estimated 20 000 individuals were exposed to high levels of aluminium and other chemicals including lead and sulphate. The first epidemiological study was some four months after the incident when 500 questionnaires were sent to households within the exposed area and 500 to controls in a non-exposed area. Even with the intense amount of public interest in the contamination incident, the response rate was only 44.6%. Half the respondents in the exposed area stated that they experienced unusual symptoms after the episode compared to 10% of people in the control sample (RR 4.2, CI 3.3-5.4). Virtually all the acute symptoms asked about in the questionnaire were reported significantly more commonly from respondents in the exposed area (headache, nausea, vomiting, diarrhoea, fever, cold, excessive tiredness, blurred vision, painful mouth/tongue, bad breath, earache, difficulty passing urine, chest pain, stomach ache, painful joints, pain in muscles and skin rashes). It was also noted that significantly reported illness levels in the exposed area were higher even before the contamination incident ($p<0.001$).

A few years later Owen and Miles reported a longer term follow-up of hospital discharge rates after the incident (In the UK discharges from, rather than admissions to, hospital are recorded).[3] The authors compared standardised hospital discharge rates in the Camelford population with that of the whole of the Cornish population. The discharge rate from Camelford rose from being below average in the year 1987-1988 to being above average in the year 1989-1990 and remained high up to at least the year 1992-1993. No single specific diagnosis was responsible for this increase.

The health implications of this incident have been open to significant debate and some dispute between health professionals and the local population. The publication of the second of two official reports published into this incident[4,5] stimulated an editorial in the British Medical Journal.[6] In that editorial the difficulty of conducting an unbiased epidemiological study after such incidents were raised. The author suggested that the situation was similar to that of a doctor trying to reassure an anxious patient. This led to a brisk complaint that the health services had not adequately investigated the incident.[7]

David and Wessely, two London psychiatrists, suggested that much of the increased reported illness following the Camelford incident was due to heightening perception of otherwise benign symptoms by both subjects and their medical attendants which were then erroneously attributed to the contamination incident.[8] David and Wessely further suggested that such attribution can be due to one of a number of reasons such as:

1. The attribution of the normal level of somatic symptoms in a community to the contamination event.
2. Symptoms due to anxiety generated by the event, possibly made worse by alarmist and irresponsible media coverage.[9]
3. The potential for compensation claims and associated litigation.
4. Somatoform disorders (description of somatoform disorders will be given below).

Further support for our hypothesis comes from the another water related incident, this time in Worcester.[10] The incident was detected following the receipt of complaints of taste problems in drinking water. The taste problems were traced to contamination of the River Severn with 2-ethyl 5,5-dimethyl 1,3-dioxane and 2-ethyl 4-methyl 1,3-dioxolane. An epidemiological survey conducted in the affected area and two control areas found increased reporting of a variety of symptoms in exposed population. Symptoms were strongly correlated with reported amount of water consumption. However, subsequent analysis showed that symptoms were correlated with belief in exposure rather than actual exposure.

3 SOMATOFORM DISORDERS, PSYCHOLOGICAL PROCESSES, AND ENVIRONMENTAL EVENTS

From records of medical consultations and hospitalisations following environmentally hazardous events, including contamination, there is evidence of an increase in diagnosed somatoform disorders. These disorders include multiple syndromes - somatisation disorder, conversion disorder, autonomic response syndrome, body dysmorphic disorder and hypochondriasis.[11] Each disorder is conceptualised as a different variation in disruptions occurring between psychological processes and physiological ones, and involve differences in presentation. As a group they share certain features. First, they are defined by the patient in terms of pervasive and disabling somatic symptoms in excess of what might be predicted from medical investigations or examinations (i.e. while they may result from disruptions in physiological functions, they are not caused by identifiable pathophysiological mechanisms). Across the spectrum of disorders, symptoms may be focused on one or a few typical locations (e.g. sensory loss or muscular pain in a limb or body region), or diffuse, involving different sites that change over time (e.g. nausea, urinary problems, headache, loss of energy, sexual dysfunctions).

Patients present for medical consultation in hope of finding out what is causing their symptoms, and believe they result from some diagnosable abnormality or disease. Further, their symptoms – including fears about their meaning and prognosis - may themselves create anxiety, which in turn can increase symptom intensity, especially when symptoms prove difficult to diagnose and medical encounters only frustrate their need for explanation. Psychological factors are believed to be fundamental to the aetiology of these disorders because of the temporal relationship between the initiation or exacerbation of symptoms and the presence of stressors, conflicts, or concurrent psychological and problematic needs.[12,13]

It does not require a big leap to predict that warnings of a potentially serious health threat would contribute to the development of illness concerns in individuals predisposed towards 'somaticising' sources of distress. For others, they might be a generating factor in the development of such a disorder, when they occurred within a context of other stressful life events, or when other psychological conflicts stimulating the perception of personal threat or loss were active at the same time.

Research studies containing examples of psychological factors implicated in cases of 'environmental illnesses' are numerous. Somatisation traits have been empirically associated with self-reported poor health when individuals believe their symptoms are caused by environmental toxins or chemical insensitivities.[14] In some cases, symptoms are chronic and specific, as in psychogenic pseudoseizures.[15] In a multi-incident study where the psychological effects of exposure to potentially hazardous materials was measured, elevated somatisation scores including symptoms of anxiety, headaches, diarrhoea, trouble concentrating, and shortness of breath were found for fourteen per cent of over three hundred individuals, between 8 and 40 days, post-incident.[16] Individuals with high sensitivity to multiple chemical substances (i.e. demonstrating illness-type responses to environmental perceptions) have been more likely to be diagnosed with somatoform disorders,[17] combinations of comorbid depressive, anxiety, and personality disorders,[18] or show a tendency to exhibit externally – orientated coping strategies.[19] Personality characteristics such as high vulnerability and anger reactivity, coupled with perceived situational danger, have been experimentally associated with an increase in illness reports.[20] The incidence of prior traumatic life events and concurrent psychosocial stress has been shown to be significantly higher in patient populations reporting environmentally-induced illnesses,[21] lending to the hypothesis that contamination events may add to the complexity of factors influencing a person's somatic perceptions at any one time.

Psychological symptoms also increase following a potentially hazardous environmental event. Negative psychological impacts (including increased stress and anxiety and depressive symptoms) following exposures to toxic materials have been well-documented even after a single event.[22-24]

Altogether these findings indicate a more pervasive and generalised phenomenon; namely, that environmental concerns reaching public awareness may act as a stimulus to increase people's awareness of their own bodily functioning and health concerns. Within the individual, events trigger response tendencies formed by pre-existing traits and prior histories with danger, as well as other sources of stress, and are processed according to established health and illness beliefs. This is further supported by findings that the decision to present symptoms for diagnosis are often found to be accompanied by pervasive illness concerns, including the presence of serious disease. These events will almost certainly raise questions in people's minds about their ability to keep themselves healthy.

4 IMPLICATIONS FOR COMMUNICATION STRATEGIES

If one accepts our hypothesis that a major component of the adverse health effects following a contamination event can be attributed to psychological rather than direct toxicological mechanisms, this raises the possibility of reducing such health effects by improved communication strategies. It seems reasonable that the severity of any psychologically mediated effects would be related to the anxiety caused by the contamination event. In this context, the purpose of communication with the public during an event is to give sufficient information to enable people to know to what risks, if any, they have been exposed and what actions they need to take to protect themselves without causing undue anxiety. There has been considerable research into what aspects of risk cause particular anxiety. There are known as the fight factors.[25] Risks are more worrying if they are perceived to be

- Involuntary
- Inequitably distributed
- Inescapable
- Unfamiliar
- Man-made rather than novel
- Cause hidden and irreversible damage
- To pose particular danger to small children or pregnant women
- To cause a particularly dreadful injury or form of death
- To be affect known rather than anonymous victims
- Be poorly understood by science
- Be subject to contradictory statements from the experts.

One of the main problems for public health during a contamination event is that there is often considerable uncertainty. Members of the public and experts may be uncertain about who were actually exposed. They may also be uncertain about the likely biological effects of exposure and especially whether there is a latency period between exposure and the onset of disease. Members of the public may also be uncertain about the motives of people communicating with them about the risks. These uncertainties are likely to be even greater when the contamination event is the result of deliberate and possibly terrorist action.

It seems to us that there are a number of problems facing the public health team responsible for communicating with the public during contamination events. The first problem is that information available during a contamination event is often incomplete and experts often do disagree about the key issues. Especially early in the course of a contamination event, full extent of the contamination may not be known. Indeed, it may not be entirely clear what contaminants are involved.

The second problem is that members of the general public are often very bad at comprehending risk.[26] In particular people tend to over-estimate risk of rare events and underestimate risk from common events.

The third problem is that people cope with perceived threats differently. While some may approach the potential problems actively, others may seek to avoid thinking about them or their consequences, especially if their experience of perceived threat is too painful. Avoidance coping, including distraction, repression and denial may inadvertently lead to careless behaviour,[27] and continued reliance on these strategies have been implicated in cases of somatisation.[12,13]

Finally, different actors in the management of such events may have very different objectives for communication. Even if all people agree that the primary objective is the protection of public health concerns about adverse publicity or the desire to promote ones own agency can lead to subtle, or at times, not so subtle differences in emphasis that could be picked up by journalists and the public.

5 AN APPROACH TO COMMUNICATION

The first question is to decide whether or not to tell the public about the event. In any well managed system, minor contamination events that have little if any relevance to public health are likely to occur much more frequently than events that are likely to lead to a real health risk. There are very good reasons for not going public on such issues such as not wanting to cause unnecessary anxiety or not wanting to cry wolf so often that when real emergencies occur people will not longer believe them. On the other hand not informing the public about a potential risk until it is too late puts people at risk and can lead to severe criticism from pubic, press and government itself. Indeed the communicator in this instance can be subject to criticism whatever they do.[28,29] The issue of what and when to tell the public is particularly difficult when terrorism is concerned and there has been much speculation about whether warnings about terrorist threats are truly designed to protect the public or encourage public acquiescence in legislative change or decisions to go to war.[30-32]

Cutlip, Center and Broom described seven key principles for communicating to an audience:[33]

- Credibility of the person giving the message
- Context in which the message is given
- Content
- Clarity
- Consistency
- Channels
- Capability of the audience

The most important aspect is credibility as people usually first judge messages based on the perceived trustworthiness of the source.[34] Once a decision is made to go public on an issue the question is raised about who should present the information to the public.

Given the importance of avoiding conflicting messages, it is usually the case that a single source should be the point of contact with the media. Again this is an issue around which the conflict between the various actors involved in incident management can develop. In this context it is worth noting that, in general, most people trust health professionals as sources of risk information above any other group, even above scientific experts in the field and consumer groups.[35] In contrast very few people trust government or industry. It follows that, in most circumstances the preferred communicator would be a health professional. Even if other agencies, such as the water company, believe they need to get their own message over in most circumstances it would be best to leave this till after the event as been controlled.

Unclear, confusing or inaccurate messages early in the delivery of news about the event have been shown to be associated with increased fear, distrust, anxiety, depression, and anger.[22,23] As a consequence, high levels of fear accompanying health warnings are likely to have a negative impact on appropriate evaluative behaviour and health practices. Therefore, warnings should offer a realistic appraisal of the risks and potential impacts of contamination, while attempting to avoid provocation of widespread alarm.

Offering the public some mechanism of control is likely to reduce anxiety and other psychological impacts, especially they offer alternative courses of action or information-seeking. Delineating specific procedures gives people a course for action, rather than leaving them without anything they can do to help themselves reduce the potential hazard. Feeling able to influence impacts of uncontrollable events can increase individuals' sense of participation in ensuring their future health and well-being. This might take one of several forms, and permit community members a role in helping public health officials with their tasks. The importance of health-maintaining behaviours, especially if there are specific ones that may have particular relevance to the crisis (e.g. advice on boiling water or using bottled waters) can be stressed.

Contamination events might offer good opportunities for teaching individuals who do develop medically unexplained symptoms about the normalcy of selective perception, and how events may unconsciously trigger deeper concerns about illness and vulnerability. The benefits of such therapeutic goals have been shown in cases of somatisation[36] and could be considered in joint public health initiatives following a contamination event as well.

In addition, it is also important to stress the value of empathy.[37] People respond better to individuals who are clearly concerned about their predicament and who are obviously concerned about resolving the problem. It is particularly important to beware of implicitly criticizing or penalising the public for their concerns. Even in cases where a prior tendency to somaticise can be found, psychologists have warned that it is not ethical – or justifiable – to attribute patients' reports of illness to psychological factors alone, as living or working within contaminated environments is likely to have deleterious psychological consequences for some, possibly producing a cascade of impacts including subsequent unhealthy behaviours that are nonetheless stress-relieving (e.g. smoking, alcohol or drug reliance). For industrial and public health officials communicating with medical officers, it is important to take care in creating an informed and empathetic approach as concerned community members begin to request consultations. Others have warned that viewing patients as chronic and problematic somatizers or hysterical characters searching for a nurturant relationship will undermine the public health or health care provider- patient relationship (Rosenberg et al., 1990).[38] In addition, patients are frequently resistant to acknowledging the possible link between symptoms and their thinking and feeling. The importance of empathy with a patients' concerns becomes therefore fundamental in treating them.

Finally, once a problem is in the public domain, silence is rarely an option.[37] In such circumstances silence speaks of ignorance, guilt, cover-up or arrogance. None of these reduce public anxiety. Positive suggestion could be powerful device, if used to construct better avenues of understanding and response. Positive imagery relating to a future of a safer environment (i.e. beneficial outcomes to an aversive event) could be evoked by the careful use of language and metaphor, and offer avenues by which the public can return to a realistic sense of security and perceive commitment by health officials over their experiences and concern.

6 CONCLUSIONS

We have argued that in a proportion of contamination events, psychological mechanisms may be more important in causing ill health then direct toxicological effects of the contaminating agent. We have further argued that good risk communication during such events can reduce the psychological impact and consequent morbidity. An example of an event which, though associated with acute symptoms did not lead to any reports of chronic illness is that of the grounding of the tanker Braer in Shetland in the early 1990s.[39,40] It could certainly be argued that the lack of chronic health effects was in part due to the proactive response of the health authorities and the strengthened impact of reassurances as a results.

Communication following a deliberate contamination event such as a terrorist attack should in theory be no different than in an accidental contamination event. Although there have been several important studies into the psychological effects of terrorist action, few have really address the impact of somatisation-type disorders and none have stressed the importance of communication in helping to reduce this outcome.[41,42] However, the greater uncertainty around the nature and extent of contamination and associated health effects will make such events considerably more frightening to the public. Furthermore, the increased number of agencies involved, especially the police and armed forces, each with their own communication agendas make getting the health message through in a way that will not adversely affect public health especially problematic. As ever for disaster management, good communication depends on good planning. It is important to ensure that communication priorities and strategies are agreed well before such events happen. Especially important is that all agencies involved in management of such events agree to who will take the lead on communication. These agencies also need to agree what the primary objectives of communication will be. The early hours of response to such events are not the time to argue priorities.

7 REFERENCES

1 P.R. Hunter. *Waterborne disease: epidemiology and ecology*. Wiley, Chichester, 1997.

2 A. Rowland, R. Grainger, R, Stanwell-Smith, N. Hicks and A. Hughes. (1990) Water contamination in North Cornwall: a retrospective cohort study into the acute and short-term effects of the aluminium sulphate incident in July 1988. *J. R. Soc. Health.*, 1990 **110**, 166.

3 P.J. Owen and D.P.B. Miles. A review of hospital discharge rates in a population around Camelford in North Cornwall up to the fifth anniversary of an episode of aluminium sulphate absorption. J Public Health Med., 1995, **17**:200-204.

4 Lowermoor Incident Health Advisory Group. *Water pollution at Lowermoor, north Cornwall.* Cornwall and Isles of Scilly District Health Authority, Truro, 1989.
5 Lowermoor Incident Health Advisory Group (1991) *Water pollution at Lowermoor, north Cornwall* 2nd. edn. HMSO, London.
6 D. Coggon. Camelford Revisited. *Brit. Med. J.* 1991, **303**, 1280.
7 R. Lawson. Camelford Revisited. *Brit. Med. J.* 1991, **303**, 1480.
8 A.S. David and S.C. Wessely. The legend of Camelford: medical consequences of a water pollution accident. *J. Psychosom. Res.*, 1995, **39**, 1.
9 G. Small et al. The influence of newspaper reports on outbreaks of mass hysteria. *Psychiatr. Q.*, 1987, **58**, 269.
10 S.E. Fowle, C.E. Constantine, D. Fone, and B. McClosky, An epidemiological study after a water contamination incident near Worcester, England in April 1994. *J. Epidemiol. Commun. Health,* 1996, **50**, 18.
11 DSM-IV *Diagnostic and Statistical Manual of Mental Disorders* – version 4. American Psychiatric Association, 1994.
12 K. Monsen and O.E. Havik. Psychological functioning and bodily conditions in patients with pain disorder associated with psychological factors. *Brit. J. Med. Psychiat.*, 2001, **74**, 183.
13 H. Levitan. Onset situation in three psychosomatic illnesses. in *Psychosomatic Medicine: Theory, physiology and practice,* Vol. 1, ed. S. Cheren. International Universities Press, Madison, CT, 1989.
14 I.R. Bell, J.M. Peterson and G.E. Schwartz. Medical histories and psychological profiles of middle-aged women with and without self-reported illness from environmental chemicals. *J. Clin. Psychiat.*, 1995, **56**, 151.
15 H. Staudenmayer and R.E. Kramer. Psychogenic chemical sensitivity: psychogenic pseudoseizures elicited by provocation challenges with fragrances. *J. Psychosom. Res.*, 1999, **47**, 185.
16 D.F. Kovalchick, J.L. Burgess, K.B. Kyes, J.F. Lymp, J.E. Russo, P.P. Roy-Byrne and C.A. Brodkin. Psychological effects of hazardous materials exposure. *Psychosom. Med.*, 2002, **64**, 841.
17 J. Bailer, F. Rist, M. Witthoft, C. Paul and C. Bayerl. Symptom patterns, and perceptual and cognitive styles in subjects with multiple chemical sensitivities (MCS). *J. Environ. Psychol.*, 2004, **24**, 517.
18 D.W. Black. The relationship of mental disorders and idiopathic environmental intolerance. *Occup. Med.: State Art Rev.*, 2000; 15:667-570.
19 B. Gottwald, J. Kupfer, I. Traenckner, C. Ganss and U. Gieier. Psychological, allergic, and toxicological aspects of patients with amalgam-related complaints. *Psychother. Psychosom.*, 2002, **71**, 223.
20 M.A. Lindberg. The role of suggestions and personality characteristics in producing illness reports and desires for suing the responsible party. *J. Psychol.*, 2002, **136**, 125.
21 L. Hillert, P. Savlin, A. Levy Berg, A. Heidenberg and B. Kolmodin-Hedman. Environmental illness – effectiveness of a salutogenic group-intervention programme. *Scand. J. Public Health*, 2002, **30**, 166.
22 A. Baum, R. Fleming, and L. Davidson. Natural disaster and technological catastrophe. *Environ. Behav.*, 1983, **15**, 333.
23 R.M. Bowler, D. Mergler, G. Huel and J.E. Cone. Psychological, psychosocial, and psychophysiological sequelae in a community affected by a railroad chemical disaster. *J Traumatic Stress,* 1994, **7**, 601.

24 H.H. Dayal, T. Baranowski, Y. Li and R. Morris. Hazardous chemicals: psychological dimensions of the health sequelae of a community exposure in Texas. *J. Epidemiol. Commun. Health,* 1994, **48**, 560.
25 P. Bennett. Understanding responses to risk. in: *Risk Communication and Public Health* ed. P. Bennett. Oxford University Press, Oxford, 1999, p3
26 P. Slovic, B. Fischhoff and S. Lichtenstein S. Facts and fears: understanding perceived risk. in: *Societal Risk Assessment: how safe is safe enough?* Plenum: New York 1980, p. 181.
27 J.L. Singer and J.B. Sincoff. Beyond repression and the defenses in *Repression and dissociation: Implications for psychological theory, psychopathology, and health.* ed. J.L. Singer. University of Chicago Press, Chicago, 1990, p. 471.
28. L.K. Altman. A Public Health Quandary: When Should the Public Be Told? The New York Times. 2005, 15th February.
29. P.M. Sandman and J. When to Release Risk Information: Early — But Expect Criticism Anyway. 2005, http://petersandman.com/col/early.htm
30. Anon. Terror probe raises issue of when to alert public. 2001. http://archives.cnn.com/2001/US/11/04/inv.warnings/
31. Anon. Blair denies terror hype claims. 2004 http://news.bbc.co.uk/1/hi/uk_politics/4036843.stm
32. G. Avery. Bioterrorism, fear, and public health reform: Matching a policy solution to the wrong window. *Public Admin. Rev.,* 2004, **64**, 275.
33. S.M. Cutlip, A.H. Center and G.M. Broom. *Effective Public Relations*, Prentice Hall, Englewood Cliffs, New Jersey. 1985
34 P. Bennett, D. Coles and A. McDonald. Risk communication as a decision process. in: *Risk Communication and Public Health* ed. P. Bennett. Oxford University Press, Oxford, 1999, p. 207.
35 S. McKechnie and S, Davies. Consumers and risk. In: *Risk Communication and Public Health* ed. P. Bennett. Oxford University Press, Oxford, 1999, p. 170.
36 R. Kellner. Psychotherapeutic strategies in the treatment of psychophysiologic disorders. *Psychother. Psychosom.,* 1979, **32**, 91.
37 S. Lang, L. Fewtrell and J, Bartram. Risk Communication. in Water quality: Guidelines, standards and Health. IWA Publishing, London, 2001, p. 317.
38 S.J. Rosenberg, M.R. Freedman, K.B. Schmaling and C. Rose. Personality styles of patients asserting environmental illness. *J. Occup. Med.,* 1990, **32**, 678.
39 D. Campbell, D. Cox, J. Crum, K. Foster, P. Christie and D. Brewster. Initial effects of the grounding of the tanker *Braer* on health in Shetland. *Brit. Med. J.,* 1993, 307, 1251.
40 D. Campbell, D. Cox, J. Crum, K. Foster and A. Riley. Later effects of grounding of tanker *Braer* on health in Shetland. *Brit. Med. J.,* 1994, 309, 773.
41 D.A. Alexander and S. Klein. Biochemical terrorism: too awful to contemplate, too serious to ignore: subjective literature review. *Brit. J. Psychiat.*, 2003, 183, 491.
42 A.L. Hassett and L.H. Siga. Unforeseen consequences of terrorism: medically unexplained symptoms in a time of fear. Arch. Intern. Med. 2002, 162, 1809.

COMMUNICATION OF TAP WATER RISKS – CHALLENGES AND OPPORTUNITIES.

Miguel França Doria, Nick F. Pidgeon, and Kat Haynes

Centre for Environmental Risk, University of East Anglia, Norwich NR4 7TJ, UK

1 INTRODUCTION

The communication of risks during a water contamination emergency can be a hard and ungrateful task. It condenses all difficulties of normal day-to-day communication with the hardship of communicating bad news to a diverse audience in a very limited period of time. The risks to be communicated may be technically complex and estimates may involve a high level of uncertainty. Under such difficult circumstances, the message can be easily distorted, misinterpreted, or even ignored. However, communicating about potential risks may help to minimise health and social impacts. Furthermore, in the absence of official messages people are likely to seek information from alternative sources. These can eventually offer ambiguous and distorted messages, aggravating a situation that is already complex. Thus, communicating and interacting with the public is often better than passive approaches.

During the last two decades, research has shown that communicating risks is not a straightforward process and can produce mixed outcomes[1]. Nonetheless, well-designed strategies can lead to far better results than poorly conceived communication procedures. Several steps can help to improve risk communication, from 'golden rules', to exploring the challenges that communicators may face, and mental models approaches[2, 3, 4].

This paper starts by exploring some of the potential purposes of risk communication during water emergencies and by clarifying the role of communication. It will then discuss the relevance of communicators and communication channels, and partially address the content of warning messages. Finally, the crucial but difficult issue of the timing of message released is presented. The solutions presented in this paper are suggestions that may have to be adapted accordingly to the circumstances.

2 WHY COMMUNICATE?

Risk communication can address a variety of purposes. During most water-related emergencies, the main goal of risk communication is likely to be the protection of public health. Several major water management frameworks, such as the IWA Bonn Charter for Good Safe Drinking Water[5] and the Canadian Multi-barrier approach[6], acknowledge the role of the general public in the management of water related risks. The underlying principle is that consumers should be involved as a legitimate partner in the decisions that may affect them, and their involvement can improve management efficiency.

In a review of about one hundred water-borne outbreaks[7], identified risk communication failure as one of several events that precede outbreaks. From both a technical and ethical perspective, outbreaks can be avoided if people are persuaded to adopt protective behaviours during a contamination event. Emergency communication does however represent a means of last resort, when all approaches to protect water quality fail. It requires that contamination can be properly identified and that those who detected it quickly inform those who are responsible for communicating with the public. Therefore, Risebro[7] also stresses the importance of communication within water companies and externally with other relevant groups.

Risk communication may also seek to minimise community stress during potentially traumatic events, such as terrorist attacks or natural disasters. The release of warning messages that offer information about hazards and adequate protective behaviours is also a way of empowering those who are affected, raising their control over a difficult situation.

Other additional objective of risk communication is to maintain or to protect a good relationship between water suppliers and consumers[1]. Informing the public of potential risks during a contamination event is a sign of openness and may contribute towards maintaining public trust (the alternative, of not communicating, can be followed by adverse media coverage and public criticism). Among other factors, trust in institutions and companies is composed by perceptions of expertise, competence, care and openness[8, 9]. In this context, public communication can be interpreted as a sign of care and openness. In contrast, the lack of official communication is a message *per se* and may seriously erode public trust[10]. People may assume that the problem was undetected or ignored, compromising perceptions of expertise and care, and may even interpret it as an attempt to undercover a hazardous situation, giving rise to conspiracy theories. Once lost, trust is very difficult to regain.

3 WHAT IS COMMUNICATION?

Risk communication encompasses a wide variety of procedures, that range from technically-oriented approaches (e.g. information) to democratic-oriented approaches (e.g. partnerships). Although the ideal procedure may vary accordingly to the circumstances, a combination of technological and democratic approaches is often advantageous[2]. One of the most frequently used definitions of risk communication was proposed by the US National Research Council (1989), which defines it as "an interactive process of information and opinion [exchange]. It involves multiple messages about the nature of the risk and other messages, not strictly about risks, that express concerns, opinions or reactions." Risk communication is not a one-way message delivery and the public should be regarded as a valuable partner during a water contamination emergency. Pre-crisis evaluation and interaction with the public helps to improve crisis communication by anticipating barriers and identifying opportunities. Getting information from the public about the way they perceive water risks and how water is used on households is important for the success of the message.

The importance of continuous interaction with the public in the context of water resources has been long acknowledged[11]. A good knowledge of the target audience is also crucial to the success of communication. It is important to note that the audience is not fixed and may vary from all water consumers in a region to a relatively small set of vulnerable groups. The public is heterogeneous and no "one message" will work for everyone[1].

4 WHO COMMUNICATES?

In democratic societies, virtually everybody can communicate about their views, concerns and values. However, not everybody can efficiently attract the attention of the public to his or her message. An old rule of rhetoric states that the value of the message largely depends on the prestige of the communicator. Modern research on risk communication further developed this rule, pointing out that the attention and importance given to a message depends on the trust the audience has on the communicator[12]. As previously discussed, perceived trust depends on several factors, including perceived expertise, competency and care. Therefore, messages from groups whose motivations or competence appears to be suspicious are likely to be ignored. Motivations can be clarified by showing that the audience concerns were addressed, by offering consensual solutions, by listening to the audience, and by offering complete messages[2].

Table 1 shows whom the public expects to communicate in case of a serious water contamination event due to an accident or to a terrorist attack. This table shows that water companies are the group that people mostly expect to issue a warning in an event of accidental water contamination. However, in the case of a terrorist attack, people expect to get information in the first place from the government or their local authority.

Expected communicator	**Accident**	**Terrorism**
Water Companies	1st	2nd
Health Officers	2nd	3rd
Government and Local Authorities	3rd	1st
Environmental Groups	4th	4th

Table 1 *Who do you expect to launch an alert in the case of a serious water contamination event by accident or by a terrorist attack?*
(n=165; ranks are based on percentages of respondents who selected a specific group; data collected by the authors in 2004 using a mail survey with phone follow-up).

Although a wide dialogue and debate among different groups is usually beneficial, it is important to avoid contradictory messages from different credible sources during a crisis event. The different groups involved should coordinate and, if possible, designate a single place to deliver information to the media and the public (National Research Council, 1989).

5 HOW TO COMMUNICATE?

Communication efforts can be undermined if the message does not reach the intended recipients or if it is ignored. Exposure to risk warnings and the attention they receive is largely influenced by the channels used to spread the message. Information provided via unexpected channels may fail to get proper attention. Table 2 shows the channels expected to deliver risk warning in the case of a serious water contamination event. The mass media, particularly TV and radio, are the channels through which people mostly expect to hear a warning message about a water contamination event. On the other hand, information by post or from friends and neighbours are relatively less expected. Nonetheless, different people may rely on different channels and the use of multiple sources to spread risk warning is advisable. A review of public awareness of a boil water notice issued during a cryptosporidium outbreak substantiates the survey results presented on table 2, having found that about 55% of respondents heard about it from the media, 25% by word of mouth, and only 20% by the distributed leaflet[13]. (For ways of interacting with the media see the special issue of *Water Science and Technology*, 45(8)).

Information source	Accident	Terrorism
TV (national or regional)	1st	1st (=)
Radio (national or local)	2nd	1st (=)
Street Notices	3rd	3rd
Newspapers	4th	2nd
Loudspeakers	5th	4th
Leaflets by post	6th	5th
Friends	7th	6th(=)
Neighbours	8th	6th(=)

Table 2 *Where do you expect to hear a warning in case of a serious water contamination event by accident or by a terrorist attack?*

(n=165; data collected by the authors in 2004 using a mail survey with phone follow-up).

Apart from traditional communication channels, there are several complimentary ways of warning the public during water emergencies. For example, warnings may be transmitted by phone, using an automatic system with recorded messages to target vulnerable groups (e.g. hospitals, schools) or small groups of consumers in a limited region. As risk perception of drinking water is largely influenced by water colour[14], the water can also be dyed with a food colorant (e.g. red). This latter option can be difficult to implement but may deserve further research. For example, rainwater and recycled water are sometimes dyed as a way of preventing consumption.

6 WHAT TO COMMUNICATE?

The content of risk communication messages should be adapted to the specificities of particular water related emergencies. However, giving the urgency of communication during emergencies, messages should be drafted and tested in advance. Although there is no single standard formula that can be applied to all cases, a number of aspects should be considered when messages are designed.

One of the most widely used approaches in risk communication is based in mental models, which refers to the way new risk information is combined with previous beliefs [15]. If the audience does not have any prior beliefs about the topic, the message may be incomprehensible; if there are erroneous beliefs, the message may be misinterpreted. Thus, messages should take into consideration the way water risks are perceived by the public, their mental models, and how these differ from the mental models of the experts responsible to draft risk communication messages[3].

Risk perception and consumer satisfaction surveys show than most people consider their tap water to be consistently safe[14]. It is therefore important to stress that this situation has temporarily changed, naming the problem and pointing out what may have caused it. For example, the message may say that there was an "accidental contamination by agricultural chemicals near one of the water supply wells". As the message may reach unintended recipients, it is also important to include information about the geographic area affected, adding maps if appropriate. To clarify that the situation is abnormal, the message may include information about the efforts being done to restore water safety and state how long the problem is likely to last.

The message may also specify the specific magnitude and the nature of health risks. However, research on this topic strongly suggests that it is extremely hard to efficiently communicate about risk probabilities and specific consequences[16, 17].

People should be clearly informed about the required behavioural changes, stressing what should and what should not be done. For example, can the water be used to drink, to cook, and to wash vegetables? Can it be used for hygiene purposes, to water plants, for house and clothes cleaning, to give to pets, to use in aquariums? If the water cannot be used for some of these purposes, information about provisional solutions should be included, clarifying appropriate behaviours (e.g. boil water) and/or alternative water sources that may be temporarily used. If suitable, in order to simplify the message, a checklist with do's and don'ts can be presented, matching water uses with provisional solutions.

A final section of the message should include contact details where more information can be provided and to receive any pertinent information that people may want to share. People should also be informed how further information will be provided and how can they know that the problem is solved.

7 WHEN TO COMMUNICATE?

Perhaps the most crucial and difficult aspect of risk communication in the context of water emergencies is to decide when specific warnings should be issued. It is extremely hard to find objective criteria to base such a decision on. For example, should the public be informed when a certain substance exceeds a guideline? How about if the problem can be quickly solved, before the warning is likely to reach the public? Or should a risk-management approach be used, warning the public when there is a certain probability that a specific number of people may be affected? For example, this approach may refer to the chance of >1% of consumers getting diarrhoea or to the chance of >.00001% consumers being seriously affected. The actual risks take into consideration the concentration of hazardous substances, the number of people that use the water supply, and several other factors. When serious contamination is suspected, should the precautionary principle be applied by immediately informing the public? For example, EPAL (the main Portuguese water company) issued an emergency warning after one of their main reservoirs in Lisbon was briefly invaded by an unidentified group of people in October 2001, one month after September 11 (the case was classified as vandalism a few days later). Extending the precautionary principle to another example, should a warning be issued when a certain number of people are hospitalised with symptoms that raise the suspicion of a water-borne outbreak, even if no causal link can be established? The DPHM[13] proposed a protocol of procedure for situations when high cryptosporidium concentrations are detected, which includes criteria to issue warnings "if the risk assessment indicates there is a continuing risk to public health that outweighs the risks of a boil water notice". Whatever criteria are used, it should not lead to the emission of numerous futile messages that may undermine the impact of warnings during serious contaminations.

Although it is extremely hard to establish specific criteria to decide when emergency communication strategies are set in place, there are some advantages if voluntary codes of conduct are available to base the decisions in. Pre-established criteria can enable quicker and more efficient decisions, which may ultimately protect the health of consumers. Since in many cases the precise risks can be relatively uncertain, such codes can represent the best possible solutions and protect water companies from hard decisions. Finally, codes of conduct can be reviewed and improved to accommodate past experience and new developments.

8 CONCLUSIONS

Risk communication may not be able to solve all problems but can be advantageous in terms of protecting public health, minimising community stress and safeguarding trust. There is no unique and universal solution to communicate risks during water contamination emergencies. Several factors should be considered when risk communication strategies are designed, including how the message should be disseminated, what should be communicated, and when should warnings be issued. Different approaches can be used and it is important to draft and test strategies in advance.

References

1 Parkin, R.T., Embrey, M.A. and Hunter, P.R. (2003) Water-Related Health Risk Communication: Lessons Learned and Emerging Issues. *Journal of the American Water Works Association*, 95(7), 58-66.

2 Rowan, K.E. (1994) Rules for risk communication are not enough: problem-solving approach to risk communication. *Risk Analysis*, 14(3), 365-373.

3 Owen, A.J., Colbourne, J.S., Clayton, C.R.I, and Fife-Schaw, C. (1999) Risk communication of hazardous processes associated with drinking water quality - a mental models approach to customer perception, Part 1 - A methodology. *Water Science and Technology*, 39(10), 183-188.

4 Morgan, M.G., Fischhoff, B., Bostrom, A. and Atman, C. (2002). *Risk Communication: A Mental Models Approach*. Cambridge: CUP.

5 IWA (2004) *The Bonn Charter for Safe Drinking Water*. (available on-line from: www.iwahq.org.uk/template.cfm?name=bonn_charter).

6 CDW-CCME (Federal-Provincial-Territorial Committee on Drinking Water and the Canadian Council of Ministers of the Environment Water Quality Task Group) (2004) *From Source to Tap: Guidance on the Multi-Barrier Approach to Safe Drinking Water*. Publication number: 1334. Manitoba: Canadian Council of Ministers of the Environment.

7 Risebro, H.L., Doria, M.F., Hunter, P.R., Andersson, Y., Medema, G., Osborn, K., and Schlosser, O. (*in prep*) Development of fault tree methodology for novel application in the analysis of enteric waterborne disease outbreaks. *Paper to be presented at* IWA

Water Micro 05 (13th International Symposium on Health-Related Water Microbiology), Swansea (UK), 4-9 Sept., 2005.
8 Johnson, B.B. (1999) Exploring dimensionality in the origins of hazard related trust. *Journal of Risk Research*, 2(4), 325-354.
9 Poortinga, W. and Pidgeon, N.F. (2003) Exploring the dimensionality of trust in risk regulation. *Risk Analysis*, 23, 961-972.
10 Wolfe, P. (2004) Lead problem difficult to solve. *Water and Waste Water International*, 19(8), 3.
11 Sadler, B.S. (Ed.) (1987). *Communication Strategies for Heightening Awareness of Water*. UNESCO.
12 Hovland, C.I., Janis, I.L., and Kelley, H.H. (1963). *Communication and Persuasion. Psychological Issues of Opinion Change*. New Haven: Yale University Press.
13 DPHM (Department of Public Health Medicine) (2000) *Report of an Outbreak of Cryptosporidiosis during August and September 2000 in the Lisburn, Poleglass and Dunmurry Areas of the Eastern Board*. Eastern Health and Social Services Board.
14 Doria, M.F., Pidgeon, N, and Hunter, P.R (2004) Perception of Tap Water Risks and Quality: a Structural Equation Model Approach. *Proceedings of the 4th International Water Association World Water Congress* (CD), paper ID 116797, 10pp.
15 Gentner, D., and Stephens, A.L. (Eds) (1983) *Mental Models*. Hillsdale: Erlbaum.
16 Stone, E.R., Yates, J.F., and Parker, A.M. (1994) Risk communication: Absolute versus relative expressions of low-probability risks. *Organizational Behaviour and Human Decision Processes*, 60(3), 387-408.
17 Yamagishi, K. (1997) When a 12.86% mortality is more dangerous than 24.14%: implications for risk communication. *Applied Cognitive Psychology*, 11(6), 495-506.

IMPROVING COMMUNICATION OF DRINKING WATER RISKS THROUGH A BETTER UNDERSTANDING OF PUBLIC PERSPECTIVES

C.G. Jardine[1]

[1]Dept. of Rural Economy, University of Alberta, Edmonton, Alberta T6G 2N1 Canada

1 INTRODUCTION

Safe drinking water is now routinely available in most developed countries, and indeed is usually considered a public right by the citizens of these countries. However, in recent years the public has shown increasing distrust of publicly supplied tap water. A poll of 500 Canadian adults conducted in 2001[5] found that 46% lack confidence in the water coming into their homes and 68% claim to use bottled water at least sometimes[1]. Although figures on national bottled water use vary, one recent report claimed that up to 17.5% of Canadians and 20% of Americans now get their drinking water exclusively from bottled sources[2]. The decision to drink bottled water is commonly attributed to either organoleptic (taste and odour) factors and/or concern about the potential health risks associated with tap water[3,4,5]. Demographic variables and trust in drinking water utility companies are also known to influence drinking water source decisions[6].

Water contamination emergencies are one factor that may undermine public trust in the drinking water supply and in drinking water utilities. Understanding how perceptions of drinking water safety change after a contamination episode provides important information on how to establish effective public communications about drinking water safety. This research was designed to investigate if the occurrence of two high profile water contamination events in Canada influenced how people think about the potential risks associated with drinking tap water. The underlying premise was that these events may have increased public perception that there might be a health risk associated with drinking tap water, and that drinking bottled water might be a "safer" alternative. It was also hypothesized that these episodes may have undermined public trust in the capability of Canadian drinking water utilities to handle emergency contamination issues. Developing a better understanding of public perspectives of drinking water risk (in the context of other risks people may be affected by in their lives) and of trust in water utilities, provides important insights on developing and maintaining effective communications on drinking water risks.

[5] Accurate within ±4.5% at the 95% confidence level for a 50% response level

2 BACKGROUND

In 1994, a telephone survey designed to assess different aspects of health risk perception among the residents of Alberta, Canada was conducted under the auspices of the Eco-Research Chair in Environmental Risk Management at the University of Alberta[7,8]. The survey was conducted in conjunction with the 1994 Alberta Survey - an annual provincial survey administered by the Department of Sociology, University of Alberta, through its research facility, the Population Research Laboratory (PRL). The questions posed were based in part on a previous Canadian health risk perception survey designed and conducted in 1992 by Decision Research and Goldfarb Consultants, under the supervision of the Department of National Health and Welfare[8,9,10]. The purpose of the 1994 survey was to assess different aspects of health risk perception, including public perception of drinking water quality and related health risks[11].

Subsequent to this survey, two high profile water contamination events occurred in Canada. In May 2000, a waterborne outbreak of *E. coli.* O157:H7 in Walkerton, Ontario resulted in 7 deaths and over 2300 cases of illness. A year later, in May 2001, an outbreak of *Cryptosporidium* in the drinking water of North Battleford, Saskatchewan resulted in ~7000 cases of illness (no deaths). A second similar survey was conducted in 2005 to assess how these visible and dramatic national events may have changed how Albertans view various health risks, how they now view health risks from drinking water, and how well they think Alberta water utilities are capable of handling an emergency situation. This survey was again conducted as part of the 2005 Alberta Survey administered by the PRL, thus providing for consistency between survey mechanisms and results.

3 METHODS

3.1 1994 Alberta Survey

The 1994 Alberta Survey was conducted between February 20 and April 11, 1994. The questionnaire was pre-tested by professional interviewers on 32 Edmonton area households. Based on the findings of this pretest, modifications were made to the questionnaire prior to conducting the survey. The target participant population for telephone interviewing was all persons 18 years of age or older who, at the time of the survey, were living in a dwelling unit in Alberta that could be contacted by direct dialling. From this population, three samples were drawn to cover Alberta, including Edmonton metropolitan area, Calgary metropolitan area, and the rest of the province ("other" Alberta). Approximately 400 people were sampled from each of these areas. A Random-Digit Dialling approach was used to ensure that respondents had an equal chance to be contacted whether or not their household was listed in a telephone directory. The survey selectively targeted equal numbers of males and females.

A total of 1259 interviews were completed during this period, with a response rate of 73%. The estimated sampling error, at the 95% confidence level, assuming a 50/50 binomial percentage distribution was ±2.7%. Survey estimates for the area sub-sample of 400 are estimated to be within ±5%, at the 95% confidence level.

The Alberta Survey routinely includes questions designed to determine the general demographic characteristics of the survey respondents. These include age, household income, education, employment status and religion. Demographic questions were asked at the beginning and end of the survey.

The survey was administered through a multi-station CATI (Computer-Assisted Telephone Interviewing) system installed on a local area network at the PRL. The Ci3 Wincati System is a PC-Windows based product of Sawtooth Software, Northbrook, Illinois. This system facilitates the exchange of information among interviewing PC stations and supervisor stations linked using a file and database server during the data collection period. Supervisors monitor call dispositions, field edit, validate data and generate progress reports. The data were tabulated and cleaned using the SPSS for Windows statistical package (a product of SPSS Inc., Chicago, Illinois). The data cleaning process included wildcode, discrepant value, and consistency checks. To analyze and interpret the results of the province-wide survey, the samples were combined as a single sample. The three sample areas were therefore weighted in proportion to the Alberta population each represents.

Survey questions were asked in six major areas: (1) attitudes, opinions and worldviews; (2) individual exposure to environmental health risks; (3) risk perception; (4) sources of information; (5) reliability of information; and (6) status of health risks. The questions on individual exposure to environmental health risks and risk perception were repeated for the 2005 survey, and are described in more detail below.

3.2 2005 Alberta Survey

The 2005 Alberta Survey was conducted between March 3 and 17, 2005. The survey was pre-tested with 20 individuals. The methods used to collect and analyze the data were essentially the same as those used in the 1994 Alberta Survey.

The response rate was 39.9%, calculated based on the number of people participating in the survey divided by the number selected in the sample. The estimated sampling error, at the 95% confidence level, for an area sample of 1208 households and assuming a 50/50 binomial percentage distribution is ±2.8%.

Question #1 of the 2005 survey, repeated directly from the 1994 survey was *In the last three years, have you (and your family) been directly affected by any particular environmental health risk? If yes, what kind of environmental health risk was this?* These questions were designed to determine personal exposure to environmental health risks, and to determine the types of environmental exposures considered by the respondent to represent risks to the health and well-being of themselves and their families. This section of questions was given before more specific questions on health risks to avoid any prompting effect arising from naming specific health risks. The conditional open ended question was coded by the researchers using the codes initially developed for the 1994 survey to provide a basis for response comparison.

Question #2 asked *For each of the following, please tell me whether you think there is "almost no health risk", "slight health risk", "moderate health risk", OR "high health risk" to the Alberta public as a whole: (1)high voltage power lines; (2) chemical pollution in the environment; (3) cigarette smoke; (4) tap water; (5) drinking alcoholic beverages; (6) depletion of the ozone layer; (7) bottled water; (8) suntanning; (9) stress; and (10) sour gas wells.* This question was also directly repeated from the 1994 survey, but for few potential risks (the 1994 survey included an additional 12 risks not included in this survey). This approach to measuring health risk perception was initially designed for the "intuitive toxicology" research studies [11,12].

Question #3 was *In your opinion have health risks to the Alberta public from drinking tap water "increased a great deal". "increased somewhat", "stayed about the same", "decreased somewhat", or "decreased a great deal" in the past ten years? How about for you and your family?* This question is a modification of a question asked in 1994 (which asked about health risks in general).

Finally, Question #4 was *Drinking water utilities sometimes have to respond to emergency situations, such as* E. coli *contamination or spills. In your opinion, are the abilities of Alberta drinking water utilities to handle an emergency situation better than 10 years ago, worse than 10 years ago, or unchanged in the last 10 years?* This was a new question designed to specifically investigate public opinions of drinking water utility abilities following an emergency situation.

4 RESULTS

4.1 Individual Exposure to Environmental Health Risks

In both the 1994 and 2005 surveys, the majority of Albertans stated that they had *not* been directly affected by any particular environmental risk in the last three years (84.4% and 84.2% respectively). The percentages of those who felt they *had* been affected by an environmental risk were essentially unchanged from 1994 to 2004 (14.7% and 15.1%).

The respondents who responded positively to this question (185 people in 1994 and 183 people in 2005) were further asked what type of risk they had been affected by. The 2005 responses to this question were analyzed using the codes developed for the 1994 survey. The risk categories fell into approximately three general areas: (1) air and water concerns; (2) chemical and waste disposal concerns; and (3) lifestyle concerns. Each of the risks cited in the latter two categories accounted for less than 10% of the responses. Given this and the subject focus of this research, this paper will focus on comparing the risks related to air and water concerns. Figure 1 compares the results between the 1994 and 2005 surveys for these two risk types. A significantly higher proportion of respondents in the 2005 survey felt they had been affected by general air related risks than in 1994 (24.6%2.4%± compared with 6.9%±1.4%). Conversely, significantly fewer of the 2005 respondents felt they had been affected by any type of water related risk than in 1994 (3.3%±1.0% compared with 8.7%±1.6% for total water risks).

4.2 Risk Perception

In this section of the survey, respondents were asked to indicate the degree of health risk they associated with each of ten hazards known to be of concern to the public, medical community and government agencies. All of these hazards were included in the 1994 Alberta Survey, and nine were taken from the 33 hazards used to determine risk perception in the 1992 Canada Survey. The addition item, "sour gas wells" was added as a hazard of specific concern to Albertans in both the 1994 and 2005 survey. The items were selected to cover a wide range of hazards, including risks from technology (e.g. "sour gas wells", "high-voltage power lines"), lifestyle (e.g. "stress", "cigarette smoking"), pollution (e.g. "chemical pollution") and common substances (e.g. "tap water", "bottled water"). The ten hazards repeated in this survey were selected on the basis of continued assumed relevance for Albertans.

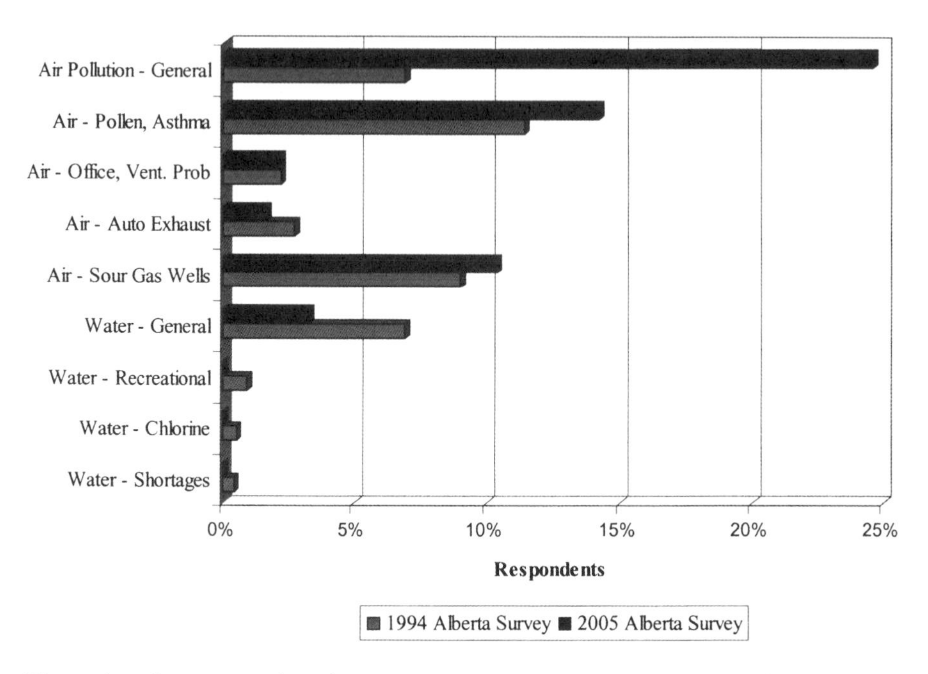

Figure 1 *Comparison between 1994 and 2005 of health risks related to air and water concerns incurred by Albertans in the last ten years (based on affected respondents only).*

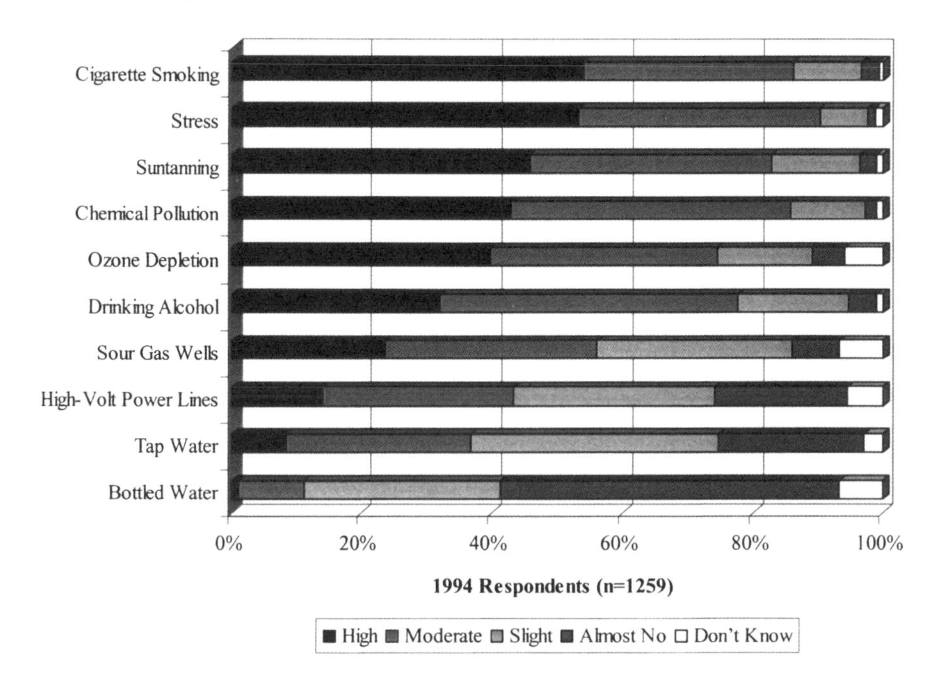

Figure 2 *Perceived health risk to the Alberta public (1994).*

The results for this question for the 1994 and 2005 surveys are shown in Figures 2 and 3, respectively. Figure 4 compares the results of high health risks for the two surveys. The most interesting change from 1994 to 2005 was that "stress" has overtaken "cigarette smoking" as having the highest percentage of responses in the high risk category. The relative ranking of the other hazards remained the same in both surveys. As in 1994, the three highest perceived risks (stress, cigarette smoking and suntanning) were all related to lifestyle hazards, as opposed to technology or pollution hazards. In 2005, a significantly higher number of people ranked stress and cigarette smoking as a "high" health risk than in 1994 (59.7%±2.8% and 64.6%±2.7% compared with 54.0%±2.8% and 53.3%±2.8%, respectively). A significantly lower number of people ranked suntanning, chemical pollution and ozone depletion as a "high" health risk in 2005, as compared with 1994 (39.7%±2.8%, 32.5%±2.6% and 30.7%±2.6%, compared with 45.8%±2.8%, 43.0%±2.7%and 39.7%±2.7%, respectively).

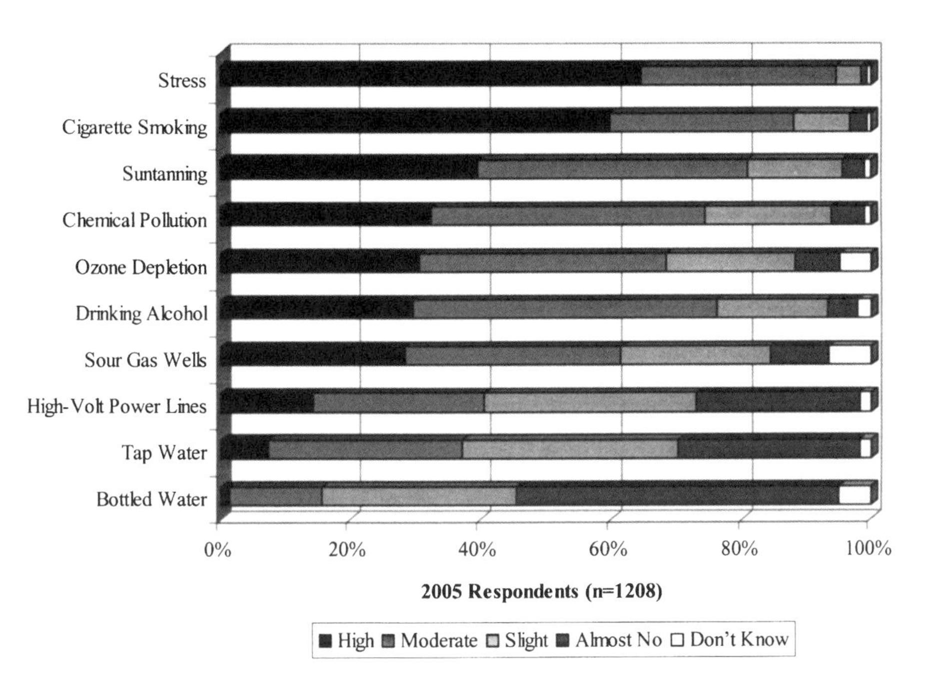

Figure 3 *Perceived health risk to the Alberta public (2005).*

"Tap water" and "bottled water" continued to be ranked low as a health risk relative to the other hazards given. Figure 5 directly compares the health risk perception for these two hazards for the two surveys. There was no significant difference in the proportion of respondents who saw tap water as a moderate to high health risk between 1994 and 2005 (36.8%±2.7% compared with 37.2%±2.8%). However, there was a small, but significant increase in the proportion of Albertans who saw bottled water as a moderate to high health risk in 2005 as compared with 1994 (15.6%±2.1% compared with 11.2%±1.8%).

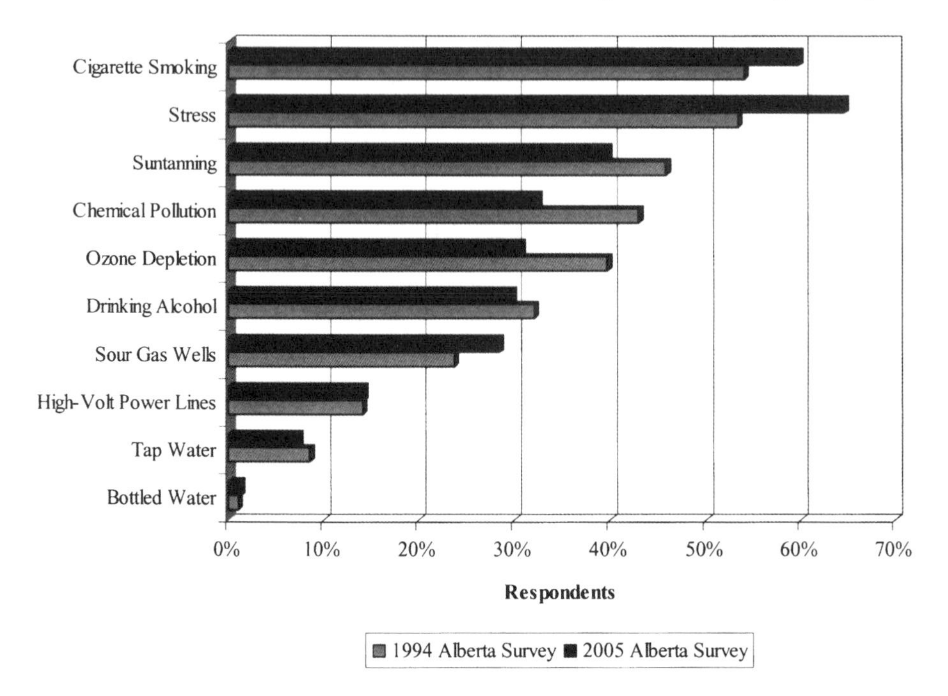

Figure 4 *Perceived high health risks to the Alberta public (1994 and 2005).*

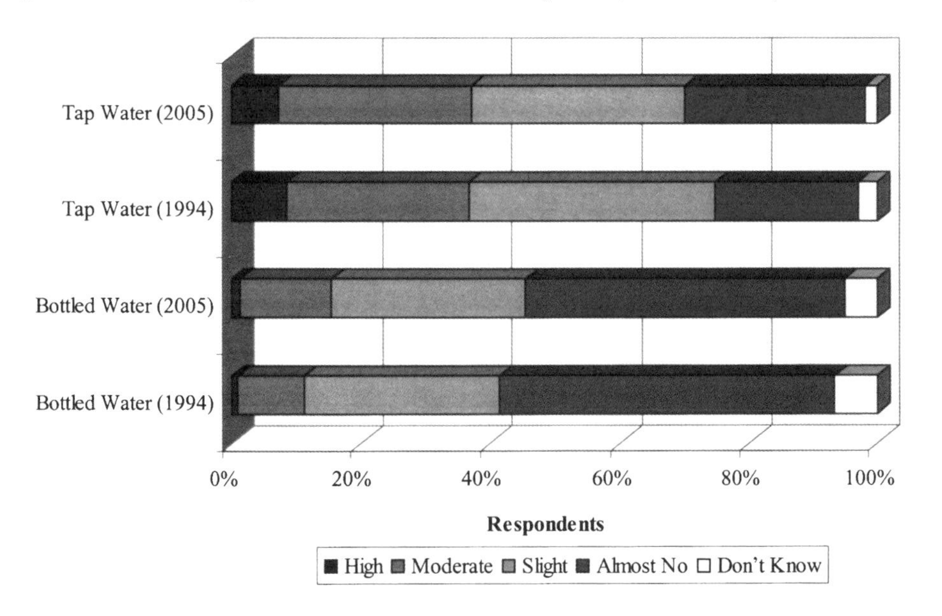

Figure 5 *Perceived health risk of tap and bottled water to the Alberta public (1994 and 2005).*

In 2005, a significantly higher proportion of females than males (63.6%±2.7% compared with 36.4%±2.7%) ranked tap water as a high health risk. A significantly lower proportion of respondents from Calgary ranked tap water as a high health risk than did respondents from Edmonton (5.7%±1.3% versus 8.7±1.6%). Similarly, a significantly lower proportion of respondents from Calgary ranked bottled water as a high health risk than did respondents from Edmonton and other areas of Alberta (0.2%±0.5% versus 1.7%±0.7 and 2.0%±0.8%)

4.3 Health Risks from Drinking Water

Figure 6 shows the perceived change in health risk from drinking tap water in the last ten years, as measured in the 2005 survey. The highest proportion of respondents in both surveys felt that the health risk from drinking tap water in the last ten years had stayed about the same for both the Alberta public and for themselves and their families, although this proportion was significantly higher in the personal context (52.3%±2.8 compared with 40.6%±2.8%). A significantly higher proportion of respondents felt that the risk had either increased somewhat or decreased somewhat for the Alberta public than for themselves or their families (21.4%±2.3% and 19.7%±2.2% versus 15.2%±2.0% and 14.0%±2.0%, respectively).

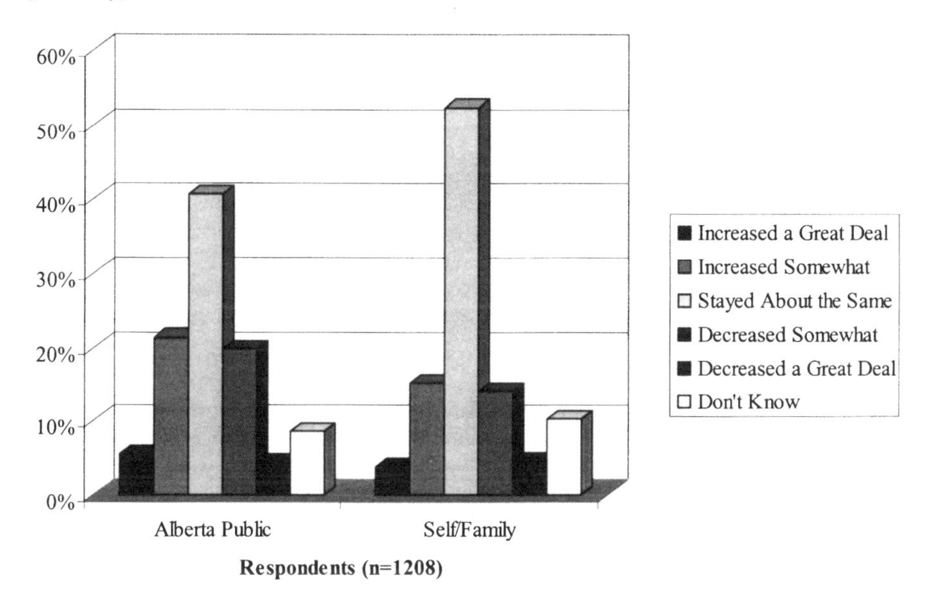

Figure 6 *Perceived change in health risk from drinking tap water in the last 10 years.*

4.4 Ability of Water Utilities to Handle an Emergency Situation

As shown in Figure 7, the majority of respondents (56.7%) felt that the ability of water utilities to handle an emergency water contamination situation is better than it was 10 years ago. Only 4.1% felt that this ability was worse then ten years ago, with 23.9% stating it was unchanged. Quite a few people (15.2%) stated they didn't know if it was better or worse.

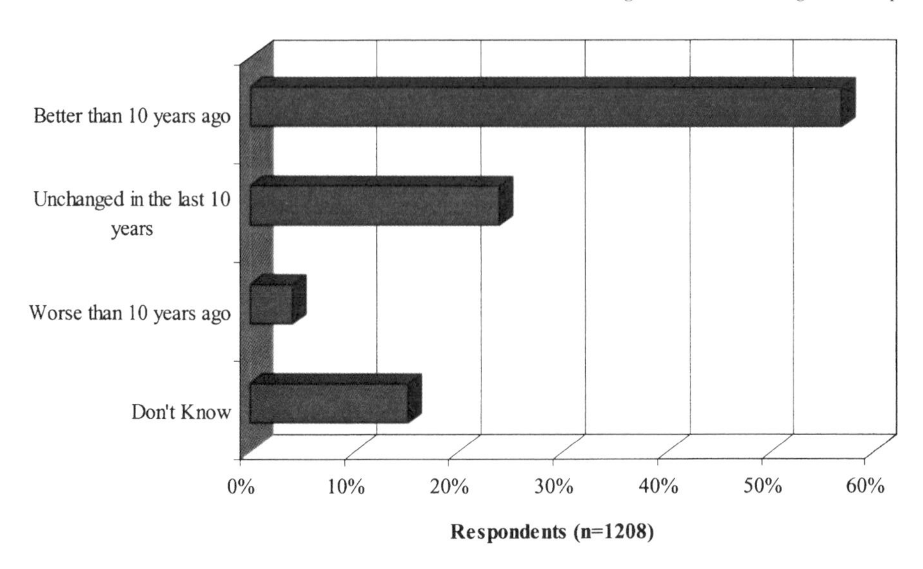

Figure 7 *Change in ability of drinking water utilities to handle an emergency situation in the last ten years.*

5 DISCUSSION

Drinking water risk perception is known to be generally affected by three factors[13]: (1) the consumer's level of awareness of a problem; (2) the presence or absence of drinking water contamination problems, and (3) the chronicity of the problem. It has also been shown to be affected by external information and past health problems[6]. On this basis, it was expected that the drinking water risk perception of Albertans would be affected to some extent by awareness of two fairly high profile contamination events in Canada, even if they were not directly affected. However, the perceived risk of drinking tap water has not been significantly changed in Alberta by the water contamination events occurring in the ten years between surveys conducted in 1994 and 2005. Contrary to expectation, the proportion of Albertans who saw bottled water as a moderate to high health risk actually increased slightly (but significantly) from 1994 to 2005. These results would seem to indicate that the impact of the drinking water contamination events elsewhere in the country did not result in either any sustained change in perceived health risk of tap water or in an increased perception that bottled water is a "safer" alternative. This is reinforced by the fact that a significantly lower proportion of respondents claiming to have been affected by a health risk cited water concerns as the reason in 2005 than in 1994.

This result is in part explained by the finding that the majority of Albertans feel that utilities are better able to handle a water contamination emergency than they were ten years ago. Although not explicitly investigated, it is hypothesized that this response might actually be predicated on the events at Walkerton and North Battleford, as people may feel that utilities have improved their practice to ensure similar events won't happen in Alberta. Regardless of reason, this trust in the water utilities has undoubtedly contributed to a lack of change in perceived risk from tap water.

In 2005, 37.2% of Albertans saw tap water as a moderate to high health risk as compared with 15.6% for bottled water. However, only about 0.5% of total respondents (3.3% of those who had directly experienced a health risk) felt they or their families had directly experienced a water-related health problem in the past three years (compared with 1.0% of total respondents or 6.9% of those who had directly experienced a health risk in 2005). This discrepancy is consistent with the results of the 1994 survey. These findings suggest there is a large difference in perspective between potential concerns about drinking water health risk and actual experience.

The extent of perceived tap and bottled water risk was influenced by gender and region, with a significantly higher proportion of females and residents of Edmonton ranking these risks as high. Previous studies on perceptions of environmental health risks have also found gender and sociopolitical factors (such as power, status and trust) are strong determinants of people's perceptions and acceptance of risks[14] and can be specifically related to perceptions of drinking water quality[15].

The perceived risks of drinking tap and bottled water continued to be small relative to other types of risks (particularly lifestyle risks such as stress, smoking cigarettes and suntanning). Indeed, the Alberta public now feels that stress and smoking cigarettes pose an even higher health risk than they did in 1994. This ranking of risks belies the commonly held view of many risk managers that the public is overly concerned about relatively minor risks over which they have little control. Instead, the Alberta public appears to be very aware of the risks associated with lifestyle choices that are generally judged to be serious by health and risk professionals.

Albertans were more likely to feel that the health risk for the public has changed (either increased or decreased somewhat), whereas personal and/or family risk was seen to have stayed the same. Again, this points to a difference in perspective between potential public health risks and personal experience. It also illustrates a classical response to population and personal health risks, where people seek to interpret population based health assessments in terms of their personal likelihood of incurring the risk[16].

6 SUMMARY AND CONCLUSIONS

Based on these results, the impact of water contamination events on perceived risk of drinking water safety appears to be mitigated by several factors, consistent with those expostulated by Anadu and Harding[13]. First, the *extent of direct impact* would seem to be very influential. Albertans do not feel that the risk of drinking tap water has changed because they have not been directly impacted in recent years. Second, the *duration of time without another reinforcing high profile event* appears to affect perceptions. Four years have elapsed since the last major event at North Battleford, and the public profile of these events has faded. Third, the *perceived ability of water utilities to learn from mistakes and made improvements* seems to favourably influence perception of drinking water safety. Previous research on trust of information about food-related risks demonstrated that sources which have moderate accountability are seen to be the most trusted[17].

Much of the information from this study can also be applied to better understanding communication needs for drinking water risks. Both the 1994 and 2005 surveys demonstrated that members of the general public generally make rationale decisions about the magnitude of various types of risks in their lives. Communications that attempt to put drinking water risks into "perspective" by comparing them with other types of lifestyle risks are thus misguided and likely to cause more alienation and distrust than enlightenment about the "magnitude" of the risk[18].

The survey results also showed that people's concerns about potential risks are generally higher than warranted by their personal risk experiences with exposure or health impacts, and that people are more likely to think risks are different for the public than they are for themselves or their families. These findings seem to indicate that people make distinctions between population and individual risk estimates. Confusion will occur if population risk is not distinguished from individual risk in communications. Understanding the different personal decision processes used in estimating individual risk will enhance understanding of risk information and thus increase communication effectiveness[16].

The high confidence in the increased abilities of drinking water utilities to handle an emergency suggests that people are reassured by the idea of the water industry learning from its mistakes. The best way to restore public confidence would thus seem to be ongoing, regular communication on what utilities have learned from the contamination event and what they are doing to make sure it never happens again. Other studies have shown that one-time or infrequent communications on drinking water quality often account for lack of change in negative public risk attitudes[19]. However, it is not sufficient for utilities to simply tell people what they are going to do – true accountability is only reinforced through actual practice. Nor is it adequate to limit communications to comparing drinking water quality with established standards, as this also does not reinstate trust[20,21]. In addition, it must be remembered that it is unrealistic to expect immediate changes based solely on initial communications and provision of information. Trust is only earned through time and consistency of action[22].

This research indicates that the occurrence of major water contamination events does not necessarily result in an irrevocable loss of public trust in the safety of their tap water and the capability of the drinking water utility. As in most cases involving risk and different public perspectives, situational factors, coupled with agency accountability and vigilance, are critical to determining if and when public trust will be restored.

7 ACKNOWLEDGEMENTS

The questions posed on the 1994 Alberta Survey were initially conceived by Drs. Steve E. Hrudey and Harvey H. Krahn. The survey was supported by the *Eco-Research* Chair in Environmental Risk Management at the University of Alberta. The Office of the Vice President (Research) at the University of Alberta made available SSHRC lapsed funding for support of the questions on the 2005 Alberta Survey. The author would also like to acknowledge the contributions of the Population Health Research Laboratory at the University of Alberta who conducted both the 1994 and 2005 telephone surveys.

References

1 COMPAS Inc. Poll, http://www.compas.ca/data/010511-WaterQuality-PC.pdf, May 2001.

2 T. Clarke, *Inside the Bottle: An Expose of the Bottled Water Industry*, Polaris Institute, 2005

3 C.G. Jardine, N. Gibson and S. Hrudey, *Water Sci. Technol.*, 1999, **40(6)**, 91.

4 P. Levallois, J. Grondin and S. Gingras, *Water Sci. Technol.*, 1999, **40(6)**, 135.

5 B.B. Johnson, *Risk: Health, Safety & Environment*, 2002, **13**, 69.

6 M.F. Doria, *Journal of Water and Health*, In Press

7 C.G. Jardine, H. Krahn and S.E. Hrudey, *Health Risk Perception in Alberta*, 1995, *Eco-Research* Chair in Environmental Risk Management Research Report, **95-1**.

8 P. Slovic,., J. Flynn, C.K. Mertz and L Mullican, *Health-Risk Perception in Canada*, 1993, Prep. for the Dept. of National Health and Welfare by Decision Research. Rept. 93-EHD-170.

9 D. Krewski, P. Slovic, S. Bartlett, J. Flynn and C.K. Mertz, *Human Ecol. Health Risk Assess.*, 1995, **1**, 53.

10 P. Slovic, T. Malmfors, D. Krewski, C.K. Mertz, N. Neil, and S. Bartlett. *Risk Analysis*, 1995, **15(6)**, 661.

11 C.G. Jardine, S.E. Hrudey and H. Krahn, *Proc. Western Canada Water and Wastewater Assoc. 50th Annual Conference and Canadian Public Works Assoc. 20th Annual Conference*, 1998.

12 N. Kraus, T. Malmfors and P. Slovic. *Risk Analysis*, 1992, **12(2)**, 215

13 E.C. Anadu and A.K. Harding, *J. Am. Water Works Assoc.*, 2000, **92(11)**, 82.

14 J. Flynn, P. Slovic and C.K. Mertz. *Risk Analysis*, 1994, **14(6)**, 1101.

15 S. Turgeon, M.J. Rodriguez, M. Thériault and P. Levallois, *J. Environ. Mgmt*, 2004, **70**, 363.

16 D. Powell and W. Leiss, *Mad Cows and Mother's Milk – The Perils of Poor Risk Communication*, McGill-Queen's University Press, Kingston, Ontario, 1997.

17 L.J. Frewer, C. Howard, D. Hedderley and R. Shepherd, *Risk Analysis*, 1996, **16(4)**, 473.

18 P. Slovic, N. Kraus and V.T. Covello, *Risk Analysis*, 1990, **10**, 389.

19 B.B. Johnson, *Risk Analysis*, 2003, **23(5)**, 985.

20 A.J. Owen, J.S. Colbourne, C.R.I. Clayton and C. Fife-Shaw, *Water Sci. Technol.*, 1999, **39(10-11)**, 183.

21 B.B. Johnson and C. Chess, *Risk Analysis*, 2003, **23(5)**, 999.

22 P. Slovic, *Risk Analysis*, 1993, **13(6)**, 675-682.

UK WATER INDUSTRY LABORATORY MUTUAL GROUP: PROGRESS AND ACHIEVEMENTS

Steve Scott[1] and K. Clive Thompson[2]

[1] Consultant, E-mail spscott@btinternet.com

[2] Chief Scientist, ALcontrol Laboratories, Rotherham, South Yorkshire, S60 1BZ, United Kingdom, E-mail clive.thompson@alcontrol.co.uk

ABSTRACT

The Water Industry Laboratory Mutual Aid group was set up in 1995 and has continuously evolved since that time. It is an informal group that was set up for laboratories involved in the analysis of emergency incidents relating to drinking waters, rivers, effluents and adverse effects upon sewage treatment works. Although water companies had set up groups to deal with emergency incidents these was not any group that dealt with this aspect with respect to laboratories. The group's main objectives are: -
To continuously improve the response capability of laboratories carrying out emergency incident analysis
To share information;
To initiate mutual new aid developments / initiatives
To share positive and negative experiences.
To identify and adopt best practices
To provide reassurances to the water industry that the laboratory services can cope with emergencies including CBRN.
To provide contingency event cover in the event of a major catastrophe occurring on one of the group member laboratories

Keywords: - mutual aid, incidents, drinking water, rapid screening tests, chemical, radiological and ecotoxicological analysis

1 INTRODUCTION

In 1995 the Water Industry Laboratory Mutual Aid group was set up as a discussion forum to review the analytical issues relating to a water incident. The forum was to cover both the drinking and waste water aspects of the industry. To achieve this it was essential that there was good liaison between the laboratory, emergency planners and the operational sections of the water company. Another aspect to be considered was the government direction that *'requires the industry to work together, including laboratories, to plan for a worst case scenario'*. It is important to note that not all water industry laboratories are an integral part of a water company, some of them operate independently in a fully commercial environment. To improve the depth of knowledge available to the group, a variety of support laboratories and organisations were invited to participate, e.g. Defence Science and Technology Laboratory (Dstl) and Drinking Water Inspectorate (DWI). The initial network of contacts gradually expanded over several years and now includes representatives from all UK water company laboratories.

Four sub-groups have been set up: -
Organics
Laboratory Environmental Analysis Proficiency (LEAP) Emergency Scheme
Radioactivity
Laboratory emergency response general issues (including ecotoxicity.)

2 EMERGENCY RESPONSE CAPABILITY STATEMENTS

During the start-up process it was recognised that it was key to standardise the response to a major incident which may even involve the loss of the company's laboratory facilities. In order to achieve this it was agreed to set up and maintain, an up to date database of Emergency Response Capability Statements. These statements would list concisely, the contact details and the analytical capabilities of the laboratory. This would allow other members to identify supportive facilities if needed. The mutual aid would be provided on a reasonable endeavours basis and would not be contractual. Several laboratories are commercial competitors therefore a ground rule that was established early on was the agreement to operate under Chatham House rules. These rules allow information to be freely exchanged within the meeting but may not be utilised without the express agreement of all interested parties. This ensures that sensitive and pertinent data may be discussed within all meetings. Table 1 lists the information that is supplied via the capability statements. These are regularly updated on a formal basis by a volunteer member of the group. All emergency work undertaken would be on a reasonable endeavours basis. Some adjacent area water companies have set up localised sub-groups with a more detailed capability statement that also specifies the maximum workload of all relevant parameters that could be undertaken at short notice. This would involve ceasing some less essential routine work for their own water company. This reduction has been agreed with the relevant water company. This would also cover a contingency event where a laboratory suffered a catastrophe and either ceased to operate (e.g. a major fire) or had key sections shut down (e.g. a localised fire)

Company Name	Date
Laboratory Name	
Address	
Prime contact details	
Other contact details	
Tel & Fax No. - Normal & Out of Hours	
Does the Laboratory carry out the routine analysis of the parameters in the Drinking Water Directive to the standard required by DWI	Y / N
Is the Laboratory UKAS accredited	Y / N
Specialist Analysis Areas:	
Radioactivity (gross alpha and beta)	Y / N

Radioactivity gamma ray spectrometry	Y / N
Specialist Microbiology	
Regulatory Cryptosporidium	Y / N
Giardia	Y / N
PCR capability	Y / N
Flow cytometry	Y / N
Class pathogens	Y / N
Virology	Y / N
Outline of specialist microbiology capability	
Typical capacity for extracting and running GC-MS samples per day	Number
Low resolution GC-MS	Y / N
High resolution GC-MS	Y / N
LC-MS	Y / N
LC-DAD	Y / N
Purge & Trap GC-MS	Y / N
Odour identification	Y / N
Gas identification	Y / N
Air monitoring sampling	Y / N
Air monitoring analysis	Y / N
Ecotoxicity testing	Y / N
Outline of specialist organic and ecotoxicity testing services	

Table 1: Emergency Response Capability Statements Outline

3 LABORATORY CHALLENGE

The water industry laboratory provides several services to a water company. One of its major tasks is to monitor the quality of water at each stage of the water cycle i.e. raw water (ground or surface), drinking water purification processes, the distribution system, effluents and waste water treatment plants. To meet this demand the laboratory has adopted an approach which utilises automated equipment using standardised procedures.
The challenge arises when an incident occurs, which, by its very nature, is an unplanned event occurring at any time, night or day, workday, weekend or holiday. In the most straight forward case the incident may involve a known contaminant where the laboratory is required to monitor the water to ensure it is of an acceptable quality. Or, it may involve unknown pollutants of indefinite concentration which might or might not be present. These incidents will arise following a variety of situations e.g. a fish kill, an operational plant failure, a taste and odour complaint, an illness linked to water quality or a security alert. These types of incidents require specialist staff, not only to use analytical techniques that may require modification to suite particular circumstances but to interpret complex data. This data interpretation is heavily dependent on the specialist's experience and knowledge of the company's waters and processes. Trying to prove a negative (i.e. absence of significant contamination) under pressured emergency incident conditions is completely different from carrying out targeted routine analysis. Unfortunately, the emphasis on high throughput analysis and cost reduction has seen a corresponding reduction in these specialist staff. It is with this challenge in mind that the companies meet

annually and participate in the various sub-groups in order to share information about new developments, experiences and to identify and adopt best practices.

4 JUDGEMENT

There are several mechanisms to measure how well laboratories are rising to this challenge. The laboratories are audited annually and may be inspected by the DWI and United Kingdom Laboratory Service (UKAS). As part of the audit process the laboratories' performance in external proficiency schemes are reviewed. Both Aquacheck and the Central Science Laboratory (CSL) provide suitable samples to test the laboratory's ability to identify unknown compounds. The CSL *'Laboratory Environmental Analysis Proficiency (LEAP) Emergency Scheme'* or more fondly known *'Mystery Exercise'*, has provided good training and feedback on the different initiatives and best practices that have arisen from the Mutual Aid meetings. This scheme actually records how long each laboratory takes to provide a provisional and a final result for each substance detected from the time of receiving the sample. In addition the accuracy of the estimated concentrations found is also reported. This information is considered essential for this type of proficiency scheme.

5 MYSTERY EXERCISES

Since its conception in 1997, when Alcontrol laboratories and Yorkshire Water established the Laboratory Environmental Analysis Proficiency (LEAP) Emergency Scheme, the exercise has been run nine times. It started with 5 laboratories and has since grown to include 22 laboratories. In 2000 it transferred to CSL to ensure that it was seen as independent of the water industry. Its main objective is:

... *'to test the ability of a laboratory to identify unknown contaminants in a water sample by the most efficient and timely means and to provide responses to questions posed by an incident manager'*...

Table 2 provides a short list of some of the results from the early days of the scheme. These results indicate that the laboratories are very good at the high throughput target analyses e.g. pentachlorophenol, solvents and metals but struggles when there are polar components present e.g. acidic herbicides (2,4-D and 2,4,5-T) and ethylene glycol.

Exercise	Test Substance	Conc. Spike (μg/l)	No of Labs	% of Labs detected substance present
1	Pentachlorophenol	1200	5	80
	Tributyltin chloride	3000	5	100
2	Benzene	4400	4	100
	Ethylene glycol	55000	4	25
3	Thallium	200	4	75

	2,4-D	50	4	50
	Mecoprop	100	4	50
	Pirimicarb	50	4	100
6	Dimethoate	20	15	93
	1,4 Dioxane	5150	15	53
	2,4,5-T (free acid)	20	15	40
	Phenyl mercuric acetate	10	15	6
	Mercury (as PMA)	5.96	15	100
	Methanol	144000	15	87
	Ethyl Acetate	16000	15	87

Table 2 Summary of the Results from Four LEAP Scheme Exercises

6 RAPID ANALYSIS DEVELOPMENTS

As a result of increased terrorist activities in recent years and the need to have robust analytical techniques to handle the low probability but high impact incidents it was agreed to set up individual sub-groups. These groups would review current capability, identify best practice and, based on this information, would develop more rapid tests. These groups involved organic chemistry, radiochemistry and ecotoxicity. The remit was to produce methodologies using, wherever possible, current equipment and to be capable of processing 100 samples in 2 hours.

6.1 GC-MS

Prior to the development work an unknown sample for organic analysis using GC-MS would take approximately 2 hours from receipt of sample to producing the first interpretation. It was agreed that speed was more important than ultimate sensitivity and a target limit of detection for drinking water samples of between 1 - 10 μg/l would be adopted.

The final method[1] used 100 ml of water spiked with a mixture of isotopically labelled compounds which would be used to give an indication of concentration for any substance detected. The sample was shaken with 4 ml of dichloromethane (DCM) for 30s and allowed to separate for 60s before transferring 2.5 ml of extract to another tube containing anhydrous sodium sulphate to remove the water. Transfer the dry DCM extract to a GC vial where, if necessary, concentrate x10 using a stream of dry N_2. The 1 ul of extract is injected onto a standard GC-MS system with a 30 m analytical column. After the elution of the solvent, the oven temperature is ramped rapidly. The same extract, after 20 ul of 0.2 M trimethylsulphonium hydroxide (TMSH) in methanol has been added, can then be injected onto another GC-MS using a hot splitless injector. This should methylate any polar compounds present. The data is analysed using the NIST AMDIS programme which automatically extracts pure (background free) component mass spectra from highly complex GC-MS data files and uses these purified spectra for a search in a mass spectral library. These libraries may be commercially provided or custom built. Using this technique a sample may be analysed within 20 minutes of receipt. If 5 extracts are combined, with a subsequent reduction in sensitivity, 15 samples / hour / instrument can be analysed.

The technique was tested using Mystery Sample 6. The results can be seen in Figure 1 where the upper trace is the same extract as the lower one after TMSH has been added. All the organic components were detected, apart from the methanol.

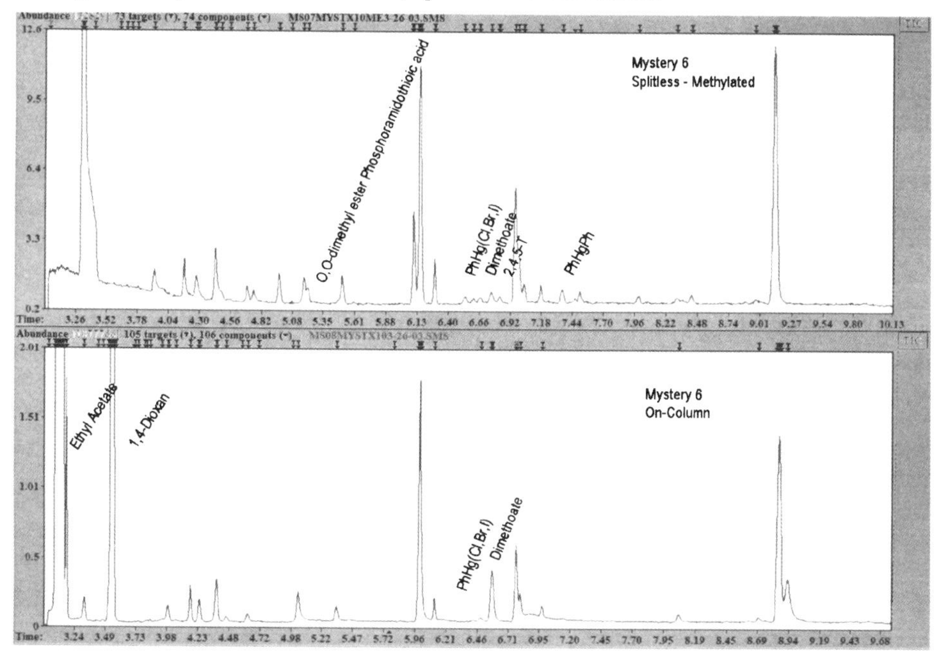

Figure 1: GC-MS traces of extracts from Mystery Sample 6

6.2 Radioactivity

A similar approach was adopted for the radioactivity analysis. A rapid screening method was developed by South West Water and Thames Water which has been validated for soft and hard waters[2]. The methodology can process ten samples in the first three hours followed by ten samples every 30 minutes. There is an informal proficiency scheme between South West, Thames, United Utilities and Alcontrol which is also open to other UK water laboratories. Further background information on radioactivity is given elsewhere in this book.[3]

6.3 Ecotoxicity

The third sub-group focussed on ecotoxicity testing which is complementary to the chemical tests. It will never replace chemical testing but the holistic approach will be run in parallel. Currently a 1h and 24h *Thamnocephalus platyurus* and a 30min Microtox® test are recommended to be run in all emergencies. The 24h *Thamnocephalus platyurus* test has a good sensitivity profile to a wide range of toxicants. Further information regarding ecotoxicity is given in this volume[4-5]

7 CURRENT DEVELOPMENTS

These sub-groups are currently involved in developing further screening tests utilising UV scanning and LC-MS techniques. Although GC-MS is a powerful technique it is important to remember that without complex derivitising procedures, it is only capable of detecting compounds that are relatively non-polar and thermally stable. LC-MS techniques do not suffer from these drawbacks but the instrumentation has only recently become widely available within the industry and libraries of MS data for them are not yet widely available. Also the mass spectra do vary from system to system and the operating conditions, much more so than GC-MS spectra. The other major initiative being undertaken is that a new sub-group has been set up that meets after every LEAP exercise to review, learn and develop best practice. The Mutual Aid group has produced a draft protocol entitled '*Guidance for the Analysis of Potable Water Samples from a Suspected Contamination Incident*'. This should be published in early 2006[6].

8 REALITY

Although progress has been and continues to be made, the reality of the situation is that there are still limitations to what a water laboratory may achieve. It is important to remember that:

- The laboratory will not be capable of identifying every contaminant,
- Most routine methods currently used within the water industry will not detect biotoxins, chemical warfare agents or pharmaceuticals,
- Detecting one contaminant does not preclude the presence of others,
- Rapid analysis is not instantaneous – operational staff need to be aware of a credible timeline.
- Prioritisation of samples for analysis by Operations will ensure that the key samples are analysed before less important samples

It is important to understand that the identification of unknowns is not an exact science. There should be no expectation that any combination of technology or analytical capability will always guarantee success. Therefore the prioritisation of submitted samples is essential. On the positive side, the vast majority of samples will prove not to be contaminated.

9 FUTURE DIRECTIONS

What is needed for the future are rapid, automated analytical techniques that are capable of detecting very large numbers of unknown compounds. In an ideal world these would be multifunctional requiring low maintenance and minimal training. The new LC-MS instruments are heading in the right direction but reliable and robust water industry databases will need to be developed if they are to be fully utilised.

10 CONCLUSIONS

The group which was set up to bring together water industry laboratory personnel in a common forum has been successful. Its major achievements have been:

- An annual review meeting which all UK Water Industry laboratories are represented.
- Four sub-groups have also been set up covering specialist areas to improve rapid screening capabilities
- A database of Emergency Response Capability Statements is maintained and distributed.
- An industry proficiency scheme (LEAP) has been set up and developed to meet the needs of the laboratories.
- Rapid methods have been developed to meet the challenges of the modern world.
- A draft protocol has been produced giving guidance on the analysis of potable water samples from a suspected contamination incident.

ACKNOWLEDGEMENTS

The authors fully acknowledge the efforts of the Water Industry Laboratories Mutual Aid Group and the LEAP proficiency Scheme

REFERENCES

1. Scott, S., Rapid Analysis (Organics), UKWIR 2003
2. Bell, I., Frewin, P., Cornwell, D. and Owen, J.,Rapid Analysis (Radioactivity), UKWIR, 2003
3. Wilkins, B. T., Incidents involving radionuclides, this volume, chapter 32
4. Thompson, K. C. and Scott, Analysis methods for water pollution emergency incidents, this volume, chapter 29
5. Persoone, G., Recent Advances in rapid ecotoxicity screening, this volume, chapter 26
6. Scott, S and Thompson, K. C. , Guidance for the Analysis of Potable Water Samples from a Suspected Contamination Incident'. This should be published by UKWIR in early 2006.

RECENT ADVANCES IN RAPID ECOTOXICITY SCREENING

G. Persoone[1,2]

1 Laboratory for Environmental Toxicology and Aquatic Ecology, Ghent University, Belgium
2 MicroBioTests Inc., Belgium

1 INTRODUCTION

Providing safe drinking water is to date a continuously increasing qualitative as well as a quantitative problem for the water industry, of which the consumer is in most cases not aware, nor concerned about, when turning on the tap at home.
Besides transforming raw water into potable water with all the inherent quality control analyses, the water industry has to be constantly alert for accidental or (even worse) deliberate malicious contamination of water supplies.
As emphasized in previous papers on the problematics of water contamination[1-2], a strictly chemical approach is confronted with the huge number of contaminants which may be present in the water samples; a full suite of analyses not only takes much time and efforts but also requires sophisticated and costly equipment which is not available everywhere.

Taking into account that potable water contamination needs to be detected very rapidly (i.e. in a matter of hours) in order to take appropriate remedial action whenever needed; toxicity testing has received considerable attention over the last few years, as a complementary technique to chemical analyses[3].

It was indeed discovered that aquatic biota can react to chemical stress in a time span of minutes but in turn the effects measured mostly don't give any information on the actual nature of the contaminants. The detection thresholds for biological reactions furthermore are "compound dependent" and may range from sub-ppb concentrations for highly toxic chemicals to hundreds of ppm for relatively harmless products. As with chemical analyses the potential as well as the limitations of toxicity tests have to be taken into account when considering water contamination issues.

2 THE FIRST RAPID TOXICITY TEST

From the 1970s a "very rapid" biological test became available, which is to date used worldwide for rapid toxicity analyses of water[4]. The "bacterial luminescence inhibition assay" with the marine bacterium *Vibrio fischeri* , commercialised under the name "Microtox®" with its DeltaTox® field version, now has various competitors such as e.g. Lumistox®, ToxAlert®, Checklight® ToxScreen® and Biotox®.

Besides their rapidity (each assay only takes 15-30 minutes) a major asset of these specific bacterial tests is their independence of the tedious and costly culturing/maintenance of live stocks of test biota. The bacteria are lyophilized (freeze dried) and hence available "anytime and anywhere" for immediate analyses of suspected water samples.

Toxicity is "species" specific as well as "chemical" specific and the extensive data base on the bacterial luminescence tests clearly shows that this assay is very sensitive to many chemical compounds but less sensitive to others. Furthermore, this specific bioassay is not sensitive to toxins of biological origin (biotoxins); prokaryotes do not have nerve cells so they cannot not react to nervous system poisons.

3 RAPID TOXICITY TESTS WITH HIGHER (EUCARYOTIC) ORGANISMS

A substantial number of bioassays with a variety of test species have been developed over the last 40 years. The majority of these are "acute tests" in which the biota are exposed to chemicals or contaminated waters for one or a few days, with measurement of mortality as the effect criterion. In view of the need for rapid toxicity screening, bioassays with a shorter duration of exposure (hours) have now also been worked out based on "sub-lethal" criteria, with measurement of enzymatic, behavioural or physiological endpoints.

3.1 Enzymatic tests

A 1h microbiotest with aquatic crustaceans under the name "Fluotox", was developed in the Laboratory for Biological Research in Aquatic Pollution (LABRAP) at the Ghent University in Belgium[5-7]. This assay is based on the visual observation of decreased fluorescence resulting from the inhibition of galactosidase activity in stressed organisms. The principle of this assay rests on the enzymatic splitting of a non-fluorescent complex substrate, composed of a fluorescent part and a galactoside. If the galactosidases are inhibited during the short exposure of the organisms to toxicants, the ingested substrate is not cleaved and the fluorescent part is not set free, so the biota do not become fluorescent.

The 1h Fluotox assay is now available commercially in the USA (independently of the Ghent Laboratory) under the name Daphnia IQ test®, for specific application with the crustacean *Daphnia magna*,. In 2003 this enzymatic assay has been evaluated along with 7 other commercially available technologies for rapid water toxicity testing. This comparative exercise has been performed by Battelle on request of the USEPA in the framework of the Environmental Technology Programme (ETV) on Anti-Terrorism Water Monitoring Technologies. Along with the chemiluminescence assay Eclox®, the Daphnia IQ test® was the only "non-bacterial" toxicity test involved in the USEPA ETV exercise in which the sensitivity of all seven rapid tests was determined on five chemical compounds (aldicarb, colchicine, potassium cyanide, dicrotophos and thallium sulphate) and four biotoxins/nerve agents (botulinum toxin, Ricin, Soman and VX). The purpose of the programme was to compare the detection thresholds of each test with the human lethal doses for each compounds or biotoxin. The detection threshold of the enzymatic *Daphnia* assay for the five chemicals ranged from 250 ppb (for cyanide) to 240 ppm (for thallium sulphate). These thresholds were either superior, equal or inferior to those of the microbial assays, depending of the chemical and the type of bacterial test. In turn, none of the bacterial bioassays, nor the chemiluminescence assay could detect all four biotoxins whereas the rapid Daphnia IQ test was quite sensitive to all four, with detection thresholds in the range 0.3 ppb to 15 ppb, i.e. from 10 to 1000 times below the human lethal doses. As indicated above, the latter finding is not that surprising since the crustacean was the only multicellular test organism of the seven assays, and hence provided with "a brain" (albeit a primitive one) capable to react to toxins acting on nerve cells.

3.2 Behavioural tests

Toxicity tests with behavioural endpoints were developed from the early 1970s and "fish monitors" based on the negative "rheotaxis" of the organisms have been used for several decades, for on-line water quality monitoring. Commercial Daphnia Toximeters are now used in different countries for continuous on line monitoring of surface waters or drinking waters, with automatic measurement of the behaviour of the crustaceans by computer-aided video image capture[8-9].

Recently a fully automated on-line biomonitoring system based on real time image analysis of movement and behaviour of a flagellated protozoan (*Euglena gracilis*) has also been developed at the Friedrich-Alexander University of Erlangen in Germany[10-12].

3.3 Physiological tests

Inhibition of food ingestion under stress is a well-known physiological phenomenon and has been explored with filter-feeding organisms for ecotoxicological applications. The concept has already been applied successfully with various species of protozoans, rotifers and crustaceans[13-18]. In most of these "rapid" assays, the decrease or absence of feeding (or ingestion of stained particles) is analysed visually under the microscope after exposure of the test organisms to chemicals or contaminated waters for a brief period.

4 CULTURE/MAINTENANCE FREE MICROBIOTESTS

With the exception of the bacterial luminescence inhibition tests mentioned above, virtually all the other rapid bioassays referred to are dependent on the culturing/maintenance of live stocks of the selected test species. The biological, technical and cost burdens inherent to year-round continuous availability of test organisms is a substantial drawback which automatically restricts toxicity testing to a few specialised laboratories.

This burden has, as of the early eighties, triggered research in LABRAP at the Ghent University, to look for alternatives which would bypass the dependency on the culturing and maintenance of live stocks of test biota. A new concept was conceived and explored, based on "dormant or immobilised" stages of selected test species providing the live biota at the time of performance of the assays.

The fundamental research performed over the last 20 years in LABRAP and presently in the spin-off company MicroBioTests, eventually led to the controlled mass production of dormant or immobilised stages of micro-algae, protozoans, rotifers and crustaceans, with additional focus on the storage conditions of these "special" life stages and on their specific hatching or reactivation conditions.

A whole battery of "culture/maintenance free" bioassays was gradually developed, which were miniaturised and incorporated in user-friendly "Toxkit microbiotests"[19-28].

The battery of acute and short-chronic Toxkit microbiotests also comprises a "Daphtoxkit" with the aquatic crustacean *Daphnia magna* and an "Algaltoxkit" with the micro-algae *Selenastrum capricornutum* (presently named *Pseudokirchneriella subcapitata)*, two species which are the most used organisms for toxicity testing on chemicals and effluents worldwide The standard operational test procedures of the former two Toxkit microbiotests follow the methodology prescribed by OECD and ISO for bioassays with these crustacean and algal test species.

Extensive in-house comparison of the former microbiotests with the "conventional" assays with *Daphnia magna* and *Selenastrum capricornutum* revealed that the test biota obtained from the dormant crustacean eggs or the immobilised stages of the micro-algae have the same sensitivity for chemical compounds and contaminated waters, as the Daphnids and micro-algae obtained from laboratory cultures[24-25]. This finding has been confirmed over the last five years by several comparative studies and intercalibration exercises in different countries[29-33].

Since the commercial launching of the first Toxkits in the early nineties, the international interest for the culture independent and user-friendly microbiotests has gradually been growing worldwide with daily applications in research and toxicity monitoring as shown by the large number of scientific publications (For further information see website www.microbiotests.be).

5 THE RAPIDTOXKIT MICROBIOTEST

During the last few years, the question was raised in MicroBioTests Inc. whether it would not be possible to combine the advantages of Toxkits with the need for a practical and low cost "rapid" toxicity test. This thought was also inspired to some extent by the fact that scientists from several countries had discovered that some Toxkits, and in particular the Thamnotoxkit microbiotest, were quite sensitive to the biotoxins produced by blue-green algae during massive algal blooming in surface waters and water reservoirs[34-36]. The 24h mortality test with the aquatic crustacean *Thamnocephalus platyurus* was ear-marked as an interesting alternative to the mouse bioassay for routine screening of suspected algal blooms[37-40].

Research was therefore initiated on a "Rapidtoxkit, with selection of the most appropriate test species and a practical "sub-lethal" test criterion. The first achievements were reported two years ago during the "International Symposium on Water Contamination Emergencies"[2], highlighting that the aquatic crustacean *Thamnocephalus platyurus* was a good candidate for a rapid toxicity test based on the physiological criterion "absence of particle uptake" under toxic stress, within a time span of 1 hour.

The 1h Rapidtoxkit microbiotest has in the meantime been completed, and the test procedure is explained briefly hereunder and visualised schematically in Figure 1.

- After hatching of the dormant eggs (which takes about 24h), the larval crustaceans are transferred into a test tube containing the suspected water sample, in parallel to a tube with clean (control) water.
- During the 1h exposure period (which can even be shortened in case of real emergency) the organisms are either or not stressed, depending of the presence or absence of toxicants.
- A small amount of "red beads" (coloured microspheres) is then added to the test tubes, which the larval crustaceans will ingest (in the controls) or possibly not ingest (in a contaminated water sample).
- After 15 minutes the test organisms are killed with a few drops of fixative and transferred in shallow cups of a transparent observation chamber for analysis under a dissection microscope.
- The percentage of organisms with "coloured" intestinal tract is determined and compared to that in the control.

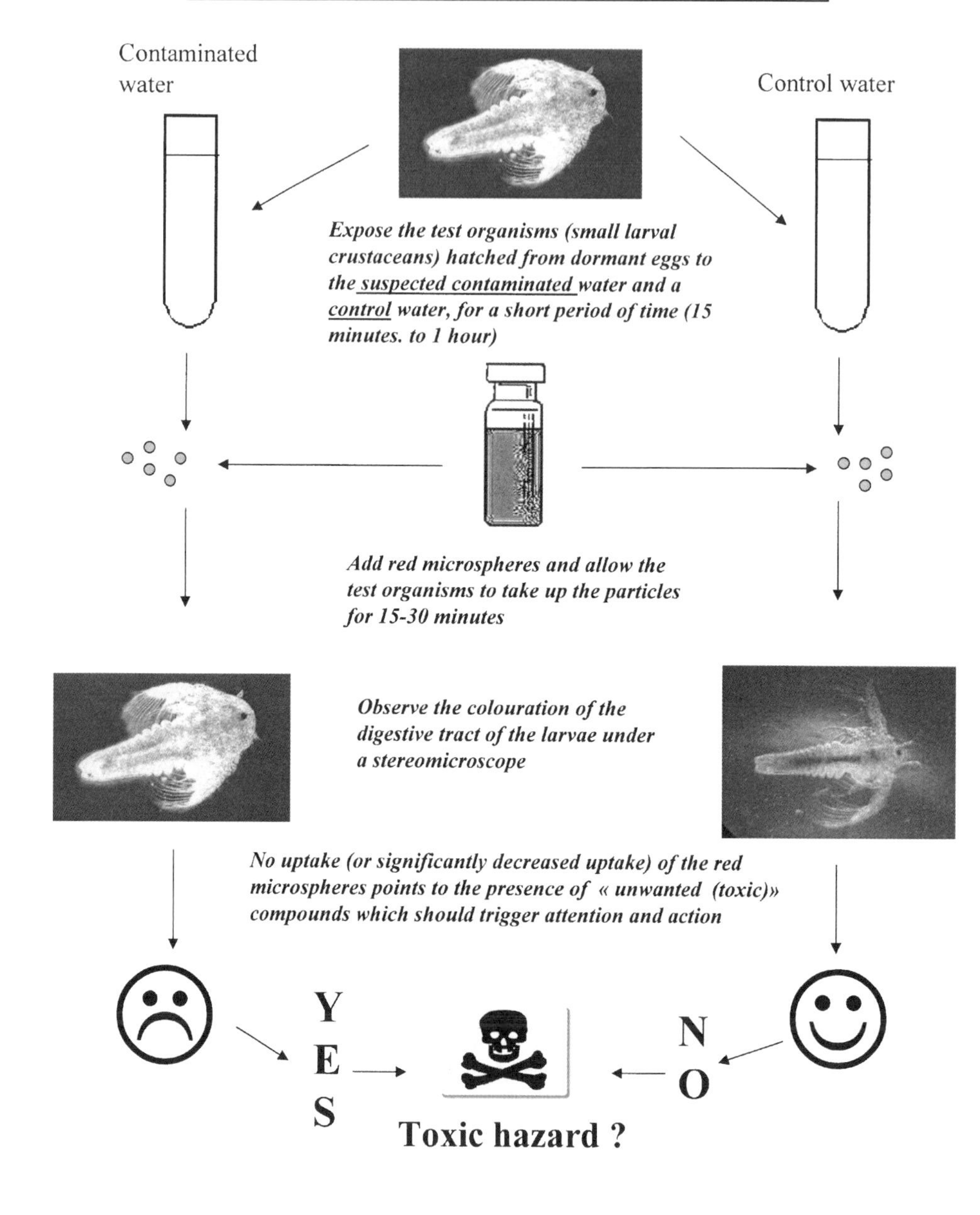
Figure 1 The RAPIDTOXKIT microbiotest in a nutshell
Contaminated water
Control water
Expose the test organisms (small larval crustaceans) hatched from dormant eggs to the suspected contaminated water and a control water, for a short period of time (15 minutes. to 1 hour)
Add red microspheres and allow the test organisms to take up the particles for 15-30 minutes
Observe the colouration of the digestive tract of the larvae under a stereomicroscope
No uptake (or significantly decreased uptake) of the red microspheres points to the presence of « unwanted (toxic)» compounds which should trigger attention and action
YES
NO
Toxic hazard ?

- The percentage of inhibition of particle ingestion is then calculated for the suspected water sample, as the effect criterion for the presence of toxic substances.

The Rapidtoxkit is a very cost-effective microbiotest since the commercial kit contains all the materials needed to perform bioassays on 21 to 45 water samples (depending of the number of replicates used for each sample). Unlike the Daphnia IQ test referred to above, this simple and very practical assay is totally independent of the continuous culturing/maintenance of live stocks. Nevertheless one needs to take into consideration that the hatching of the dormant eggs takes one day to obtain the larval crustaceans; in practice this means that one always has to anticipate one day in advance on the need for live (fully active) test organisms! In order to bypass this handicap a programmable MiniToxkit Incubator has recently been developed which provides the test organisms exactly at the time of performance of the rapid assays.

6 FIRST APPLICATIONS OF THE RAPIDTOXKIT

In the context of the practical application of the Rapidtoxkit to water contamination problems, it was important to compare the sensitivity of the sub-lethal effect parameter with the "lethal" effect levels after a longer period of exposure. Mortality is indeed "the milestone" effect used in most acute toxicity tests with invertebrates. Such a sensitivity comparison had already been made by Janssen et al. (1993) for the 1h enzymatic Fluotox assay referred to above. In the latter study 1h EC50s of the Fluotox were compared with 24h EC50s for *Daphnia magna* (EC50 is the "immobility" effect criterion used for acute Daphnia assays). Comparison of the data pairs of 20 inorganic and organic compounds revealed an excellent correlation ($r^2 = 0.95$) between the effect levels found with the rapid assay and the standard 24 hour test respectively.

A first comparative study of the Rapidtoxkit and the Thamnotoxkit was recently performed by Nalecz-Jawecki and Persoone[41] on 28 pharmaceuticals belonging to 5 different categories (non steroidal anti-inflammatory agents, biocides, cardiovascular compounds, nervous system drugs and purine alkaloids). The results of this study first of all showed very large differences in toxic effects among the 28 drugs, ranging from virtually not toxic to very toxic. For the majority of the pharmaceuticals with low toxicity, the median effect levels of the Rapidtoxkit are higher than those of the Thamnotoxkit, indicating a more intense (rapid) sub-lethal reaction of the crustaceans than a lethal one. The correlation coefficient for the effect data pairs for all the drugs is very high ($r^2 = 0.93$) and confirm that the physiological effect criterion applied with the Rapidtoxkit is, like the enzymatic Fluotox effect criterion mentioned above, a good (rapid) "predictor" of important biological effects (mortality) after more prolonged exposure of the test species to toxicants.

A second comparative study of the Rapidtoxkit and the Thamnotoxkit microbiotests has been performed in 2005 by Törökne[42], on biotoxins produced by blue-green algae. As already indicated above, *T. platyrus* is very sensitive to the hepatotoxins, neurotoxins and cytotoxins synthesized by particular cyanophycean species in surface waters. The outcome of this study performed on the extracts of 15 samples of algal biomass known to contain various cyanotoxins, showed that the 1h EC50 effect levels of the Rapidtoxkit were quite near the 24h LC50's of the Thamnotoxkit. According to Törökne these findings indicate the promising potential of the new microbiotest for very rapid detection of the biotoxins produced by blue-green algae.

7 APPLICATION OF THE RAPIDTOXKIT TO CONTAMINATED WATERS

The very first confrontation of the new rapid microbiotest with "real world samples" was made during the "Workshop on Biodefense" at the University of Pecs in Hungary at the end of August 2004 and to which laboratories and commercial companies were invited to apply their respective (rapid) toxicity tests on 30 "blind" water samples.

Eventually 15 different types of assays were tried out by the participants, based on various effect endpoints (toxicity, mutagenicity, endocrine disruption, etc).

The testing on blind water samples was a unique occasion to determine the potential of the Rapidtoxkit with "real world samples" and to compare its sensitivity with that of the other assays applied during this Workshop, also including the 24h Thamnotoxkit. The sub-lethal and lethal effects of the 30 waters were determined with both microbiotests using a tenfold dilution series and the EC50's and LC50's calculated, with subsequent transformation in Toxic Units (TU) for more easy comparison.

The water samples were then classified into four toxicity classes on the basis of the following arbitrary subdivision of the TUs : not toxic (0 Tu); toxic (0.1-1 TU), very toxic (1-10 TU) or extremely toxic (10->100 TU).

The nature and the chemical composition of the 30 samples were only disclosed at the end of the Workshop and revealed that 13 samples waters were "natural" (surface waters, groundwaters, tap water or sediment pore waters); the others were surface and groundwaters spiked with one or more chemicals, in different concentrations.

The chemicals compounds used for the spiking were mercuric chloride, potassium cyanide, citric acid, estradiol, DES (diethyl stilbestrol), coumesterol or MNNG. (N-Methyl-N'-Nitro-N-Nitrosoguanidine)

Since the detailed results of this Workshop will be published elsewhere, only the major findings obtained with the Rapidtoxkit and the Thamnotoxkit will be briefly commented upon hereunder.

None of the four sediment pore waters gave a toxic signal with either the Rapidtoxkit or the Thamnotoxkit, but in turn only three of the other natural water samples were found not toxic to both microbiotests. No inhibition of particle uptake was noted after 1h exposure of the six natural waters which induced mortality in the crustaceans within 24h. The chemical composition of these six waters clearly showed that each of them contained concentrations of particular compounds which were above the lethal thresholds.

None of the waters spiked with ppb concentrations of estradiol, DES or coumesterol, or with low concentrations of KCN (0.4 ppm) or MNNG (0.1 ppm) gave any effect with the two microbiotests. In turn all the natural waters spiked with higher concentrations of the chemicals listed above were toxic to extremely toxic to the crustaceans.

Overall, the intensity of the toxicity signal (i.e. the toxicity class) was the same for 21 of the 30 water samples (i.e. 73%). For the other waters, the 24h mortality test in most cases gave a somewhat higher toxicity signal (higher toxicity class) than the sub-lethal test.

This very first "real world" study undoubtedly shows the potential of the Rapidtoxkit for a quick determination of the toxic hazard of suspected waters. It is worth mentioning that in the framework of Anti-Terrorism Water Monitoring Technologies, the USEPA has now also requested Battelle to perform an analogous ETV study with the Rapidtoxkit (with the same chemicals and biotoxins as in the first ETV study referred to above), to determine the usefulness of this new microbiotest for rapid detection of water contamination.

8 CONCLUSIONS

A variety of "rapid" tests with bacteria, unicellular organisms and crustaceans are now available for either manual analysis, or for on-line monitoring of surface waters or drinking waters. Some of these assays, such as the bacterial luminescence inhibition tests and the Rapidtoxkit microbiotest are fully independent of the culturing/maintenance of live stocks of the test species, and hence particularly attractive for rapid testing of suspected samples.

However, as emphasized in multiple scientific publications, no single test species is "most sensitive" for all the toxicants which may be present in water. Consequently, and in order to avoid "false negatives" it is always preferable, not to rely on one single test for rapid determination of toxic hazards (with whatever type of test species) but to apply two assays with organisms from different phylogenetic groups.

References

1 G. Persoone, Microbiotests for rapid and cost-effective hazard assessment of industrial products, effluents, wastes, waste leachates and groundwaters. In *Rapid detection assays for food and water,* eds. S.A. Clark, K.C. Thompson, C.W. Keevil and M.S. Smith, Royal Society of Chemistry, Cambridge 2001. Special Publication No 272, 109-115.

2 G. Persoone, K. Wadhia and K.C. Thompson, Rapid Toxkit microbiotests for water contamination emergencies. In *Water contamination emergencies – can we cope ?* eds. J. Gray and K.C. Thompson, Royal Society of Chemistry, Cambridge 2004. Special Publication 293, 122-130.

3 K.C. Thompson and S.Scott, Analysis methods for water pollution emergency incidents. In *Water Contamination Emergencies – Enhancing our Response*, Royal Society of Chemistry, Cambridge, 2005. (this volume)

4 A.A. Bulich, Use of luminescent bacteria for determining toxicity in aquatic environments. In *Aquatic Toxicology* ASTM 667, eds. L.L. Markings and R.A. Loùerme, 1979, 98-106.

5 C.R. Janssen and G. Persoone, Rapid toxicity screening tests for aquatic biota. I. Methodology and experiments with *Daphnia magna. Environ. Toxicol. Chem.*, 1993, **12**, 711-717.

6 C.R. Janssen., E.Q. Espiritu and G. Persoone, Evaluation of the new enzymatic inhibition criterion for rapid toxicity testing with *Daphnia magna*. In *Progress in standardization of aquatic toxicity test,* eds. A.M.V.M. Soares and P. Calow, 1993, Chapter 5. 71-80.

7 E.Q. Espiritu, C.R. Janssen and G. Persoone, Cyst-based toxicity tests VII. Evaluation of the 1 hour enzymatic inhibition test (Fluotox) with *Artemia* nauplii. *Environ. Toxicol. Water Qual.*, 1995, **10**, 25-34.

8 M. Lechelt, W. Blohm, B. Kirschneit, M. Pfeiffer, E. Gresens, J. Liley, R. Holz, C. Lüring and C. Moldaenke, Monitoring of surface water by ultra-sensitive *Daphnia* Toximeter. *Environ. Toxicol.*, 2000, **15**, 5, 390-400.

9 U. Green, J.H. Kremer, M. Zillmer and C. Moldaenke, Detection of chemical threat agents in drinking water by an early warning real-time biomonitor. *Environ.Toxicol.*, 2003, **18**, 6, 368-374.

10 H. Tadehl and D.P. Häder, Fast examination of water quality using the automatic biotest ecotox based on the movement behaviour of a freshwater flagellate. *Wat.Res.*, 1999, **33**, 2, 426-432.

11 H. Tadehl and D.P. Häder, Automated biomonitoring using real time movement analysis of *Euglena gracilis. Ecotoxicol.. Environ. Saf.*, 2001, **48**, 161-169.

12 C. Streb, P. Richter, T. Sakashita and D.P. Häder, The use of bioassays for studying toxicology in ecosystems. *Current Topics in Plant Biology*, 2002, **3**, 131-142.

13 M.D. Ferrando and E. Andreu, Feeding behaviour as an index of copper stress in *Daphnia magna* and *Brachionus calyciflorus. Comp. Biochem. Physiol.*, 1993, **106C**, 327-331.

14 M.D. Ferrando, C.R. Janssen, E. Andreu and G. Persoone, Ecotoxicological studies with the freshwater rotifer *Brachionus calyciflorus* : The effects of chemicals on the feeding behaviour. *Ecotoxicol. Environ Saf*, 1994, **26**, 1- 9.

15 C.M. Juchelka and T.W. Snell, Rapid toxicity assessment using ingestion rate of cladocerans and ciliates. *Arch. Environ. Contam. Toxicol.*, 1995, **28**, 508-12.

16 G. Bitton, K. Rhodes and B. Koopman, CerioFAST : an acute toxicity test based on *Ceriodaphnia dubia* feeding behaviour. *Environ. Toxicol. Water. Chem.*, 1996, **15**, 123-125.

17 K. Jung and G. Bitton, Use of CerioFAST for monitoring the toxicity of industrial effluents : comparison with the 48h acute *Ceriodaphnia* test and Microtox. *Environ. Toxicol. Chem.*, 1997, **16**, 2264-2267.

18 S.I. Lee, E.J. Na, Y.O. Cho, B. Koopman and G. Bitton, Short term toxicity test based on algal uptake by *Ceriodaphnia dubia. Water Environ. Res.*, 1997, **69**, 1207-1210.

19 T.W. Snell and G. Persoone, Acute toxicity bioassays using rotifers. I. A test for brackish and marine environments with *Brachionus plicatilis. Aquat. Toxicol.*, 1989, **14**, 65-80.

20 T.W. Snell and G. Persoone, Acute toxicity bioassays using rotifers. II. A freshwater test with *Brachionus rubens. Aquat. Toxicol.*, 1989, **14**, 81-92.

21 G. Persoone, Cyst-based toxicity tests : I. A promising new tool for rapid and cost-effective toxicity screening of chemicals and effluents. *Z.Angew. Zool.*, 1992, **78**, 2, 235-241.

22 M.D. Centeno, L. Brendonck and G. Persoone, Cyst-based toxicity tests : III. Development and standardization of an acute toxicity test with the freshwater anostracan crustacean *Streptocephalus proboscideus*. In *Progress in Standardization of Aquatic Toxicity Tests*, eds. A.M.V.M. Soares and P. Calow, 1992, 37-55.

23 M. Van Steertegem and G. Persoone, Cyst-based toxicity tests : V. Development and critical evaluation of standardized toxicity tests with the brine shrimp *Artemia* (Anostraca, Crustacea). In *Progress in Standardization of Aquatic Toxicity Tests*, eds. A.M.V.M. Soares and P. Calow, 1992, 81-97.

24 G. Persoone, Development and first validation of a "stock culture free" algal microbiotest : the Algaltoxkit. In *Microscale Aquatic Toxicology, Advances, Techniques and Practice,* eds. P.G. Wells, K. Lee and C. Blaise, 1998, Chapter 20, 311-320.

25 G. Persoone, Development and validation of Toxkit microbiotests with invertebrates, in particular crustaceans. In *Microscale Aquatic Toxicology,*

Advances, Techniques and Practice, eds. P.G. Wells, K. Lee and C. Blaise, 1998, Chapter 30, 437-449.

26 W. Pauli and S. Berger, A new Toxkit microbiotest with the protozoan ciliate *Tetrahymena*. In *New Microbiotests for routine Toxicity Screening and Biomonitoring*, eds. G. Persoone, C. Janssen and W. De Coen, 2000, 169-176.

27 B. Chial and G. Persoone. Cyst-based toxicity tests, XII. Development of a short chronic sediment toxicity test with the ostracod crustacean *Heterocypris incongruens* : selection of test parameters. *Environ.Toxicol.*, 2002, **17**, 6, 520-527.

28 B. Chial and G. Persoone, Cyst-based toxicity tests. XIII. Development of a short chronic sediment toxicity test with the ostracod crustacean *Heterocypris incongruens* : methodology and precision. *Environ.Toxicol.*, 2002, **17**, 6, 528-532.

29 P. Fochtman, Acute toxicity of nine pesticides as determined with conventional assays and alternative microbiotests. *In New Microbiotests for routine Toxicity Screening and Biomonitoring,* eds. G. Persoone, C. Janssen and W. De Coen, 2000, 233-242.

30 L. Ulm, J. Vrzina, V. Schiesl, D. Puntaric and Z. Smit, Sensitivity comparison of the conventional acute *Daphnia magna* immobilisation test with the Daphtoxkit F™ microbiotest for household products. *In New Microbiotests for routine Toxicity Screening and Biomonitoring,* eds. G. Persoone, C. Janssen and W. De Coen, 2000, 247-252.

31 M. Latif and A. Zach, Toxicity studies of treated residual wastes in Austria, using different types of conventional assays and cost-effective microbiotests. In *New Microbiotests for routine Toxicity Screening and Biomonitoring*, eds. G. Persoone, C. Janssen and W. De Coen, 2000, 367-384.

32 R. Baudo, A. Sbalchiero and M. Beltrami, Test di tossicita acuta con *Daphnia magna. Biologi Italiani. Higiene dell'ambiente e del Territorio.*, 2004, **6**, 62-69.

33 M. Daniel, A. Sharpe, J. Driver, A.W. Knight, P.O. Keenan, R.M. Walmsley, A. Robinson, T. Zhang and D. Rawson, Results of a technology demonstration Project to compare rapid aquatic toxicity screening tests in the analysis of industrial effluents. *J. Environ. Monit.*, 2004, **6**, 855-865.

34 B. Marsalek and L. Blaha, Microbiotests for cyanobacterial screening. In *New Microbiotests for routine Toxicity Screening and Biomonitoring,* eds. G. Persoone, C. Janssen and W. De Coen, 2000, 519-526.

35 M. Tarczynska, G. Nalecz-Jawecki, M. Brzychcy, M. Zalewski and J. Sawicki, The toxicity of cyanobacterial blooms as determined by microbiotests. In *New Microbiotests for routine Toxicity Screening and Biomonitoring,* eds. G. Persoone, C. Janssen and W. De Coen, 2000, 526-532.

36 A. Törökne, The potential of the Thamnotoxkit microbiotest for routine detection of cyanobacterial toxins. In *New Microbiotests for routine Toxicity Screening and Biomonitoring* eds. G. Persoone, C. Janssen and W. De Coen, 2000, 532-540.

37 A. Törökne, A new culture-free microbiotest for routine detection of cyanobacterial toxins. *Environ. Toxicol.*, 1999, **14**, 5, 466-472.

38 A. Törökne, E. Laszio, I. Chorus, J. Fastner, R. Heinze, J. Padisak and F.A.R. Barbosa, Water quality monitoring by Thamnotoxkit F including cyanobacterial blooms. *Water Sc. Technol.*, 2000, **42**, 1, 381-385.

39 A. Törökne, E. Laszio, I. Chorus, J. Fastner, R. Heinze, J. Padisak and F.A.R. Barbosa, Cyanobacterial toxins detected by the Thamnotoxkit (a double blind experiment). *Environ.Toxicol.*2000, **15**, 5, 549-553.

40 M.N. Reskone and A. Törökne, Toxic *Microcystis aeruginosa* in Lake Velencei. *Environ.Toxicol.*, 2000, **15**, 5, 554-557.

41 G. Nalecz-Jawecki and G. Persoone, Toxicity of selected pharmaceuticals to the anostracan crustacean *Thamnocephalus platyurus* : comparison of sub-lethal and lethal effect levels with the 1h Rapidtoxkit and the 24h Thamnotoxkit microbiotests. *Environ.Sci. Pollut. Res.*, 2005, (in press).

42 A. Törökne, Rapid test for detecting cyanobacterial toxins. In *Proceedings of the 12th International Symposium on Toxicity Assessment*, Skiathos, Greece, ed. A.Kungolos, 2005, 9.

A WATER COMPANY PERSPECTIVE

J. Gary O'Neill

Yorkshire Water Services, Bradford, Yorkshire, UK.

1 INTRODUCTION

Possibly the main problem with establishing a protocol for handling water contamination emergencies is their infrequency.

The Regulator has a rigorous, prescriptive approach to the quality of treated water and even more so to the routine sampling and analysis. The Regulator also has an expectation that Water Companies should have appropriate levels of security against deliberate contamination and an analytical capability for possible candidates. However, this does not have regulatory enforcement and there is a large degree of variability amongst UK Water Companies on attitudes and approach to these issues. Even wider variability is apparent around the world. This paper can only outline the approach and potential development of that approach in one company – Yorkshire Water Services (YWS). YWS supplies 4.5M customers from 81 Water Treatment works using rivers, upland reservoirs and groundwater as sources.

This approach to field and laboratory analysis for detection of unknown deliberately released contaminants in Yorkshire Water in conjunction with its provider of analytical services, ALcontrol Laboratories, was outlined at the previous water contamination conference in 2003[1]. Results were given of the evaluation of a series of test kits for field use, specifically for arsenic and cyanide, and a generic test assessing cholinesterase inhibition to cover organophosphate and carbamate pesticides which would also cover some nerve gases. Non-specific toxicity testing devices using bacteriological and enzymatic luminescence were evaluated against a range of toxicants. Consideration was also given to instruments measuring photo-ionisation, radioactivity and more conventionally, pH, turbidity, conductivity, and to the use of a UV spectrophotometer for broad based analysis of organic compounds. The only field kit tested in relation to microbiological analysis involved ATP measurement.

The conclusion was essentially that although each of the kits had its value and scored over the laboratory in terms of travel time the overall coverage was limited compared with the laboratory. The laboratory also maintained a toxicity biomonitoring capability (using the crustacean, *Thamnocephalus*)[2] superior to any of the field toxicity tests.

2 DEVELOPMENTS SINCE 2003

The number of field kits involved rapidly precluded the back of a van / boot of a car approach and the kits were eventually deployed in a mobile laboratory; an adapted vehicle with benching, a sink and a generator providing 250V. The latter is essential for the UV spectrophotometer which otherwise could not be used in the field. Some other tests have been evaluated, in particular a more sensitive assay for cholinesterase inhibition, but none have proved practicable.

The advantage of the mobile laboratory is that it can be deployed on site and hence removes the time delay in transport to the main laboratory. However, although it provides a range of analytical capability it could never approach that of the main laboratory. In practice, in the two years that it has been available the opportunity for use has been very limited and. the main laboratory with its more comprehensive capability and 24 hour availability has been the mainstay of emergency analysis. Maintaining the mobile laboratory is now likely to be discontinued.

3 SAMPLING AND LABORATORY REQUIREMENTS

In the event of a suspected contamination incident, a water company would ideally want to have samples taken and delivered to the laboratory immediately. At best even with a 24-7 capability this may often provide the longest delay in the process. In Yorkshire the best approach we have found is to have specific sampling kits available at specific locations and in the possession of the standby scientists. In the event of intelligence or initial observation giving a strong indication of toxic substances, precautions will be necessary during sampling.

The laboratory is notified that samples would be arriving with up to eight scientists called out if it is outside normal working hours. Analysis would be expected to commence as soon as the sample(s) arrives. A more comprehensive description of methodology is given elsewhere in these Proceedings[3] but briefly: appearance and organoleptic characteristics are an obvious first step. A UV scan will give an immediate indication of organic contamination with substances containing an aromatic ring. Gross toxicity will be apparent within an hour from the Microtox and Thamnocephalus results, although the latter needs 24 hours for completion. A 1 -2 hour non-mortality version of this test is described elsewhere in these proceedings. Rapid radioactivity testing is carried out; GC-MS, HPLC/MS/MS will cover a wide range of organic compounds; and ICP-MS covers metals analysis. Specific tests for paraquat, diquat and fluoroacetate are carried out in parallel. The analysis would be expected to be completed within three hours. Microbiological and the final *Thamnocephalus* tests will be completed the next day.

4 SPECIALISED LABORATORIES

In the event of suspected deliberate contamination, rapid extensive testing for a wide range of specific pathogens and toxic substances is available to all water companies through a consortium of specialised laboratories on a 24-7 basis.

5 ON-LINE MONITORING

The value of on-line monitoring lies in the transmission of data to a point where any problems can be immediately recognised and the benefits far outweigh any protocol involving grab samples and carriage to the laboratory because of the time delays. The limitations of on-line monitoring lies in its scope.

As a matter of routine, on-line data at treatment works for turbidity, pH, chlorine, flow and selected metals are monitored through a SCADA system. In Yorkshire Water the data in telemetered to a central data storage system. Since 1998 particular emphasis has been placed on turbidity as a surrogate for Cryptosporidium. In order to maintain the company treated water standard of 0.1 FTU, a great deal of maintenance, scrutiny and interpretation on the output is required on the 400 turbidimeters deployed at the different stages of the process for 32 treatment works.[4]

This process, described in the company as 'Filter Management' would not provide protection in the event of contamination but does give the fundamental basis and ethos for on-line systems. All critical instruments must provide credible data at all times. Problems with any specific instrument must rapidly become apparent and be remedied as soon as soon as practicable. Instruments must telemeter to a central location in addition to any SCADA system. Alarms must be set realistically and must be credible and a response procedure must be in place on a 24-7 basis. Any novel instrumentation must follow the same pattern. As is apparent elsewhere at this conference a range of instrumentation is commercially available or being developed, e.g. Einfeld[5] but consideration of purchasing such instrumentation needs to be done with full knowledge of resources required for day to day use, maintenance and interpretation of output.

5.1 Yorkshire Water Pollution Monitoring System (YW-PMS)

The company has eight direct river abstractions with limited raw water storage capacity and has had in place since 1985 a scanning UV spectrophotometer system on the river water for pollution detection. This broad screen approach to detection has been in use as there are no readily identifiable individual potential pollutants on the largely agricultural based catchment. In 2004 these were replaced by new UV based instrumentation developed in house using a solid state diode array, the Yorkshire Water Pollution Monitoring System (YW-PMS). These are modular systems needing very little maintenance and score over the previous systems as they can operate under high river turbidities and still maintain their sensitivity to pollutants by using software which automatically compares the current scan with the most relevant stored river spectra. The limitations of the device are that it will only detect organic compounds which display a UV spectrum, essentially with double bonds. This includes most pesticides but not diesel fuel. The limits of detection will depend on the actual compound but values in the 0.1 – 1 mg/l are common and alarms are set accordingly on the eleven devices in place.

A UV probe produced in Austria by S::CAN[6] similarly uses UV diode array technology and is described elsewhere at this conference[7]. This has not yet been evaluated in YWS but had there had been an awareness of it in 2002 then the company may not have embarked upon its own development of a diode array system.

5.2 Biomonitors

As part of the company's ongoing evaluation of sensors for river abstractions two biomonitors are currently being assessed. Firstly a German system - the Daphnia Toximeter from bbe Moldaenke[8]. This records behaviour of Daphnia and gives an integrated toxicity signal. Secondly a Dutch system - TOXcontrol from MicroLAN[9], which assesses toxicity by inhibition of light from light emitting bacteria. Both have been successfully used in Europe as described elsewhere at this conference, [10] and are currently undergoing trials within YWS.

5.3 Electronic nose

Various versions of the electronic nose have historically been developed but the one currently under development with Manchester University[11] uses metal oxide semiconductor sensors that operate at high temperature and exhibit good immunity to environmental factors (temperature / humidity). The primary development is for assessment of sewage quality but its sensitivity suggests that there could be drinking water applications. In use on a river abstraction it has been found to be robust in a water treatment works environment with low maintenance requirements. It is an intelligent system that develops fingerprints for known contaminants; early results show good sensitivity to diesel.

5.4 CENSAR

The CENSAR[12] system which owes some of its development to YWS is essentially a multi-parameter microchip based system used in distribution mains under pressure routinely used mainly for changes in turbidity conductivity and chlorine for operational purposes. Over 200 are in use in Yorkshire; some are fixed but the primary use is in control of resuspended material in water mains during valving operations. Changes in turbidity are monitored as on-line output on a lap top computer as a valve is turned providing the opportunity to slow the turning if turbidity starts to increase. Small changes in temperature are monitored simultaneously and can indicate movement of water across a valve.

6 DEVELOPMENT OF ON-LINE MONITORING FOR DETECTION OF CONTAMINATION OF TREATED WATER

As outlined above the company has wide experience in development and use of non-specific quality monitors on rivers and also routine measurement of conventional drinking water determinand in the network. No systems are in place specifically for detection of deliberate contamination, although the screening systems in place at river abstractions would provide a degree of cover.

Protection of treated water becomes more problematic. Monitoring of a treatment works outlet or single point in the network would not be too difficult. A range of systems could be used but YWS is likely to use its own experience. The YW-PMS together with a CENSAR would provide a capability for a UV screen together with conductivity and chlorine. This combination would provide cover for a range of contaminants. In the network CENSARs would be used together with pressure sensors. The CENSAR is highly portable and the YW-PMS relatively so and these could be important factors in some situations. The S::SCAN also uses a UV scan and, as with the CENSAR, can be used in pressurised mains and may be superior to the YW-PMS for such developments.

Temperature measurement provides an interesting area of development. CENSAR provides very precise measurements and detection of temperature differences of 0.01°C are feasible. It can be envisaged that any direct contamination of the network could be picked up by slight temperature changes which, together with pressure measurements, could provide the basis for whole system monitoring requiring numerous units and a period of assessment to determine normal or operational changes. Network SCADA systems together with telemetry would complete whole system monitoring.

The Biomonitors: the Daphnia toximeter and the ToxControl could also be feasibly used in treated water if any chlorine present is neutralised.

References

1 G.O'Neill, C.Ridsdale, K. Clive Thompson and K. Wadhia. 'Field and Laboratory Analysis for detection of unknown deliberately released contaminants' in *Water Contamination Emergencies: Can We Cope*, eds, KC Thompson and J Gray. International Conference, Kenilworth, UK, 2004.
2.G Persoone, K. Wadhia and K. C. Thompson. Rapid Toxkit Microbiotests for Water Contamination Emergencies in *Water Contamination Emergencies: Can We Cope*, Eds, KC Thompson and J Gray. International Conference, Kenilworth, UK, 2004.
3 K. C. Thompson and S. Scott. Analysis methods for water pollution emergency incidents in *Water Contamination Emergencies: Enhancing our response*, Eds, KC Thompson and J Gray. International Conference, Manchester, UK, 2004.
4 A.Wetherill and J.G. O'Neill in *Advances in Rapid Gravity Filtration in Water and Wastewater*. CIWEM International Conference, London. 2001
5 W. Einfeld. Faster, smaller, cheaper. Technical innovations for next-generation water monitoring in *Water Contamination Emergencies: Enhancing our response*, Eds, KC Thompson and J Gray. International Conference, Manchester, UK, 2004.
6 B. Tangena. Be prepared, the approach in the Netherlands. *ibid*
7 S::CAN.- scan Messtechnik GmbH, Herminengasse 10, A-1020 Wien, Austria http://www.s-can.at
8 Daphnia Toximeter - bbe MOLDAENKE, GmbH, Wildrosenweg 3, D-24119 Kiel-Kronshagen, Germany
cmoldaenke@bbe-moldaenke.de
9 TOXcontrol - Microlan bv, PO Box 644, 5140 AP Waalwijk, The Netherlands
Joep.appels@microlan.nl
10 J. Appels. Development of an on-line biological warning system, presented at *Water Contamination Emergencies: Enhancing our response*, International Conference, Manchester, UK, 2004.
11 Electronic Nose - Dr Peter Wareham, School of Chemical Engineering & Analytical Science, The University of Manchester, Sackville St, Manchester, M60 1QD. p.wareham@manchester.ac.uk
12 CENSAR - CENSAR Technologies Ltd, Unit G2, Ground Floor, 6 Whittle Road, Ferndown Industrial Estate, Wimborne, Dorset, BH21 7RU

RAPID DETECTION OF VOLATILE SUBSTANCES IN WATER USING A PORTABLE PHOTOIONIZATION DETECTOR

P. J. Bratt[1], K. C. Thompson[2] and P. Benke[2],

[1]RAE Systems UK Limited, PO Box 490 Abingdon, Oxon, OX14 2WW
E-mail pbratt@raeeurope.com
[2]ALcontrol Laboratories, Rotherham, South Yorkshire,
S60 1BZ, United Kingdom, E-mail clive.thompson@alcontrol.co.uk

1 INTRODUCTION

Although many rapid screening techniques have been developed for dealing with water contamination emergencies, most are laboratory based, require a highly trained operative and may need a fairly complex calibration and QA/QC procedures. The photoionisation detector (PID) system for the detection of volatile organics is simple, robust, can be operated by non-scientific personnel after basic training and only needs a very simple QA/QC procedure.

A PID uses an electrodeless ultraviolet (UV) light source to photoionize a gas sample or a vapour sample and detect its concentration. Ionization occurs when a molecule absorbs the high energy UV light, ejecting a negatively charged electron and forming of positively charged molecular ion[1]. These charged particles produce a current that is easily measured at the sensor electrodes. Only a small fraction of the VOC molecules are ionized. Therefore, PID measurements are non-destructive and vapour samples can be bagged and used for further analysis if required.

The PID order of sensitivity (10.6 eV source) is: - aromatics, iodine compounds > olefins, ketones, ethers, amines, sulfur compounds > esters, aldehydes, alcohols, aliphatics > chlorinated aliphatics > low molecular weight aliphatics such as propane, ethane and methane (no response). Typically a high sensitivity PID system will have a useable operating range from 10 ppbV - > 2,000 ppmV

One of the main advantages of a PID system is that it can readily be transported and used at the site of an incident. A complete system will fit inside a small briefcase, and is powered by re-chargeable batteries. Up to a few years ago, a weakness of the technology was the lifetime of the light sources. Although this still applies to some degree for the 11.7 eV light sources where a lithium fluoride window is required to transmit wavelengths below 160 nm, the 10.6 eV lamps are constructed from fused silica and have a guaranteed lifetime of three years. These lamps are suitable for the detection of a significant number of the substances of interest with respect to a water contamination incident. An instrumental sensitivity check can simply be carried out by monitoring the response from a gas sample taken from a small gas cylinder containing 100 ppmV of isobutylene in air. This key system suitability check takes less than one minute.

A summary of the advantages and disadvantages of the PID are given below:-

Advantages.

. Simple and robust.
. Can be operated by trained field (non-laboratory) staff.
. Top of the range P1D systems are very sensitive.
Can detect down to – 20 µg/litre diesel in potable water.
. Can screen up to 60 samples/ hour for volatile organics.
. For a known (single) volatile pollutant in potable water, it is possible to calibrate the system and obtain a semi-quantitative result.

Disadvantages.

. Will only detect non polar volatile substances that will transfer to the headspace from a water sample. Volatile polar substances such as methanol and acetone are only weakly detected because they have low Henry's law constants and remain in the aqueous phase.
No indication of the identity of any pollutants detected.
Need to avoid condensation on lamp surface.

2 SAMPLE PRE-TREATMENT AND MEASUREMENT PROCEDURE

The recommended simple sample pre-treatment procedure is to use one litre glass bottles with PTFE inserts filled to about the 500 ml mark (± 100 ml) with the sample, the bottle is then shaken vigorously for 20-30 s. Then after a 5 s delay, the cap is removed and immediately a small piece of aluminium foil is tightly placed over the mouth of the bottle. The probe from the high sensitivity RAE ppb PID detector is inserted through the aluminium foil and the measurement taken. The system is shown in Figure 1. A similar system has been used for measuring VOCs in soils[2]. It would be possible to improve the stability of the signals if a sample line were taken out the exit of the instrument back into the bottle. For this work this modification was not used.

Figure 2 shows the response from 1 ppmV of diesel in various river waters and effect of 5 mg/l sodium dodecylbenzenesulphonate (SDS), a surface active substance. It can be seen that the response is similar in deionised/RO (reverse osmosis) water and the four different river samples. However, in the presence of 5 mg/l SDS there was a significant decrease in response. This was attributed to the increased solubility of diesel in SDS-containing water.

Figure 3 shows the effect of temperature on the response of 1 ppmV of diesel in deionised/RO water over the temperature range 5 – 40°C, the increase in response observed at higher temperatures was not large and of no great benefit. (The increased diesel vapour pressure is almost cancelled out by the increased solubility.) It is not recommended to heat the sample above ambient temperature because it increases the risk of water condensation on the PID source window and sensor, which will lead to an erratic response.

3 APPLICATIONS

The system can be used both in the laboratory and also in the field at the site of an incident. In the laboratory it is possible to screen up to 60 samples/hour to indicate samples which have detectable levels of volatile organics. These then can be prioritised for conventional GC-MS headspace analysis in order to identify the pollutant. In the field the system is used for rapid monitoring at the site of an incident.

Also it can be used in a semi-quantitative mode to check for removal of a volatile pollutant from a distribution system after flushing the system through with water. This can save significant time compared with having to send samples back to a central laboratory many miles away.

An example of this was an incident where a contractor had grossly polluted the mains distribution system in a small village with a recovered, distilled mineral oil, having a boiling point of 260-330°C and containing a complex mixture of hydrocarbons.
The composition of the oil was as follows:
3% aromatics, mostly alkylnaphthalenes (C1 to C6)
Smaller amounts of fluorene, acenaphthene, anthracene, phenanthrene
42% aliphatics - mostly nC14 - nC19 alkanes, smaller amounts of branched alkanes
55% naphthenics - equal amounts of tri- and tetra-cyclic alkanes and their alkyl substituted derivatives?
(The toxicological aspects of this incident are described in the Chapter in this book by P.C. Rumsby, W.F. Young, N. Sorokin, C.L. Atkinson and R. Harrison.)
The nature of the pollutant was known, however the main laboratory was over 80 miles away. Fortunately the PID system could readily detect 0.05 ppmV of this substance in drinking water. Thus each dwelling in the village was flushed with fresh water to remove the pollutant and by using the PID device it was possible to rapidly assess when adequate flushing had been carried out for each property. The final 'clear' samples were taken in duplicate and the duplicate set of samples subsequently confirmed by the main laboratory.

4 RANGE OF DETECTABLE SUBSTANCES.

Table 1 lists a number of common substances that can be readily detected using a 10.6 eV source. Reference 1 lists a very wide range of chemicals and also includes relative response information to various chemicals

Table 1 Some common substances that can be readily detected using a 10.6 eV PID detector

Petrol
Kerosene
White spirit
Jet fuel
Diesel
Benzene
Toluene
Ethyl benzene
Xylenes
Trichloroethylene
Dichloroethenes
Ethyl acetate
Propyl acetates
Butyl acetates
Styrene

The following common substances cannot be detected using this source: - oxygen, nitrogen, carbon dioxide, carbon monoxide, methane, ethane, propane, chloroform and carbon tetrachloride. (i.e. all substances that have ionisation potentials greater than ~ 10.6 eV.)

5 ASSESSMENT TRIAL WITH RESPECT TO POTENTIAL DIESEL POLLUTION INCIDENT

Eighteen tap water samples were taken, six samples were spiked with 0.1 ppmV diesel, another six were spiked with 0.05 ppmV diesel and the remaining six were unspiked tap water samples. The 18 samples were then randomised so the analyst did not know which were the spiked samples. The analyst was then requested to detect as rapidly as possible, which samples contained diesel, using the PID. The results are summarised in Figure 4. It can be seen that all contaminated samples were detected and that 0.1 ppmV diesel could be clearly distinguished from the blank unspiked tapwater samples. Using a criterion of <5 arbitrary PID units being equal to zero, all the blank samples were recorded as non-contaminated. This assessment was completed within 30 minutes.

*6 **CONCLUSIONS***

The RAE ppbPID detection system has been found to be a robust fit for the purpose of rapidly screening potable water samples for the presence of volatile non-polar substances. It can easily be used by trained field personnel as well as by scientific staff in a central laboratory. The system is portable, powered by rechargeable battery and can readily be transported in a small briefcase. Using a very simple sampling technique it is possible to screen over 60 samples an hour for diesel pollution with a detection limit better than 0.1 ppmV. A system suitability check can easily be carried out by monitoring a gas sample from a small cylinder containing 100 ppmV of isobutylene in air or nitrogen and this takes less than one minute to perform.

REFERENCES

1. Haag, W. R and C. Wrenn, Theory and Applications of Direct-Reading Photoionization Detectors. RAE Systems Inc. 2002, 175 pp.

2. Hewitt, A. D. and J. E. Lukash, PID-based estimation of the concentration of aromatic and chlorinated VOCs in soil. Field Analyt. Chem. Technol. 1999, 3(3):193-200

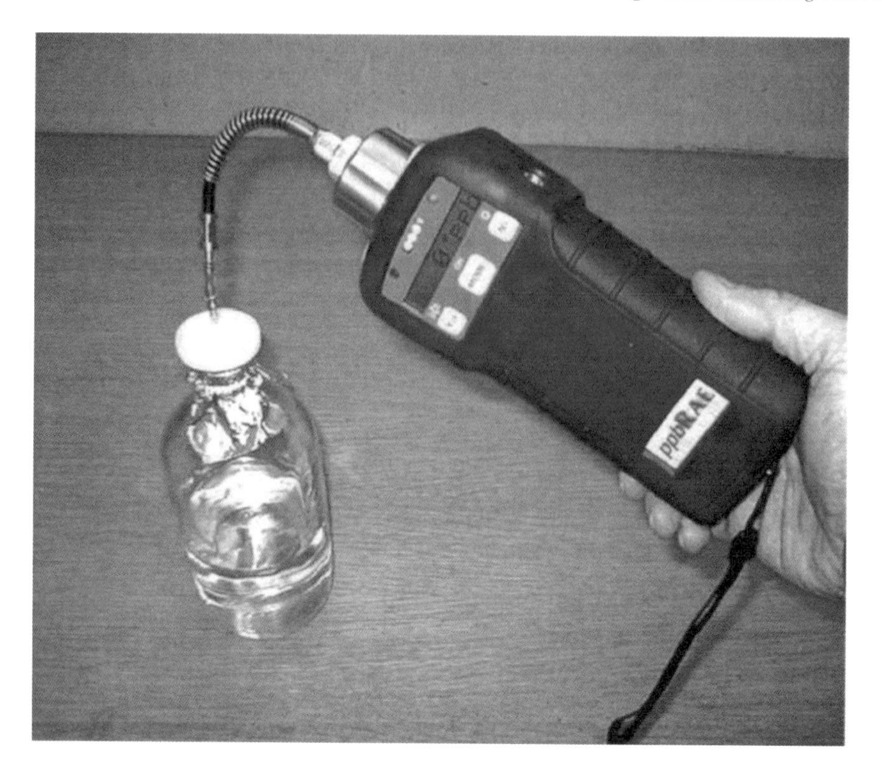

Figure 1 *PID Detection Set-up One litre glass bottle, aluminium foil and ppbRAE PID detection system*

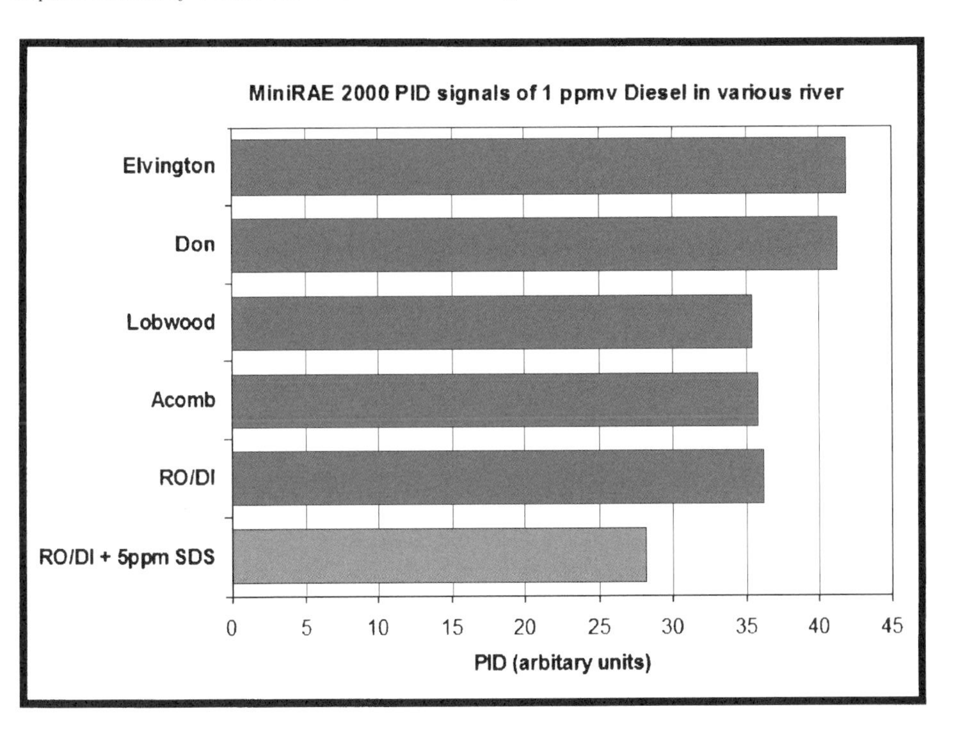

Figure 2 *Response from 1 ppmV of diesel in various rivers and effect of 5 mg/l sodium dodecylbenzenesulphonate (SDS)*

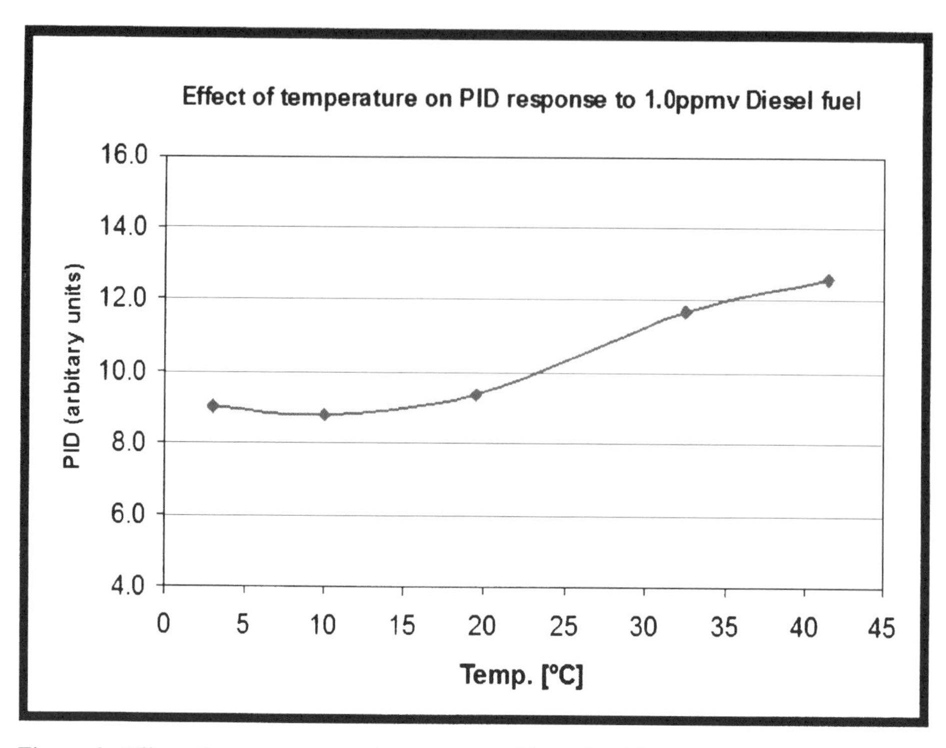

Figure 3 *Effect of temperature on the response of 1ppmV of diesel in deionised/RO water*

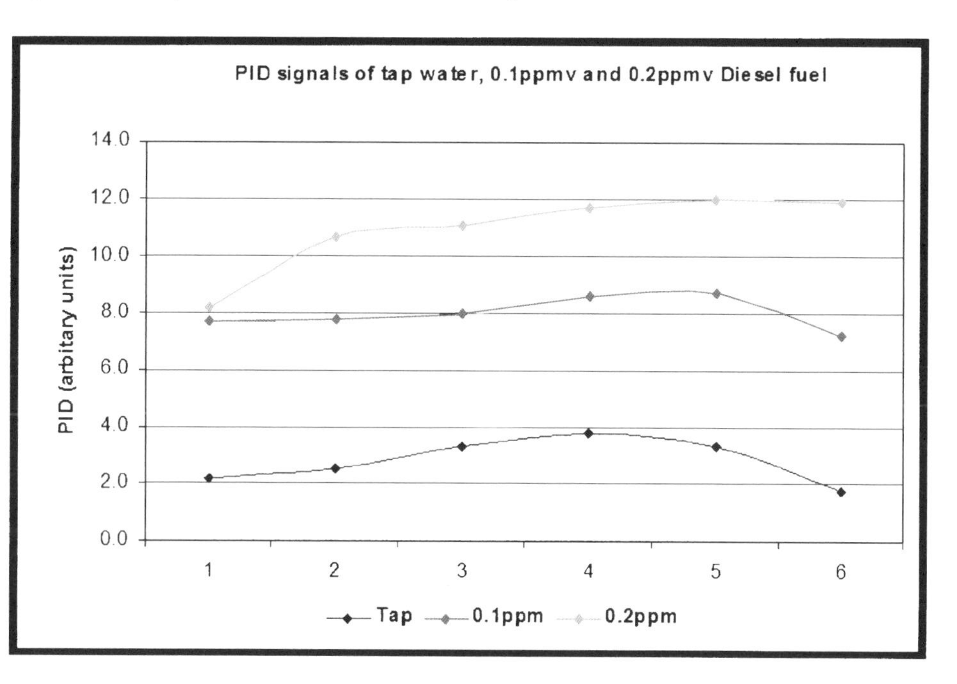

Figure 4 *Randomised Unpiked and Spiked Tap Water Sample Results*

ANALYSIS METHODS FOR WATER POLLUTION EMERGENCY INCIDENTS

K. C. Thompson[1] and S. Scott[2]

[1] Chief Scientist, ALcontrol Laboratories, Rotherham, South Yorkshire, S60 1BZ, United Kingdom, E-mail clive.thompson@alcontrol.co.uk

[2] Consultant, E-mail spscott@btinternet.com

ABSTRACT

Planning for the analysis of high impact very low probability events is very difficult. This is particularly true when dealing with the analysis arising from potable water emergency pollution incidents. The main issues are: - how to rapidly detect when significant contamination has occurred; to identify the cause or convincingly prove a negative in the absence of contamination and finally maintain an efficient and effective 24h/365d response system on a long-term basis for very low frequency events. This paper considers water pollution emergency incident analysis issues with respect to water laboratories. Other key issue are how to assess the emergency response performance on a regular basis and the need to minimise operational costs. Chemical, radiological and ecotoxicological screening protocols are discussed. Microbiological emergency incidents are not covered. The numerous benefits of setting up a mutual aid laboratory response scheme have been outlined in a previous presentation at this conference.

Keywords: - pollution incidents, drinking water, rapid screening tests, chemical, radiological and ecotoxicological analysis

1 GENERAL BACKGROUND

Planning for high impact very low probability events is notoriously difficult.[1, 2] This is particularly true when dealing with the laboratory analysis arising from potable water emergency pollution incidents. There are three main issues; firstly maintaining an efficient and effective 24h/365d response system on a long-term basis when there can be many months between major pollution events; secondly how to rapidly detect when potential contamination has occurred and thirdly identify the cause or convincingly prove a negative in the absence of contamination.

The Rotherham laboratory of ALcontrol is responsible for handling these issues for a population of over 8 million drinking water consumers. After a high profile contamination incident has occurred, a post-mortem is held and the response and analysis systems appropriately upgraded. Often very useful lessons are learnt from this exercise. The problem is how to retain the improvement on a long-term basis in the absence of further significant contamination incidents and the need to continually minimise operational costs. It is important that a laboratory positions itself on the risk / operational cost curve at an appropriate and agreed point. See Figure 1. Guaranteeing 365d/24h availability of sufficient suitably skilled staff is not easy and it is important that all key stakeholders are aware of this positioning. They should also be aware that to reduce the risk of producing an unfit for purpose emergency analytical response to zero for all emergency incident eventualities (chemical, radiological and microbiological) is prohibitively expensive. There is an asymptotic relationship between cost of running an emergency laboratory service and this risk.

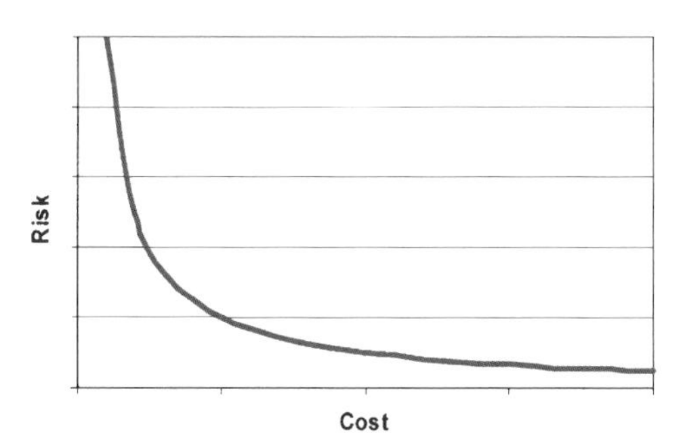

Figure 1. *Exponential risk versus cost of emergency analysis service relationship*

Ordinate. Risk of failure to provide fit for purpose analysis. **Abscissa.** Cost of maintaining the lab emergency 365d/24h system

Modern water laboratories are extremely efficient at carrying out fit for purpose high throughput targeted analysis. Some of the larger water laboratories in the UK routinely handle 500+ samples per day for a wide range of targeted parameters to comply with relevant regulations. Chemical analysis costs in real terms have reduced by well over ten times in the UK over the last 30 years. Currently the bulk of regulated drinking water chemical analysis is carried out by less than 12 highly automated laboratories. Thirty years ago this analysis was carried out by well over 300 laboratories mainly using manual techniques. Targeted and planned analysis of known substances delivered at a preset time requires much less skill than trying to identify completely unknown substances in infrequent completely random samples. These can arrive without warning on a 24h/365d basis. Equipping a changing staff base with the skill to rapidly detect and identify unknown substances on a 24h/365d basis is not an easy task.

Emergency incidents have an unnerving tendency to commence around 17-00 hours on a Friday evening often just before a Bank Holiday. Consequently, the number of staff available is limited. This creates additional logistical difficulties. Also incidents can occur in remote areas and obtaining relevant samples from natural water bodies in these remote areas can be difficult especially during the hours of darkness. Prolonged incidents, that necessitate 24 hour working, can rapidly exhaust available staff and increase the chance of analytical errors.

The issue of health and safety with respect to samplers and staff is another very difficult area. The chemical, microbiological and radioactivity risks need to be assessed. If staff are not fully appraised of the risks prior to potential major events there is a danger they may refuse to take or analyse "high profile samples". A well-documented risk assessment/safe working procedure system is a pre-requisite for any robust emergency response system.

2 TARGET VERSUS SCREENING ANALYSIS

Currently most UK major company water laboratories carry out high volume target analysis for all regulatory and operational parameters. However, this targeted regulatory analysis will not cover a significant number of potential threat agents. A completely different approach is needed for the handling of emergency incidents. As these often occur out of working hours with limited skilled staff availability, a more focussed approach is required. It is very difficult to prove a negative result, especially in the context of an emergency incident situation with intense pressure to provide rapid results. The consequences of false negative or positive results must be fully appreciated by both the laboratory and its clients. The potential public health and financial liability implications resulting from either scenario can be very significant. Much effort has been expended for the development of a number of robust rapid screening tests that will help to screen large numbers of samples quickly and hopefully allow prioritisation of the samples that require further investigation.

Any out of hours service cannot be as comprehensive as that available during working hours when many more staff are available. Thus it is important to tailor the normal out of hours guaranteed minimum response to the available limited staff resource. Rapid screening analysis allows the prioritisation of samples for more time-consuming analysis. ALcontrol has investigated a range of chemical[3] and ecotoxicological screening tests[4]. For instance any samples displaying significant toxicity would be prioritised. At ALcontrol, as soon as a potential emergency pollution sample is received, a Thamnocephalus screening test (See 4) is immediately set up to run in parallel with the normal chemical analysis.

3 CHEMICAL SCREENING TECHNIQUES

3.1 Aims and Objectives

The principal aims and objectives are considered to be: -

- Screen 100 samples in 2 hours by relevant chemical procedures
- Simplify sample pre-treatment procedures. If possible use the as received sample
- Identify best practice
- Develop more rapid robust tests
- Identify promising future developments and techniques
- Maintain the 'experts' network (See relevant previous presentation)

In addition it is important to gather as much information as possible about incident for example as a minimum: -

- Observation – sight and smell
- TOC (15 min) and diode array uv trace
- Headspace or Purge & Trap GC-MS

- Liquid / Liquid extraction (DCM and / or Hexane) GC-MS
- pH adjustment Liquid / Liquid extraction and derivitisation GC-MS

It is important to appreciate that typically there will be a two hour turnaround before first result is received

3.2 Expert Knowledge Considerations

This is considered to be a key area that requires further development. A number of key issues can be summarised as follows:

- Few experts are available and many are due to retire in the next ten years. In the authors' opinion these are not being replaced at a sufficient rate.
- MS deconvolution and interpretation software e.g. AMDIS is extremely useful when trying to interpret rapid screening chromatographic runs that contain overlapping peaks.
- An industry database is needed to store chromatographic information on the main risk threat substances, relevant reference unknowns and artefacts.
- A baseline database of chromatographic and other relevant chemical data of key uncontaminated water supplies should be set up by all water companies. This will allow any significant changes to be rapidly detected. It is important to appreciate that all waters will contain very low levels of naturally occurring (harmless) organic matter (e.g. fulvic and humic acids) and modern sensitive instrumentation will detect these substances. Consequently prior knowledge of these natural substances is essential; otherwise false alarms could be set off in an incident. The nature of the background organic matter tends to vary with the season and regular baseline scans should be run on a routine basis.
- Laboratories should have access to relevant reference publications

3.3 Metals and metalloids

Screening for metals and metalloids using induction coupled plasma mass spectrometry (ICP-MS) is a well-proven technique and scans for up to 70 elements can be rapidly carried out on potable water samples. (Typically 20 samples per hour.) Drinking water is a relatively easy matrix as the vast majority of samples contain more than 99.8% m/m water. Toxic elements such as arsenic, antimony, cadmium, lead, mercury, selenium, thallium, uranium etc. can all be readily detected at 1ug/litre levels well below the short-term acute toxicity threshold

3.4 Organic Substances

Unlike potentially toxic metals, where there are a limited finite number, there are a very large number of potentially toxic organic substances and screening for these is much more problematical. (There are over 13 million known organic substances.)

Many organisations use some form of solvent extraction and gas chromatography mass spectrometry (GC-MS) as their main screening technique. There are three problems with the approach, firstly many potential toxic substances are semi-polar or polar and will not be extracted and/or chromatographed efficiently unless they are initially derivatised; secondly the mass spectrometry library may not provide a correct identification and thirdly a number of natural harmless humic/fulvic acid derived substances are also detected with very poor or no identification.

Ways of overcoming these limitations are: -

i) Development of high pressure liquid chromatography mass spectrometry (HPLC-MS/MS) screening techniques using direct injection of the as received water sample. This avoids time-consuming sample pre-treatment (e.g. derivatisation /sample pre-concentration steps). Initial results look encouraging. It is relatively easy to build up a library of the main risk threat substances.

ii) Development of rapid GC-MS screening techniques including a derivatisation step. It has been found possible to reduce the run time from 60 to 15 min and automate the sample pre-treatment / sample pre-concentration steps. Conventional GC-MS and HPLC-MS techniques can be time consuming (up to 1 hour per sample). Thus there is a requirement to be able to screen samples quickly and prioritise the analysis so "suspect samples" are analysed first. Considerable method development work has been carried out to significantly reduce the analysis time down to 15 min for GC-MS and 5 min for HPLC/MS/MS. Another issue is that in order to run and interpret the output from GC-MS or HPLC /MS/MS screening techniques highly skilled/trained and experienced staff are required. These are becoming increasingly a "rare breed" because nowadays staff tend to be trained in carrying out high volume targeted analysis in large contract laboratories and are only infrequently involved in non-routine "one off" type analysis.

iii) Development of specialised GC-MS and HPLC/MS/MS libraries of the most likely organic substance toxic threats. There are already a number of toxic threat substance lists available in the literature and on the web [5-7].

As previously outlined, it is important for water companies to regularly screen their water sources so that naturally occurring harmless organic substances that are normally present are detected and recorded. The chromatographic software system used can then highlight these as normal background peaks. It is important to screen over the four seasons as the pattern of natural substances often change with season.

ALcontrol Laboratories have also developed a number of simple rapid screening tests such as diode array ultraviolet spectrometry (DAUVS); photoionisation detection (PID) and ion chromatographic scans. (This latter technique can be used to detect fluoroacetate.

3.5 Diode Array Ultraviolet Spectrometry (DAUVS)

Treated potable water samples give a very simple ultraviolet spectrum from 230–400 nm and using a 50mm path length silica cell it is possible to detect 10–50 µg/litre of most substances containing an aromatic ring in a treated drinking water. Uncontaminated waters give a simple spectrum with no obvious peaks. (See Figure 2). The sensitivity and discrimination of the technique can be improved further by using the first derivative. (See Figure 3). Each sample can be run in less than one minute.

This technique has been developed further as an automatic on-line monitor for Yorkshire Water Services (YWS) and is used for screening the raw water (river) inlet at major water treatment works using a 20 mm path length cell. Some specialised software has been developed that overcomes the problem of occasional high turbidity in raw waters allowing the instrument to operate at optimum sensitivity despite wide variations in light intensity received at the detector. Much effort was expended in trying to utilise on-line filter devices to remove all particulate matter, but these were found to be unreliable, relatively expensive and significantly increased the maintenance requirements. Without a pre-filter to remove suspended particulate and colloidal matter, the devices only require minimal maintenance (typically one visit per three or four weeks). The ultraviolet radiation appears to limit biofilm growth on the flowcell window. Many commercial on-line river intake devices have failed to be adopted because of the problems of reliably pre-treating the sample.

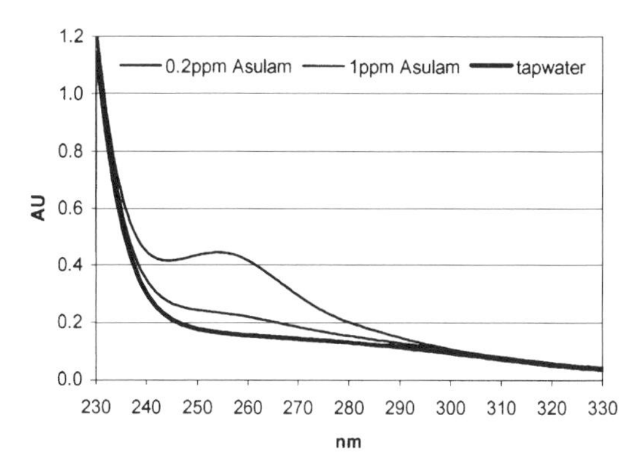

Figure 2 *UV spectra of a tapwater and a tapwater spiked with asulam (40 mm path length cell)*

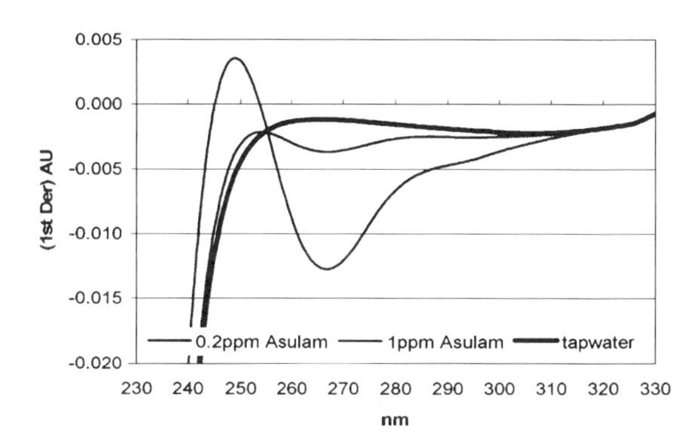

Figure 3 *1st derivative UV spectra of a tapwater and a tapwater spiked with asulam (40 mm path length cell)*

The device only utilises a crude wire mesh filter to remove gross particulate matter and the associated software can cope with high levels of turbidity. The system can be trained to recognise "normal signatures" which can be season dependent. It is programmed to alarm to a sudden step change or an abnormal change in the overall spectrum. It will respond to about 0.2 - 0.4mg/litre of most substances containing an aromatic type ring (e.g. atrazine, paraquat, diquat, phenol etc.) even in turbid river waters. The detection limit is degraded with respect to treated water because of the turbidity and natural colour of raw waters relative to treated waters and also the shorter path length cell. Very few false alarms have been observed and this is essential prerequisite for a successful on-line monitor.

Soon after this work was completed the one of the authors became aware of the S::CAN sealed unit diode array detector system[8]. This unit comprises a high energy throughput optical system with a spectral range of 190 nm to 380 nm. It has no user serviceable parts.

The lamp expected life is over 50 years. It can be used for measurements in treated and even in raw waters with high turbidity and/or high optical density. Turbidity is compensated by a mathematical model that reflects the particle distribution of the turbidity source, so no sample pre-treatment is necessary. It is claimed to be superior to fibre optic instruments with respect to energy and process stability. It utilises a pulsed Xenon arc lamp; double beam optics for very good baseline stability and optical pathlengths up to 100 mm. It can be used either totally submersed (as it can withstand pressures up to 10 Bar) or used in a conventional bypass mode. The software "learns" the shape and features of the "normal" spectrum of the water body and can alarm to user selected criteria. The S::CAN system will rununattended for very long periods. The two systems (YWS and S::CAN) are shortly to be compared as part of a full scale trial.

3.6 Photoionisation Detector (PID)

The photoionisation detector utilises a 10.6 ev electrodeless discharge lamp emitting far UV radiation which will ionise virtually all organic substances except for alkanes (e.g. methane, ethane, propane etc.). An RAE ppb high sensitivity PID has been evaluated. A 500 ml sample aliquot is placed in a one litre glass bottle which is sealed with aluminium foil. The bottle is shaken for 30 sec and the headspace gas simply introduced into the PID detector nozzle (Figure 4). With this simple system it is possible to detect down to 0.05 mg/litre diesel at an analysis rate of up to 60 samples per hour. The apparatus has also be used directly in the field to confirm adequate flushing of the water supply to all the houses in a small village over 100 miles from the Rotherham laboratory where a contractor carrying out mains relining managed to contaminate the village input main with a hydrocarbon solvent.

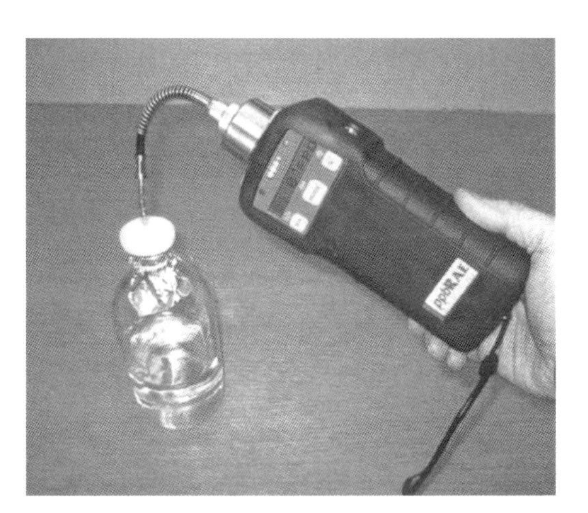

Figure 4 *Use of the RAE ppb PID detector*

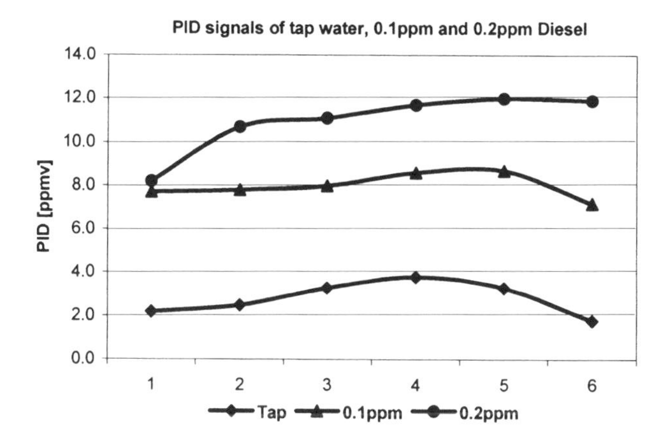

Figure 5 *Rapid detection of trace diesel levels in tapwater.*

To assess the sensitivity for rapid screening purposes, a series of 18 coded tapwater and diesel spiked tapwaters (0.1 and 0.2 mg/litre added diesel) were prepared and randomly sorted. They were then screened using the RAE PID detector by a member of the laboratory staff. The results (the 18 points) are shown in Figure 5. This clearly shows that all samples containing 0.1 mg/litre of diesel could rapidly be sorted from uncontaminated tapwater. Diesel and gasoline contamination are thought to be the most common forms of drinking water contamination. The PID responds in a more sensitive manner to gasoline than diesel owing to the significantly higher vapour pressure of gasoline relative to diesel. Also this portable battery powered device can readily be used in the field as indicated above. It is a much quicker technique than extraction into tetrachloroetylene, Freon™ or carbon tetrachloride and measurement of the appropriate region of the infrared spectra. Also it does not require the use of these environmentally unfriendly organic solvents.

3.7 Other Useful Screening Tests

Other useful screening tests include total organic carbon (TOC); pH; electrical conductivity; turbidity; free cyanide (CHEMetrics™); ammonia; parquet/diquat and ion chromatographic scan. For any tests carried out in the field, it is essential that method validation tests are carried out at the range of temperatures likely to be encountered. Some colorimetric tests may exhibit sensitivity changes at low temperatures.

4 ECOTOXICOLOGICAL SCREENING TESTS

One problem with chemical testing is that there are a very large number of potentially toxic substances and it is very difficult to prove a negative particularly under pressure in an emergency situation involving large numbers of samples. A complementary approach is to employ simple screening ecotoxicity testing. This is a holistic approach that just detects toxic effects on species employed as biological indicators in the test used.

Currently ALcontrol Laboratories routinely employs two tests for this work. The well known Microtox® test[9] and the Thamnotoxkit F® with Thamnocephalus platyurus[10]. The Microtox® test employs a marine luminescent bacteria, (Vibrio fischeri). Any toxicity is displayed as a decrease in light output from the bacteria.

$$FMNH_2 + O_2 + RCHO \xrightarrow{\textbf{Luciferase}} FMN + RCOOH + H_2O + Light\ (490\ nm)$$

The principle of the analytical bioluminescence reaction is that in the presence of oxygen and energy source, luciferase enzyme oxidizes the substrate and one of the end products is light. $FMNH_2$ = Reduced Flavin Mononucleotide, FMN = Oxidized Flavin Mononucleotide, RCHO = long chain aldehyde, RCOOH = long chain fatty acid. This test will typically detect significant organic pollution within 15min and toxic metal pollution within 30min.

The Thamnocephalus platyurus Microbiotest® employs dormant organism technology[10]. This technology avoids the need for culturing and effectively reduces the cost of ecotoxicity testing by an order of magnitude. The typical mortality endpoint method takes up to 24 hours before a negative result is confirmed, however if there is a significant concentration of toxic substance present, an effect is seen well before 24 hours. As stated earlier these tests are seen as complementary to chemical tests and effectively increase the range of toxic substances that can be detected. Table 3 gives on indication of the sensitivity of the two tests and the range of substances that can be detected.

A recent development of the Thamnocephalus test, to speed up the detection of toxicity, reduces the time for a negative result from 24 hours to ~1 hour[4, 11]. This is based on monitoring the feeding of the Thamnocephalus with highly coloured food beads after one hour. A simple analogy is that a human being before succumbing to Typhoid fever (which may takes 4 or 5 days from infection) will rapidly lose appetite well before death occurs. The Thamnocephalus cysts are stable for up to nine months if kept at 5°C and require ~24 hours to hatch. ALcontrol Laboratories keeps some cysts hatched on a 24hours/365 days basis to cope with any water pollution emergencies.

Work has also been carried out on the ECLOX® enhanced chemiluminescence light (ECL) test[12] which is based on chemiluminescence. (See Figure 6)

The enhanced chemiluminescent reaction utilises the enzyme horseradish peroxidase (HRP) to catalyse the oxidation of luminol. The result is a flash of light, and in order to prolong the emission of light an enhancer is added. Light output is measured using a luminometer.

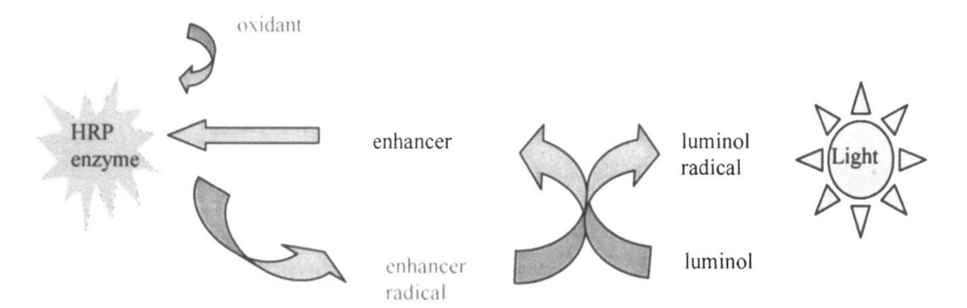

Figure 6 *Schematic diagram showing the Enhanced Chemiluminescence Light (ECL) reaction.*

Contaminants that interfere with the ECL reaction include: -

- Radical scavengers (such as antioxidants which remove oxygen from water and scavenge free radical molecules)
- Competitive enzyme inhibitors
- Non-competitive enzyme inhibitors
- Miscellaneous influences on general chemiluminescent reactions (including trace levels of manganese).

Testing performed using this system indicated lower sensitivity (with the exception of KCN) than that found with the Thamnocephalus test. (See Table 1).

5 RADIOACTIVITY SCREENING

Natural radioactivity in the UK area serves by ALcontrol Laboratories is not considered to be very significant. This is true for much of the UK. However, incidents such as Chernobyl have shown that there is a need for rapid screening methods for detecting radioactivity contamination in a large number of samples at short notice.
With the current UK water supply regulations in areas of low risk it is only necessary to monitor gross alpha and gross beta activity to demonstrate that the radioactivity levels in a given water are below 0.1 and 1 Bq/litre respectively.

However, to achieve this, it requires a sample volume of at least one litre of water to be evaporated to dryness, ashed with sulphuric acid and then counted for 18 hours. This is a very time consuming method with a low sample throughout[13]. The current regulatory limits of 0.1 and 1 Bq/litre are considered to be very conservative and for short-term emergencies situation the limits can be increased to 5 and 30 Bq/litre with negligible risk to consumers. Consequently the UK water laboratories mutual aid scheme have developed a rapid screening method involving a 100ml initial sample volume to be evaporated and ashed with sulphuric acid, followed by 30 minutes counting time. With an eight place alpha/beta counter it is then possible to screen well over 200 samples/day for gross alpha and gross beta activity assuming sufficient manual labour for the simple water evaporation and ashing steps.

Chemical	Eclox™ (IC_{50})	Microtox® (EC_{50})	Thamnotoxkit F™ (LC_{50})
	(mg/l)		
As (III)	26	0.3	1.3
KCN	0.07	3.5	0.3
Dichlorvos	354	29	19
Thallium	113	450	0.2
Paraquat	73	20	0.6
Rodenticide	>1000	507	0.4

Table 1, - *Comparison of test sensitivity*

6 NEED FOR PROFICIENCY TESTING

An emergency incident proficiency scheme has been set up by CSL-LEAP™ (the Central Science Laboratory [Ministry Of Agriculture, Fisheries and Food]). There are currently two distributions per annum. Participating laboratories receive (at the same time) a simulated pollution incident sample with an associated scenario and they are judged on how quickly and accurately they can detect and semi-quantify the toxic substances present.

A typical exercise contains 4 – 6 different substances. The laboratory staff do not know when this will be delivered. The sample is treated as per a real pollution emergency sample except that all significant results are immediately communicated to the proficiency scheme organiser rather than the relevant water company. Ten circulations have been run to date and participating laboratories have gained considerable experience in handling and analysing a wide range of completely unknown samples. The early exercises indicated problems with the identification of more polar substances and the scheme has proved a very useful learning exercise. Post mortems of the exercises are held on a regular basis. (For details of the CSL-LEAP™ proficiency schemes, see http://ptg.csl.gov.uk/leap.cfm) The subsequent presentation by Barry May will deal with issue.

7 PASSIVE SAMPLING

Often by the time samples are taken in an emergency incident situation, the peak concentration of a pollutant may be missed. It would be possible to insert simple low-cost passive sampling devices into key parts of a water distribution network that could operate for up to one month.

There would be two types, those that collect metals[14] and those that collect organics[15]. Potable water within a distribution system typically contains very low levels of toxic metals; radioactive elements or organic substances. After an incident has occurred the devices could be removed and quickly analysed. It is felt that this approach could also be very useful for detecting radioactivity incidents. If the passive devices were changed and analysed on a regular basis, it would give some indication of longer-term low level pollution.

8 CONCLUSIONS

Planning for high impact, low probability potable water pollution incidents is very difficult; once a response system is set up, maintaining it on a long-term basis needs careful thought. Staff phone numbers, e-mail addresses, screening analysis methodology can change and careful check mechanisms are essential so that the system remains robust and effective. Regular random auditing of the emergency response system should be carried out. It is important to appreciate systems that are not in routine use, rapidly deteriorate unless they are regularly tested.

A key consideration is how to rapidly detect when a significant contamination has occurred then identifying the cause of the contamination or convincingly demonstrate absence of contamination.

A key judgement is where a water company should position itself on the degree of acceptable risk/cost curve for responding to emergency incidents. To remove all risk is prohibitively expensive and probably not feasible.

Robust rapid chemical and ecotoxicological screening tests have been and continue to be developed. Ideally a good screening test should be able to screen 100 samples in two hours.

Consideration should be given to the setting up of two or three centres of excellence state of the art laboratories

The development of rapid HPLC/MS/MS methods with no or minimal sample pre-treatment look very promising.

The assistance from the UK mutual aid group with 365 days/24 hours contact numbers for specialist analysis requirements has been found to be very useful in challenging emergency situations. However these groups need managing and regular (at least annual) meetings of all members are considered essential. (A previous presentation dealt with this topic.)

Participation in a relevant emergency incident proficiency scheme is considered essential to assess laboratories on an on-going basis. In this type of scheme, emergency incident samples arrive on an unannounced basis and the receiving laboratory is judged on the speed and accuracy of its response.

Occasional completely unannounced dummy incidents need to be run to ensure that the complete emergency response system from the reporting of a potential incident, the subsequent taking of the samples; the analysis of the samples; the reporting of results and all associated communications function as intended. All shortfalls need to be fully investigated. This is expensive, but considered to be an essential requirement.

On-line monitors at water treatment work intakes are very useful, but must be reliable, run unattended for long periods and should not give a significant number of false positive or false negative results.

ACKNOWLEDGEMENTS

One of the authors (KCT) would like to thank Dr Gary O'Neill of Yorkshire Water Services for actively supporting this work.

REFERENCES

1. Eds. K. C. Thompson and J. Gray, *Water Contamination Emergencies: Can we cope?* Proceedings of RSC/SCI/IWO International Conference held on 16 – 19th March 2003 at Kenilworth. RSC Publications, June 2004. ISBN 0-85404-628-3
2. S. States, J. Newberry, J. Wichterman, J. Kuchta, M. Scheuring and L. Casson, *Rapid analytical techniques for drinking water security investigations,* Jour. AWWA, 2004, 96:1, pp 52 – 64
3. G. O'Neill, C. Ridsdale, K. C. Thompson and K. Wadhia, *Field and laboratory analysis for detection of unknown deliberately released contaminants.. Water Contamination Emergencies: - Can we cope?* Proceedings of RSC/SCI/IWO International Conference held on 16 – 19th March 2003 at Kenilworth. RSC Publications, June 2004. ISBN 0-85404-628-3, pp 100-109
4. G. Persoone, K. Wadhia and K. C. Thompson, *Rapid toxkit microbiotests for water contamination emergencies Water Contamination Emergencies: - Can we cope?*

Proceedings of RSC/SCI/IWO International Conference held on 16 – 19th March 2003 at Kenilworth. RSC Publications, June 2004. ISBN 0-85404-628-3, pp 122-130

5. http://toxnet.nlm.nih.gov/; http://www.intox.org/databank/pages/all_pims.html; http://www.atsdr.cdc.gov/; http://www.pesticideinfo.org/Search_Chemicals.jsp; http://www.intox.org/databank/index.htm; http://extoxnet.orst.edu/ghindex.html; http://chemfinder.cambridgesoft.com/; http://webbook.nist.gov/chemistry/cas-ser.html; http://www.pan-uk.org
6. USEPA, *Standardized Analytical Methods for Use During Homeland Security Events*, Revision 1.0, September 29, 2004
7. B-L True and R. H. Dreisbach, *Dreisbach's handbook of poisoning*, 13th Edition, the Parthenon Publishing Group, 2002. ISBN 1-85070-038-9
8. G. Langergraber, A. Weingartner and N. Fleischmann, Time-resolved delta spectrometry: a method to define alarm parameters from spectral data, *Water Science and Technology, 2004,* Vol 50, No 11, pp 13 – 20. (See also http://www.s-can.at/)
9. K. L. E. Kaiser and V. S. Palabrica, *Vibrio fischeri* (*Photobacterium phosphoreum*) toxicity data index, *Water Poll.* Res. J. Canada, 1991, Vol 26, No. 3, pp 361 – 431
10. (See also http://www.sdix.com/ProductSpecs.asp?nProductID=7)
11. Eds. G. Persoone, C. Janssen and W. de Coen, *New microbiotests for routine toxicity screening and biomonitoring,* Kluwer Academic/Plenum Publishers, 2000, ISBN 0-306-46406-3 (See also http://www.microbiotests.be/product.htm),
12. http://www.microbiotests.be/toxkits/rapidtoxkit.pdf
13. *Review of the enhanced chemiluminescence (ECL) test, R & D Technical Report* E28, Environment Agency, 2000. (E-mail: publications@wrcplc.co.uk)
14. *Measurement of alpha and beta activity of water and sludge samples. The determination of radon-222 and radium-226. The determination of uranium (including general x-ray fluorescent spectrometric analysis)* 1985-6, *Methods for the examination of waters and associated materials,* Her Majesty's Stationery Office, 1986, ISBN 0-11-751909-X
15. Davison, W. and Zhang, H. (1994). In situ speciation measurements of trace components in natural waters using thin-film gels. *Nature*, 367, 546-548.
16. F. Stuer-Lauridsen, Review of passive accumulation devices for monitoring organic micropollutants in the aquatic environment, *Environmental Pollution, 2005,* In press

LABORATORY ENVIRONMENTAL ANALYSIS PROFICIENCY (LEAP) EMERGENCY SCHEME

Barry May

Central Science Laboratory, Sand Hutton, York, YO41 1LZ, UK

1 INTRODUCTION

Many laboratories now participate in proficiency testing schemes that provide independent and unbiased information on the analytical performance of a laboratory. These schemes however do not test the ability of a laboratory to analyse samples in an emergency incident situation. In 1997, ALcontrol laboratories then owned by Yorkshire Water established the Laboratory Environmental Analysis Proficiency (LEAP) Emergency Scheme. The primary function of this scheme was to test the ability of a laboratory to analyse chemical samples from a simulated potable water contamination incident.

In 2000 LEAP transferred to the Central Science Laboratory (CSL), which is an executive agency of UK Department for Environment, Food and Rural Affairs (DEFRA) formerly Ministry of Agriculture Food and Fisheries (MAFF). This move allowed the LEAP Emergency Scheme to operate independently of UK water companies.

2 REQUIREMENT FOR PROFICIENCY TESTING

In the event of a potable water contamination incident water testing laboratories will be called upon to rapidly identify any contaminants that may be present. An incident manager will undoubtedly be requiring a rapid analytical result turnaround together with a comprehensive analysis of samples arriving at the laboratory. Whilst speed of reporting will be a major requirement, the quality of any results reported will also be of equal importance, as the consequences arising from not identifying a contaminant or wrongly identifying a contaminant could be grave. Clearly there is a need to test the ability of the laboratory to identify rapidly and accurately unknown chemical contaminants in an emergency contamination incident.[1]

3 SCHEME AIMS AND OBJECTIVES

A number of key aims and objectives were used in order to structure a scheme that would provide participants with the information they would require to measure their ability to produce fit for purpose timely results during an emergency incident situation.

The key aims and objectives were considered to be: -

- Test ability of a laboratory to identify unknown chemical contaminants in a potable water emergency incident situation
- Measure turnaround time for analytical results during an emergency incident
- Test ability of a laboratory to produce fit for purpose results
- Bring all participating laboratories up to same (high) standard

4 GENERAL OPERATION OF THE LEAP EMERGENCY SCHEME

4.1 Sample Preparation

The exercise takes place over a seven day period and begins with LEAP spiking a potable water sample with chemical contaminants. A combination of inorganic and organic contaminants is chosen, with a typical exercise containing four to six contaminants. Sample spiking takes place on a Monday, which is also the day of sample dispatch.

The following spiked samples are supplied: -

- 2 x 1 litre in glass bottles for organics analysis
- 1 x 1 litre in a plastic bottle for inorganic analysis
- 1 x 250ml in a plastic bottle acidified to (0.1M HNO_3) for metals analysis

From April 2005 a duplicate set of unspiked samples will be supplied. This will allow participating laboratories to compare the potable water sample supplied both before contamination and after contamination.

4.2 Sample Dispatch

The samples are dispatched on Monday for next day delivery to senior laboratory management who are requested to store all samples at 4°C on receipt.

4.3 Contamination Scenario

Included with each set of samples is a contamination scenario, which details the circumstances of the contamination. A typical example is given below: -

At 06-30 hours this morning an intrusion alarm at a major service reservoir was activated. A site inspection was promptly instigated. Padlocks to one of the inspection hatches had been forcibly removed.

A slight "organic type" odour was evident. No containers of any description were located in the vicinity of the broken inspection cover or around the reservoir. A phone call was subsequently received at 07-00 hours by a local radio station stating that the water supply has been poisoned. The caller then promptly rang off.

A grey (non-water company) van was seen by two local residents at 06-45 hours driving away from the site at high speed. Unfortunately the registration number was not recorded.

Shortly after this several taste complaint calls were received at the customer centre from consumers served by this service reservoir.

The contamination scenario asks the following questions: -

- Is there any significant contamination present
- If so, what is the approximate concentration (s) present
- What are the potential sources of contamination
- What analytical methods have been employed to detect the contaminant

4.4 Sample Analysis

In order for the participant to gain maximum benefit from participation in the exercise senior laboratory management are requested to ensure that the laboratory staff have no prior warning of delivery of samples or contamination scenario. Each participant is requested to present the samples and contamination scenario to the laboratory at 10.00 hours on Wednesday. Laboratories are also provided with a response sheet that requires the laboratory to record contaminants identified in test samples together with answers to other questions given in the contamination scenario. The laboratory is then requested to e-mail or fax their answers to LEAP as soon as any information becomes available.

4.5 LEAP Reporting

Each laboratory is informed of contaminants present by e-mail shortly after the closing date for the exercise. This allows the laboratory to investigate its performance rapidly. A full confidential report is issued approximately one month from the exercise closing date.

This report details the following information: -

- The contaminants correctly identified by each laboratory
- The time each contaminant was reported
- The potential sources of each contaminant reported
- The methods used to identify each contaminant
- Other contaminants reported which were not added i.e. false positives

4.6 User Forum Meeting

This meeting is usually held approximately one to two months after the closing date of the exercise. The meeting gives each laboratory an opportunity to discuss their performance with other participating laboratories with a view to exchanging valuable analytical information.

5 SUMMARY RESULTS FROM 2003 AND 2004 EXERCISES

5.1 Exercise 6 (March 2003)

Fifteen laboratories submitted results for this exercise. Inspection of the results (Table 1) shows that most laboratories identified dimethoate, and ethyl acetate. All participating laboratories identified mercury in the test material, which was present due to the addition of phenyl mercuric acetate. A number of laboratories also reported phenyl-mercury and organo-mercury. In addition one laboratory noted a possible contamination from latex paint phenyl-mercury acetate. Eight laboratories identified 1,4-dioxane and six 2,4,5-T. This was far lower than would have been expected. All laboratories that correctly identified 2,4,5-T carried out a derivatisation process. This sample also contained methanol at a high concentration (144 mg/l) but only four laboratories i.e. 27% of participating laboratories correctly identified this contaminant.

Contaminants present	*Concentration ug/L*	*Correct identification*	
		number of labs (max = 15 labs)	*percentage (100% = 15 labs)*
Dimethoate	20	14	93%
2,4,5-T	20	6	40%
1,4-dioxane	5,150	8	53%
Phenyl mercuric acetate	10	0	0%
Mercury	6	15	100%
Methanol*	144,000	4	27%
Ethyl acetate*	16,000	13	87%

* Present due to spike solvents

Table 1 *Contaminants identified in exercise 6 (March 2003)*

5.2 Exercise 7 (November 2003)

Eighteen laboratories submitted results for exercise 7 (Table 2).

Contaminants present	*Concentration ug/L*	*Correct identification*	
		number of labs (max = 18 labs)	*percentage (100% = 18 labs)*
Uranium	20	8	44%
Chromium (III + VI)	200	17	94%
Chromium (VI)	200	1	6%
Picric acid	1,000	1	6%
Paraquat	200	2	11%
Cyclohexanone	19,000	18	100%

Table 2 *Contaminants identified in exercise 7 (November 2003)*

All laboratories correctly identified cyclohexanone and all but one laboratory identified total chromium (III + VI), which was present as a result of the addition of potassium chromate. However, only one laboratory correctly reported that the chromium was present as chromium (VI) only, which has a far higher toxicity than chromium (III).[2] This laboratory reported using a spectrophotometric test kit to identify the chromium (VI). Eight laboratories identified uranium, which was considered poor, as this element should have been easily identified at this concentration from an ICP-MS scan. Poor results were also obtained for picric acid and paraquat with only one and two laboratories respectively identifying these contaminants. The two laboratories that correctly identified paraquat used colorimetric test kits. The one laboratory that identified picric acid reported using a diazomethane derivatisation.

5.3 Exercise 8 (May 2004)

Eighteen laboratories submitted results. Inspection of the summarised results for exercise 8 (Table 3) shows that eleven laboratories identified bromate. Five laboratories i.e. 28% identified thorium, tellurium and aldicarb. All participants who identified these contaminants used ICP-MS and GC-MS or HPLC-MS. Only one laboratory identified sodium fluoroacetate and diquat. The laboratory that correctly identified sodium fluoroacetate reported using an ion chromatographic method. The laboratory that successfully identified diquat used a colorimetric method. Ethylene glycol was present at 2.2 mg/L and was not detected by any of the participants in this exercise.

Contaminants present	*Concentration ug/L*	*Correct identification*	
		number of labs (max = 18 labs)	*percentage (100% = 18 labs)*
Sodium fluoroacetate	2,000	1	6%
Thorium	51	5	28%
Tellurium	100	5	28%
Diquat	50	1	6%
Bromate	200	11	61%
Ethylene glycol	2,220	0	0%
Aldicarb	50	5	28%

Table 3 *Contaminants identified in exercise 8 (May 2004)*

5.4 Exercise 9 (November 2004)

Twenty-two laboratories submitted results. Inspection of the summarised results for exercise 9 (Table 4) shows that twenty one laboratories i.e. 95% of participants identified phorate and thallium.

Contaminants present	*Concentration ug/L*	*Correct identification*	
		Number of labs (max = 22 labs)	*percentage (100% = 22 labs)*
Phorate	20	21	95%
Triton X-100	1,000	10	45%
Thallium	120	21	95%
Strychnine	100	16	73%
2,4,5-T	150	12	55%
Chlorate	2,000	6	27%
Methanol*	80,000	7	32%

* Present due to spike solvents

Table 4 *Contaminants identified in exercise 9 (November 2004)*

Sixteen laboratories identified strychnine. Twelve laboratories identified 2,4,5-T i.e. 55%, which was an improvement on exercise 6 where only 40% of participants identified 2,4,5-T. Most of the laboratories that correctly identified 2,4,5-T carried out a derivatisation process. Triton X-100 was identified by less than half the participants, which was surprising given that the sample had a visible froth on shaking. A low number of participants detected chlorate, which was present at 2.0 mg/l. No significant improvement in the detection of methanol was seen from that obtained in exercise 6.
Fifteen laboratories reported elevated TOC levels in this sample with three laboratories also reporting the presence of a peak on UV scan. Positive responses were also obtained for Microtox^{TM} and proprietary chemiluminescence tests.

6 FUTURE

A number of improvements to the existing scheme are planned for 2005/6. These are detailed below: -

- Blank uncontaminated sample to be supplied
- Improvements to result reporting format for participants
- Possible additional rounds for industrial pollution of sewers and rivers
- User forums to be set up to help promote quality emergency incident analysis

7 CONCLUSION

Inspection of the data reported in LEAP emergency rounds to date clearly indicates that there is a need for laboratories to continue to improve their performance in these simulated contamination incidents. The results reported in the latest round distributed by the LEAP emergency scheme, exercise 9 (November 2004), show a general improvement in the detection of contaminants present. The lessons learned from participation in the scheme coupled with the production of a draft general guidance document [3] on emergency analysis (due for final publication in 2005/6) has undoubtedly helped bring about this improvement.

Participation in simulated emergency exercises such as the LEAP Emergency Scheme offer a number of clear benefits to laboratories that may become engaged in emergency incident analysis. Some of the major benefits are listed below: -

- Allows a laboratory to demonstrate its ability to identify unknown chemical contaminants in a potable water emergency incident sample by most efficient and timely means
- Provides useful quality information e.g. testing out new screening techniques and also shows where additional development work is needed
- Will allow a laboratory to measure its performance against other laboratories engaged in emergency incident analysis
- Will serve as an ongoing test of laboratories` emergency analysis procedures
- Highlights false positive and false negative result reporting

8 ACKNOWLEDGEMENTS

The Scheme would like to thank all the participants for the provision of the data, which makes this type of proficiency testing comparison possible.

References

1 G. O Neill, C. Ridsdale, K. C. Thompson and K.Wadhia, Field and laboratory analysis of unknown deliberately released contaminants pp 100 – 109 in K. C. Thompson and J. Gray (Eds), Water Contamination Emergencies: Can we cope? Proceedings of RSC/SCI/IWO International Conference held on 16 – 19th March 2003 at Kenilworth. RSC Publications, June 2004. ISBN 0-85404-628-3.
2 http://cerhr.niehs.nih.gov/CERHRchems/chromium_VI_reviews.htm, Chromium VI – Recent Reviews.
3 Draft Guidance for the Analysis of Potable Water Samples from a Suspected Contamination Incident. Issue 3 Aug 2005. Final version due for publication in 2005 by UKWIR.

ELECTRONIC ATTACK ON IT AND SCADA SYSTEMS

Peter Davis

National Infrastructure Security Co-ordination Centre (NISCC), PO Box 832, London SW1 1BG

1 INTRODUCTION

Information Technology (IT) systems and Supervisory Control and Data Acquisition (SCADA) systems are both major elements in the control and monitoring of water system operations. If these IT and SCADA systems are tampered with or made unavailable, results would range from minor losses of visibility of the condition of the operations, to serious damage, including the production of sub-standard drinking water or the release of sewage into the general environment.

The National Infrastructure Security Co-ordination Centre (NISCC) is an inter-departmental government organisation, which has been set up to minimise the risk to the critical national infrastructure (CNI) from electronic attack. This is done in partnership with the owners and operators of CNI elements, whether they are in the public or private sectors. The CNI includes water services, and this covers both the supply of drinking water and sewage services.

2 THE PROBLEM OF ELECTRONIC ATTACK

2.1 The extent of the problem

Electronic attack on SCADA systems does not appear to be common at first sight, but some instances are recorded. CSX Transportation operates railway services in the eastern USA, and on 20th August 2003 they were forced to stop the trains because the Slammer worm had infected their networks. This caused network congestion, and signalling & dispatching systems failed. In another case, the Port of Houston had to shut down on 20th September 2001 because ship movement information was made unavailable. This information was displayed on a web site, which suffered a denial of service attack, and the web site crashed. An Englishman, Aaron Caffrey, was charged with launching the attack, but was acquitted after claiming that although the attack came from his PC, it had been created and launched by an unknown third-party.

More specifically, an example comes from Australia of an attack on water control systems. In this case, a disgruntled ex-employee of a SCADA-system supplier attacked the systems run by Maroochy Shire, a local authority in Queensland. By using his knowledge of how the systems worked and some stolen equipment, he was able to manipulate valves and release raw sewage on several occasions. Environmental effects included damage to a golf course and to an estuary which was a nature reserve.

Although the recorded number of attacks is quite low, it must be remembered that many electronic attacks go unreported. In the 2004 CSI/FBI Computer Crime and Security Survey in the USA, it was estimated that only 1 in 20 attacks is recorded. In the three cases reported above, the damage was very public, so there was ultimately no question of the attacks being unrecorded. However, it is worth noting that in the Australian case the incident was not at first recognised as an attack. It was being investigated as an ordinary equipment failure until the local police questioned the perpetrator on another matter and discovered his activities. Thus although few attacks are currently reported, there is likely to be a larger number that are either unreported or not even recognised as attacks.

2.2 The sources of the problem

IT and SCADA systems face the threat of electronic attack from a number of different sources. The nature of the source has a significant impact on how it can be stopped or mitigated. Sources considered significant for IT and SCADA systems in the CNI comprise:

- Untargeted attacks, such as viruses and worms endemic to the Internet
- "Insiders", who have inside knowledge of how the systems works
- Hackers and script kiddies, who will attack a specific target
- Organised criminals, whose aim is to make money
- Terrorists, who are ideologically motivated
- Foreign Intelligence Services, seeking militarily or commercial useful information

These sources are well known for IT systems, and they are not significantly different for SCADA systems, except that SCADA systems may be more attractive for some threat agents such as terrorists and less attractive for others such as organised criminals. All sources must be considered for their threat potential in the light of the functions of systems to be protected.

2.3 The points of attack

If a system is perfectly secure, then the threat source is not an issue as no attack will be effective. However, all systems have vulnerabilities, which the attackers will seek to discover and exploit. Vulnerabilities arise in a number of different areas of a system, and once again the different areas will have different mitigation techniques. The principal areas are:

- Design, implementation and operations errors, which relate to individual components and systems
- Protocol and standards flaws which effect all implementations
- People problems, including bad procedures and poor administration

When considering SCADA systems, it is also necessary to look at both the part of the system where an attack might start and the part that the attack might focus on. SCADA system elements are diverse, and each system element has its own characteristics which will define the scope of potential attacks and potential mitigations. A useful list of elements includes:

- Local devices, such as PLCs, RTU, and PCs
- Local SCADA systems, on a single site
- SCADA communication links
- Servers and control centres

- The Corporate network, which may be integrated with SCADA systems
- The Internet, which may be connected to the corporate and/or SCADA system

Thus to mitigate the risks to a SCADA system, the threat sources must be compared to the vulnerabilities which the different system elements might have. The matrix of threats and vulnerabilities can then be used to assess the risks the system faces by considering the potential impacts that attacks might have.

2.4 Changing risk levels

Risks identified by the considerations described above are not static. They will change as the system and its connectivity changes, as new vulnerabilities are discovered and fixed, and as threat sources change in nature or in intensity. All these factors are currently working to increase the risk levels of SCADA systems.

The systems were in the past quite isolated, with few connections, if any, to other systems. Increasingly they are now linked to corporate networks and to the Internet, which makes changes in three areas. Firstly it increases the visibility of the systems to corporate and Internet users, who may be looking for attack targets. Secondly it exposes the systems to untargeted Internet attacks such as worms and viruses. Thirdly, the new connectivity provides remote access, whereas previously attackers would have needed physical access to the systems.

The underlying platforms and protocols used by SCADA systems are also changing. In the past SCADA systems used a wide variety of platforms and network protocols, but new systems tend to use standard platforms (Windows and Linux) and standard open protocols (TCP/IP, DNP3, etc). This change exacerbates the problems of increased connectivity, as attackers are more familiar with these IT-industry standard systems, and there are a large number of known attack methods and vulnerabilities for these systems. They are also the systems which vulnerability researchers spend most time examine and testing.

Recent changes in threat sources are based on a general increase in the knowledge of what can be achieved by attacking SCADA systems. This knowledge is often the result of publicised incidents of such attacks. Insider knowledge of how such systems work is also more widespread than in the past, as more IT people become involved in the management and support of SCADA systems.

3 RISK MITIGATION

3.1 Managing risk mitigation programmes

Risk mitigation means making changes to reduce the identified risks. This is a wide-ranging task, needing changes in a number of areas, which can broadly be described as technical, personnel or management. The technical changes will effect the systems themselves or the way they are operated. The personnel changes involve recruitment screening for those in sensitive positions, and changes to system access routines. Management changes would be in constructing and implementing comprehensive security policies, procedures and controls, tailored to both IT and SCADA systems.

The wide range of the tasks makes risk mitigation difficult to manage. It will affect a number of different organisational functions and departments and is likely require assistance from outside the organisation. It therefore need clear leadership from senior management, to give it the cross-organisation priority that is required to make changes happen. This is also likely to involve budgetary powers, as significant spending may be required.

3.2 Partners in the programme

A number of different partners will need to be included in the risk mitigation process. Firstly the system owners must set the policy and procedures for running the systems and implementing technical mitigations. The manufacturers and suppliers of these technical solutions are the second key partner group. Thirdly there are the industry standards bodies on whose work many of the technical solutions are based. The activities of these three must be co-coordinated over a period of time as the changes will not happen over night.

It is also necessary to consider the system's users, who must adhere to the revised policy and procedures. Finally regulators may have a role to play, if system security is seen to be a factor in the supply of the regulated services (although this is not currently the situation in the water industry).

4 THE FUTURE

The increasing risks to SCADA systems described above are going to be key issues in the future. Interconnection of networks and standardisation of systems will continue to increase. Increasing reliance on automated systems (and the consequent de- manning and de-skilling) will increase the impact of any serious breach in security.

These trends can be countered in the short term by the increasing use of security policies and procedures to structure the defence efforts and guide the everyday practice of safer systems operation. In the longer term, improved technical solutions will help to secure SCADA systems. Developments are already underway to make communications protocols more secure. Also the awareness of security weaknesses in SCADA systems is leading manufacturers to create products such as firewalls and intrusion detection systems which will handle SCADA protocols.

Ultimately there is no such thing as perfect security for SCADA systems, and all security measures must be justified on a cost-benefit basis. However, a thorough assessment of the risks and vulnerabilities will provide the foundation for effective security measures for IT and SCADA systems.

INCIDENTS INVOLVING RADIONUCLIDES

B.T. Wilkins

Health Protection Agency, Radiation Protection Division, Centre for Chemical, Radiation and Environmental Hazards, Chilton, Didcot, Oxon, OX11 0RQ, UK

ABSTRACT

This paper provides some background information on radioactivity and radiation in the environment. It sets out the Action Levels that would be used in the United Kingdom to trigger substitution of supplies of drinking water in the event of a future accident or incident involving radionuclides. The associated screening levels based on gross measurements of activity are also given. An incident may affect a water treatment works as well as the drinking water itself. The potential doses to the operators of such works deserve further study once a current review of information is available.

1 INTRODUCTION

Under the Water Industry Act of 1991[1], water companies have a duty to maintain supplies of wholesome water. The term "wholesome" is taken to mean that the water does not contain anything at a concentration that would be detrimental to public health. Water companies spend a great deal of time collecting samples and carrying out analyses for a wide range of potential pollutants. This includes radionuclides, which under normal circumstances are present at very low concentrations. However, in the event of an incident involving radioactivity, concentrations in water and the water treatment system could be much higher. Terms such as "radiation" and "radioactivity" are emotive. Consequently, there is understandable concern amongst monitoring teams from the water industry that, if asked to attend the scene of a possible incident and collect samples, they may be exposed to radiation. The aim of this paper is to provide some basic information on radioactivity in the environment and to set out the criteria that would trigger a need to substitute water supplies to the general public. Some topics that deserve further attention are also discussed.

2 ACTIVITY AND DOSE

The amount of a given radionuclide in a particular material is usually expressed in terms of activity or activity concentration, for which the basic unit is the Becquerel (Bq). Activities and activity concentrations are measurable quantities. Radionuclides can differ markedly in the way in which they decay, in their behaviour within the human body and in their radioactive half-life. People can be exposed to radiation via external irradiation, where the source of the activity is outside the body, and via internal irradiation, where the source has been taken into the body, for example via ingestion or inhalation. Consequently, the relative importance of different routes of exposure to radioactivity is dependent on the radionuclide of interest. These different factors have to be taken into account in order to bring the effects of different radionuclides on to a common basis. This requires the calculation of a quantity usually referred to as "dose", for which the basic unit is the Sievert. So-called "dose coefficients" have been published relating intakes of activity and dose, based on the results of extensive international research.[2] In practical terms, doses received as a result of radioactivity in the environment are usually expressed in terms of milliSieverts (mSv), which is one thousandth of a Sievert, or microSieverts (μSv) which is one millionth of a Sievert.

3 RADIOACTIVITY AND RADIATION IN THE ENVIRONMENT

People have always been exposed to radiation from natural sources. Reviews of the exposure of the UK population are published every few years. Table 1 has been taken from the most recent edition,[3] and shows very clearly that on average most of the dose that we all inevitably receive is of natural origin.

Since we generally measure concentrations of radionuclides via their decay, it is possible to detect very small quantities. As an example, the testing of nuclear weapons in the atmosphere took place largely in the 1960s, but residues of Pu can still be detected in appropriate samples of soil[4] and ^{137}Cs is still detectable in milk.[5] However, in the UK the contribution from artificial sources to the average annual individual dose is extremely

Source	Average annual dose, μSv
Natural	
Cosmic	330
Gamma	350
Internal	250
Radon	1300
Artificial	
Medical	410
Occupational	6
Fallout	6
Disposals	0.9
Consumer products	0.1
Total (rounded)	2,700

Table 1 *Average annual exposure of the UK population from all sources of ionising radiation*

small, the value for weapons fallout being about 6 μSv compared with 2230 μSv from all sources of natural radiation.

Artificial radionuclides are widespread in the environment as a result of the testing of nuclear weapons in the atmosphere and major accidents notably that at Chernobyl in the Ukraine in 1986. In the marine environment, some radionuclides discharged under authorisation to sea from certain nuclear sites can become very widely dispersed, whereas the effects of corresponding discharges to atmosphere are generally only discernible within a limited area.

The operators of licensed nuclear sites in the UK are required to carry out environmental monitoring as part of the authorisation process. Independent monitoring around these sites and more generally is carried out by government agencies and the results are published annually.[6] Whether it is intended for use under normal circumstances or after an accident or incident, one primary objective of a monitoring programme is to provide the most direct estimates of doses to people. The design of a monitoring programme therefore requires the exposure pathways of importance to be identified. Both the radionuclides of interest and the usage of the area then need to be considered. As an example, the selection of sampling and measurement locations along a coastline would need to take account of where people spent their time as well as concentrations of radionuclides in materials such as sand and sediment. The locations chosen might not then be those where activity concentrations were highest. The radionuclides of interest will affect the approach adopted at a particular location. In the case of a coastal area, if the radionuclides concerned emitted gamma rays, then external irradiation would be important and *in situ* measurements of external dose rate could be the best approach. If however the radionuclides concerned were isotopes of plutonium that emit alpha particles, then the inhalation of resuspended material would be important, and this would require sampling and analysis. The amount of material that might be resuspended could be very different for a muddy area that was frequently inundated by the sea compared with a sandy area around the high water mark.

This example relates primarily to authorised continuous discharges. However, it illustrates the point that, just because radioactivity can be detected in an environmental material, it need not constitute a hazard. Rather, it is the radiation dose received by people that is important. The same principle applies in the event of incidents involving radioactivity.

4 ACCIDENTS AND INCIDENTS AND THE WATER INDUSTRY

Emergency exercises based on major releases from licensed nuclear sites rarely result in priority being given to the monitoring of drinking water except by the water industry itself. This is because exercise scenarios often give rise to widespread deposition of radionuclides on to agricultural land. In the short term priority is then given to foodstuffs such as cows milk and any above - ground vegetables that are about to be harvested, because these are expected to give the highest radiation doses. Radionuclides would also be deposited on to the surface of water bodies such as reservoirs and lakes, but monitoring of drinking water is not considered of general importance because extensive dilution is assumed to take place.

There is however the potential for incidents that involve the introduction of radionuclides more specifically into the drinking water supply, either directly or via a spillage nearby. In such cases, the consumption of drinking water would potentially become an important exposure pathway. The awareness of such possibilities has increased considerably in recent years, and this has prompted the re-examination of earlier advice and information. Some of these topics are set out in the rest of this section. Where appropriate, comparisons are drawn with the situation relating to accidents at nuclear sites.

4.1 Availability of Monitoring Resources in an Emergency

If an accident occurs at a nuclear site, some monitoring resources would be available immediately. *In situ* detectors and air samplers are likely to be situated at the site perimeter and possibly further away. In addition, mobile monitoring teams from the site should be available to carry out measurements of external dose rate and gross activity concentrations in air and to collect samples such as grass and soil. In an incident involving the potential contamination of the water supply, the monitoring capability immediately available could involve teams that are less familiar with radiation and that have more limited resources. It is for this reason that the water industry in the UK is trying to ensure that there is someone with appropriate radiological protection training within or readily accessible to each individual water company.

4.2 Radionuclides of Interest

Releases from accidents at licensed nuclear sites are likely to involve a complicated mixture of radionuclides. Many of these are very short-lived; those of principal interest include radioisotopes of iodine, caesium and strontium and these are well defined. An incident in which the water supply is contaminated specifically is more likely to involve radionuclides that are in use in industry.

Radionuclide	Half-life (days)	Main mode of exposure	Principal uses
^{60}Co	$1.92\ 10^3$	Photons	Medical sterilisation industrial radiography medical treatment
^{137}Cs	$1.10\ 10^4$	Photons	Industrial level gauges medical treatment, instrument testing facilities
^{192}Ir	$7.40\ 10^1$	Photons	Industrial radiography
^{75}Se	$1.20\ 10^2$	Photons	Industrial radiography
^{241}Am	$1.58\ 10^5$	Alpha particles, weak photons	Smoke detectors
^{90}Sr	$1.06\ 10^4$	Beta particles	Paper thickness gauging, heat/power source
^{226}Ra	$5.84\ 10^5$	Alpha particles, photons	Medical treatment
^{238}Pu	$3.20\ 10^4$	Alpha particles, weak photons	Satellite power source, heat generators, heart pacemakers

Table 2 *Some radionuclides in common use in industry*

Some of the radionuclides that are commonly used in the form of sealed sources are shown in Table 2.[7] This list is not exhaustive, but does illustrate the diversity of radionuclides that might be encountered.

In contrast to accidents at nuclear sites, only one radionuclide might be involved in an incident with industrial sources. However, prior to measurements being carried out there may be no information on which to identify the radionuclide, and this clearly inhibits any predictions of doses and environmental behaviour. In terms of assessing any immediate hazard to monitoring teams needing to work close to the scene of an incident, there is a variety of commercially available survey instruments that could be of potential use. These are normally intended for use in industrial premises rather than in the natural environment. As an example, the commercial instruments held by the Radiological Protection Division (RPD) laboratory at Chilton are shown in Table 3, together with their application.

Instrument	Photon dose rate	Beta dose rate	Alpha contamination Bq m^{-2}	Beta contamination Bq m^{-2}	Photon contamination Bq m^{-2}
Eberline RO-10	Yes	Yes	No	No	No
Minirad 1000	Yes	No	No	No	No
Mini 900D	Yes	Crude	No	No	No
Mini 900 44/42	Crude	No	No	Gross	Yes
Mini 900 EP15	No	Crude	Gross	Yes	Gross
DP2 probe	No	No	Yes	Yes	No
Berthold LB1210B	No	No	Yes	Yes	Yes
AP2 probe	No	No	Yes	No	No

Table 3 *Survey instruments held at RPD Chilton and their application*

Taking the contrasting capabilities of these instruments together with the varying characteristics of the radionuclides in Table 2, it is apparent that in the absence of any information on the radionuclide involved, specialist staff would be needed to provide suitable surveys and reliable advice.

4.3 Contamination of Drinking Water

If the drinking water supply did become contaminated, then the radiation dose that a consumer would receive would depend on the radionuclide involved, the activity concentration in the water and the amount of water that they consumed over the period that the contamination persisted. In the aftermath of an incident, the doses that people receive could be controlled by substitution with unaffected water supplies. Consequently, it is helpful to have a system in place that provides guidance on when such action should be considered.

For foodstuffs, the European Council has specified concentrations of radionuclides at which restrictions on marketing would be imposed.[8] These would apply in the event of a future nuclear accident or any other radiological emergency. In 1994, the forerunner of RPD, the National Radiological Protection Board (NRPB) issued advice for drinking water which stated that Action Levels at which substitution of drinking water supplies should be initiated should correspond to the EC values for liquid foods.[9] The relevant values are shown in Table 4. It is important to note that these are radionuclide specific. Many of the radionuclides that might be encountered in an incident involving industrial sources fall within a single broad category.

Radionuclide	**Action Levels[b] (Bq l^{-1})**	**Categorisation of radionuclides including those given in Table 1**
Isotopes of strontium, notably ^{90}Sr	125	^{90}Sr
Isotopes of iodine, notably ^{131}I	500	^{131}I
Alpha-emitting isotopes of plutonium and transplutonium elements	20	^{238}Pu, ^{239}Pu, ^{241}Am
All other radionuclides of half-life greater than 10 days, notably radioisotopes of caesium and ruthenium[c]	1,000	^{60}Co, ^{75}Se, ^{134}Cs, ^{136}Cs, ^{137}Cs, ^{192}Ir, ^{226}Ra

Table 4 *Recommended UK Action Levels for drinking water supplies[a]*

Notes:

a) These Action Levels refer to all water supplies that are intended, at least in part, for drinking and food preparation purposes.
b) It is the sum of the concentrations of all the radionuclides included within a category and detected in the water which should be compared with the Action Level.
c) This category does not include ^{14}C, ^{3}H or ^{40}K.

If the concentration of a particular radionuclide exceeds the relevant Action Level, then from the radiological protection point of view this does not mean that an alternative supply of drinking water needs to be provided immediately. As part of a handbook on recovery after an incident[10], RPD has estimated the doses that would be received if an individual drank water that contained concentrations of radionuclides equivalent to the Action Level. Values are given for infants, children and adults and cover consumption over periods of 1 week, 1 month or 1 year. A wide range of radionuclides has been considered, and the published values are reproduced in full in Table 5. The data in Table 5 indicate that, in many cases, the doses that would be received over periods of a few weeks are less than 1 mSv, notable exceptions being ^{235}U and ^{226}Ra. Generally, in an incident the activity concentrations would be expected to decrease with time, although in the case of more widespread contamination the levels in surface water bodies might remain elevated because of run-of from the surrounding catchment. Overall, if activity concentrations in drinking water were around the relevant Action Level, then substitution of water supplies over timescales of up to a few weeks would generally be sufficient. Other factors such as public concern may however mean that some action would need to be taken on a shorter timescale.

		Committed effective dose, mSv, following consumption for:								
Radio-nuclide	**Half-life[c]**	**1 week**			**1 month**			**1 year[c]**		
		1 yr old	**10 yr old**	**Adult**	**1 yr old**	**10 yr old**	**Adult**	**1 yr old**	**10 yr old**	**Adult**
^{60}Co	Long	9 10^{-2}	4 10^{-2}	3 10^{-2}	4 10^{-1}	2 10^{-1}	1 10^{-1}	5	2	1
^{75}Se	Long	4 10^{-2}	2 10^{-2}	2 10^{-2}	2 10^{-1}	1 10^{-1}	8 10^{-2}	2	1	1
^{90}Sr	Long	3 10^{-2}	3 10^{-2}	3 10^{-2}	1 10^{-1}	1 10^{-1}	1 10^{-1}	2	2	1
^{95}Zr	Long	2 10^{-2}	7 10^{-3}	7 10^{-3}	8 10^{-2}	3 10^{-2}	3 10^{-2}	1	4 10^{-1}	4 10^{-1}
^{95}Nb	Long	1 10^{-2}	4 10^{-3}	4 10^{-3}	5 10^{-2}	2 10^{-2}	2 10^{-2}	4 10^{-1}	2 10^{-1}	2 10^{-1}
^{99}Mo	Short	1 10^{-2}	4 10^{-3}	5 10^{-3}	5 10^{-2}	2 10^{-2}	2 10^{-2}	4 10^{-2}	1 10^{-2}	1 10^{-2}
^{103}Ru	Long	2 10^{-2}	6 10^{-3}	6 10^{-3}	7 10^{-2}	2 10^{-2}	2 10^{-2}	7 10^{-1}	3 10^{-1}	3 10^{-1}
^{106}Ru	Long	2 10^{-1}	6 10^{-2}	5 10^{-2}	7 10^{-1}	2 10^{-1}	2 10^{-1}	8	3	3
^{131}I	Short	3 10^{-1}	1 10^{-1}	8 10^{-2}	1	4 10^{-1}	4 10^{-1}	3	9 10^{-1}	8 10^{-1}
^{132}Te	Short	1 10^{-1}	3 10^{-2}	3 10^{-2}	4 10^{-1}	1 10^{-1}	1 10^{-1}	4 10^{-1}	1 10^{-1}	1 10^{-1}
^{134}Cs	Long	5 10^{-2}	5 10^{-2}	1 10^{-1}	2 10^{-1}	2 10^{-1}	6 10^{-1}	3	3	7
^{136}Cs	Short	3 10^{-2}	2 10^{-2}	2 10^{-2}	1 10^{-1}	7 10^{-2}	1 10^{-1}	5 10^{-1}	3 10^{-1}	3 10^{-1}
^{137}Cs	Long	4 10^{-2}	4 10^{-2}	1 10^{-1}	2 10^{-1}	2 10^{-1}	4 10^{-1}	2	2	5
^{140}Ba	Short	6 10^{-2}	2 10^{-2}	2 10^{-2}	3 10^{-1}	9 10^{-2}	8 10^{-2}	9 10^{-1}	3 10^{-1}	3 10^{-1}
^{140}La	Short	4 10^{-2}	2 10^{-2}	2 10^{-2}	2 10^{-1}	7 10^{-2}	6 10^{-2}	8 10^{-2}	3 10^{-2}	3 10^{-2}
^{144}Ce	Long	1 10^{-1}	4 10^{-2}	4 10^{-2}	6 10^{-1}	2 10^{-1}	2 10^{-1}	7	2	2
^{169}Yb	Short	2 10^{-2}	6 10^{-3}	5 10^{-3}	7 10^{-2}	2 10^{-2}	2 10^{-2}	6 10^{-1}	2 10^{-1}	2 10^{-1}
^{192}Ir	Long	3 10^{-2}	1 10^{-2}	1 10^{-2}	1 10^{-1}	5 10^{-2}	5 10^{-2}	2	6 10^{-1}	6 10^{-1}
^{226}Ra	Long	3	3	2	1 10^{1}	1 10^{1}	9	2 10^{2}	2 10^{2}	1 10^{2}
^{235}U	Long	4 10^{-1}	3 10^{-1}	4 10^{-1}	2	1	2	2 10^{1}	1 10^{1}	2 10^{1}
^{238}Pu	Long	3 10^{-2}	2 10^{-2}	3 10^{-2}	1 10^{-1}	8 10^{-2}	2 10^{-1}	1	1	2
^{239}Pu	Long	3 10^{-2}	2 10^{-2}	4 10^{-2}	1 10^{-1}	9 10^{-2}	2 10^{-1}	1	1	2
^{241}Am	Long	2 10^{-2}	2 10^{-2}	3 10^{-2}	1 10^{-1}	7 10^{-2}	1 10^{-1}	1	9 10^{-1}	2

Table 5 *Committed effective doses from the consumption of tap water[a] contaminated at the Action Levels[b]*

Notes:

a) Consumption rates for tap water: 1 year old = 172 l y^{-1}, 10 year old = 197 l y^{-1}, Adult = 391 l y^{-1}.[9] See Table 4 for Action Levels.

b) Half-life: short = < 3 weeks; long = > 3 weeks.

c) For short-lived radionuclides (half-life < 1 month) the committed effective dose after 1 year of ingestion was calculated for a period equivalent to 8 radioactive half-lives.

4.4 Contamination in a Water Treatment Plant

Section 4.3 considered radionuclides in drinking water at the point of consumption. However, an incident may well affect an earlier part of the supply chain. In such cases, the contaminated water may have to pass through a water treatment plant prior to consumption. Several different forms of treatment are available and different companies have different combinations. These treatments are designed to take out various pollutants, some of which are analogous to certain radionuclides. Some treatment plants might therefore be expected to have an effect on activity concentrations in the water passing through them. If this is the case, then the reconcentration of activity within the treatment plant also needs to be considered. As part of the development of a handbook on recovery,[10] RPD has compiled some readily - available information on the effects of various treatments on radionuclide concentrations in drinking water and this is reproduced in Table 6.

Radio-nuclide	**Aeration and granular activated carbon filtration**	**Slow sand filtration**	**Chemical coagulation, settling and/or filtration**	**Softening**		
				Precipitate	**Ion exchange**	**Reverse osmosis**
^{60}Co						
^{75}Se						
^{90}Sr		(0-5)	(0-70)	(20-97)	(90-100)	(>99)
^{95}Zr						
^{95}Nb						
^{99}Mo						
^{103}Ru						
^{106}Ru			(77-96)	(0-83)		
^{131}I	(20-40)	(50-99)	(0-44)	(0-10)		
^{132}Te						
^{137}Cs		(50)	(0-6)	(0-80)	(>99)	
^{140}Ba						
^{140}La						
^{144}Ce		(99)	(80-94)			
^{169}Yb						
^{192}Ir						
^{226}Ra	(0-20)		(0-44)	(59-96)	(81-100)	(90-99)
^{235}U			(40-85)	(80-90)	(90-99)	(90-100)
^{238}Pu		(90)				
^{239}Pu			(>95)			
^{241}Am						

Table 6 *Indication of water treatments that may be effective in removing radionuclides from drinking water*[a]

Note
(a) The values given for each entry indicate the range of effectiveness of the treatment in reducing activity concentrations in drinking water. Values are given in terms of percentage removed

The recovery handbook was compiled with a wide range of accidents and incidents in mind, and some of the radionuclides considered in Table 6 are relevant to accidents at nuclear sites rather than incidents involving industrial sources. Some of the categories used here may include several types of treatment conditions. However, the range of values is in some cases very large, while in others data are absent. Consequently, it is difficult to make general predictions about doses to treatment plant workers and to those who may have to deal with contaminated plant materials. This is a topic that deserves further research attention.

4.5 Analyses after an Incident

Many of the radionuclides in Table 2 emit characteristic photons, which means that they can be identified readily using gamma-ray spectrometry. Robust equipment is available that can be used in the field and the number of instruments currently available in the UK is increasing. The resolution of these instruments is limited, and so they would be most useful where only a few well-defined gamma-ray emissions were involved. However, each unit costs about £6000 and again specialist staff is needed to produce reliable information.

High-resolution gamma-ray spectrometry would be a valuable asset in a post-incident monitoring programme, especially where there was a complex mixture of gamma-ray emitting radionuclides. Until recently, the need for cooling with liquid nitrogen has meant that such systems have not lent themselves readily to rigorous outdoor use. Recent developments include a robust system where cooling can be provided via a car battery, although this has yet to be evaluated comprehensively under demanding field conditions. The cost of high-resolution gamma-ray spectrometry systems of any sort is high in terms of both initial outlay and on-going maintenance. This type of equipment is not therefore recommended for use as a contingency item.

The determination of radionuclides such as ^{90}Sr that emit only beta particles or alpha-particle emitters such as ^{238}Pu requires radiochemical isolation prior to measurement. The analytical methods are time-consuming and labour intensive. Even for a simple matrix such as water, at the levels normally encountered in the environment the analyses take some time to complete. This approach is not appropriate for the circumstances after an accident, and an assessment has been carried out of rapid methods of radiochemical analysis for use after an accident.[11] The equipment and expertise needed for this type of analysis is extensive and costly. The maintenance of such a capability simply as a contingency would be difficult to justify.

The water industry does however already make extensive use of gross measurements of alpha and beta activity as part of its ongoing programme. Given the cost of more sophisticated radioanalytical facilities and the possible absence of resources elsewhere, it was considered important to determine whether gross measurements could be of use after an accident. The potential for using this approach to demonstrate conformance with the Action Levels (Section 4.3) was evaluated some years ago, largely in the context of accidents at nuclear sites. As a result, the Environment Agency (EA) set out screening levels of 5 Bq l^{-1} for gross alpha activity and 30 Bq l^{-1} for gross beta.[12] The applicability of these values to radionuclides such as those in Table 5 has been considered during the development of the handbook on recovery by RPD.[10] For most of the radionuclides in Table 5, conformance with the screening levels should ensure that the radionuclide-specific activity concentrations are below the Action Levels in Table 4. The exceptions are ^{75}Se, ^{95}Nb, ^{103}Ru and ^{169}Yb. If it was suspected that these radionuclides were present then a more detailed analysis would be needed.

5 CONCLUSIONS

There is understandable concern amongst monitoring teams in the water industry that, if asked to attend the scene of an incident involving radionuclides, they may become exposed to radiation. It is important to distinguish between the presence of radioactivity in the environment, which can be measured in very small quantities, and the dose that people might subsequently receive. The circumstances after an incident involving an industrial source are very different to those encountered after an accident at a nuclear installation. Monitoring teams may not know the radionuclide of interest beforehand. Survey instruments are available to provide guidance and reassurance, but these need to be operated by specialist staff.

In the event of a future accident or incident, Action Levels have been published that provide guidance on when substitution of drinking water supplies should be considered. Screening levels based on measurements of gross activity were originally developed when the main concern was with accidents at nuclear sites. However, these can also be applied to many of the radionuclides that might be encountered in incidents involving industrial sources. Incidents may affect treatment plants as well as the drinking water itself. Information on the effects of various treatments on the radionuclide content of water seems scarce, and there is a need to consider possible doses to workers in treatment plants once a review commissioned by DWI is available.

References

1 UK Parliament, The Water Industry Act 1991, London, HMSO, 1991.
2 International Commission on Radiological Protection, ICRP Publication 72, Age-dependent doses to members of the public from intakes of radionuclides: Part 5, compilation of ingestion and inhalation dose coefficients, *Ann. ICRP*, **26** (1), 1996.
3 S.J. Watson, A.L. Jones, W.B. Oatway and J.S. Hughes, Ionising radiation exposure of the UK population: 2005 review, Health Protection Agency, Radiation Protection Division, (in press).
4 F.A. Fry and B.T. Wilkins, Assessment of radionuclide levels around the former air base at Greenham Common, Berkshire, National Radiological Protection Board, NRPB-M752, 1996.
5 D.J. Hammond, Environmental radioactivity surveillance programme: results for 2002. National Radiological Protection Board, NRPB-W47, 2003.
6 Environment Agency, Environment and Heritage Service, Food Standards Agency, Scottish Environment Protection Agency, Radioactivity in Food and the Environment 2003, (2004).
7 European Commission, Council Directive 2003/122/Euratom of 22 December 2003 on the control of high activity sealed radioactive sources and orphan sources, *Off. J. Eur. Commun. L346 of 2003.*
8 European Commission, Council Regulation (Euratom) No. 3954/87 laying down the maximum permitted levels of radioactive contamination of foodstuffs and feedingstuffs following a nuclear accident or any other case of radiological emergency, *Off. J. Eur. Commun.* L211/1, 1989.
9 National Radiological Protection Board, Guidance on restrictions on food and water following a radiological accident, *Docs. NRPB* **5,** National Radiological Protection Board, Chilton, 1994.
10 Health Protection Agency, Radiation Protection Division, UK recovery handbook for emergency response to radiation incidents, 2005.

11 N. Green, An evaluation of rapid methods of radionuclide analysis for use in the aftermath of an accident, *Sci. Total. Environ.*, 1993, **130/131,** 207-218.
12 Environment Agency, Review of alpha and beta blue book methods: drinking water screening levels, National Compliance Service Technical Report NCAS/TR/2002/003, 2002.

CBRN ISSUES[6]

A. J. Clark

Dstl, Porton Down, Salisbury

1 INTRODUCTION

Water supply and distribution systems are a critical element of the national infrastructure, and protecting quality and availability is essential for maintaining public health and confidence in our drinking water. Recently, there have been growing concerns over the possibility that water supplies might become an attractive target for future terrorist attack. Acts of deliberate contamination could involve chemical, biological, radiological or nuclear (CBRN) hazards. UK Government agencies have, for many years, provided assistance to water companies in developing new detection and protection measures in addition to providing advice and emergency support functions. This presentation considers how use of CBRN differs from other potential contamination hazards, under what circumstances it might pose a threat, the requirement for special precautions, and the nature of the protective measures already in place. Lastly, some suggestions are put forward for new scientific developments which could enhance our future capability.

While the focus here is on the possible poisoning of drinking water, other acts of sabotage on the water infrastructure cannot be ruled out, including physical destruction, disruption of power supplies or electronic attacks (often referred to as cyberterrorism) [1]. While use of conventional weapons such as explosives or firearms is still considered the most likely form of terrorist attack, there is an increasing probability that CBRN will be employed in future. This chapter will focus primarily on the threat from C and B agents, RN aspects also having been dealt with already.

2 CBRN HAZARDS – WHAT ARE THEY?

Chemical, Biological, Radiological and Nuclear (CBRN) terminology has historically been applied in a military context to threats of enemy attack from chemical, biological and nuclear weapons (which are often referred to in the media as Weapons of Mass Destruction). More recently, usage has been extended to include a wider potential threat, both in the military and civil context, from terrorist (so-called asymmetric) attack. In the case of water supplies, there has been heightened concern over CBRN agents in acts of deliberate contamination[2]. There is therefore a need for water utilities to consider the likely impact this could have on their business and how it might differ from other contamination risks.

2.1 Chemical Warfare (CW) agents

The so-called classical CW agents encompass a wide spectrum of materials (Table 1), ranging from the highly toxic nerve agents, through vesicants, and choking agents, to incapacitating and riot control agents.

Chemical Agent Types	Examples
Nerve gases	VX, sarin, soman, tabun
Blister/Vesicant agents	Mustard, lewisite
Blood poisons	HCN, arsine, cyanogen chloride
Choking/Lung/Pulmonary	Chlorine, phosgene
Vomiting	Adamsite, diphenylchloroarsine
Incapacitating/Hallucinatory	LSD, Bz
Tear/Riot control	CS, CR, CN

Table 1. *Chemical Warfare Agents*

CW agents are optimised for aerial release as the favoured method of dissemination. They range from volatile gases and vapours to persistent materials. Properties vary from high inhalation toxicity to percutaneous supertoxics. Solubility and hydrolytic stability also vary. More detailed descriptions of these materials and their properties are available elsewhere[3,4,5]. CW agents and their precursors are controlled under the Chemical Weapons Convention[6].

2.2 Biological Warfare (BW) agents

These include a range of micro-organisms and biotoxins posing a serious risk to health. Examples include bacteria such as anthrax, plague and tularaemia, a number of viruses and toxins such as botulinum and ricin. Various threat lists have in the past been proposed, with the one compiled by the US Centres for Disease Control being perhaps the best known[7]. The CDC threat list[8] is divided into Categories according to priority, those of highest concern being Categories A and B (Table 2). Category A agents have the greatest potential for adverse public health impact with mass casualties. Category B agents which include credible waterborne threats, also have some potential for large scale dissemination with resultant illness, but in general exhibit lower morbidity and mortality. Some of the agents on the threat list such as *Bacillus anthracis* spores and *Cryptosporidium parvum* are known to be resistant to chlorine. Category C covers new and emerging threat agents. A second, widely quoted list is the Australian Group List of Biological Agents for Export Control which documents a range of BW agents and materials with controlled status[9].

Biological agent	Disease
Category A	
Variola major	Smallpox
Bacillus anthracis	Anthrax
Yersinia pestis	Plague
Clostridium botulinum (botulinum toxins)	Botulism
Francisella tularensis	Tularaemia
Filoviruses and Arenaviruses (e.g. Ebola virus, Lassa virus)	Viral hemorrhagic fevers
Category B	
Coxiella burnetii	Q fever
Brucella spp.	Brucellosis
Burkholderia mallei	Glanders
Burkholderia pseudomallei	Melioidosis
Alphaviruses (VEE, EEE, WEE)	Encephalitis
Ricketsia prowazekii	Typhus fever
Toxins (e.g. ricin, Staphylococcal enterotoxin B)	Toxic syndromes
Chlamydia psittaci	Psittacosis
Food safety threats (eg *Salmonella spp, Escherichia coli* 0157:H7)	
Water safety threats (see Table 3)	
Category C	
Emerging threat agents (e.g. Nipah virus, hantavirus)	

Table 2. *Critical Biological Agent Categories as Defined by US Centres for Disease Control*

2.3 Radiological & Nuclear (RN) agents

The radiological (R) threat covers risks from Radiological Dispersion Devices (RDD), encompassing a range of potential radionuclide contamination sources[10]. The nuclear industry is tightly regulated but the use and disposal of civilian sources of radiation is more difficult to monitor. Therefore, use of an RDD would be low-cost, low-risk and relatively easy to achieve. The main threat posed by an RDD would be to create panic, as well as potential to cause long term chronic illness. Clean up operations could also be costly and difficult. A nuclear (N) threat involves an explosive release and could be manifest via an Improvised Nuclear Device which, besides causing disruption and damage, might also lead to contamination of water.

2.4 Toxic Industrial Chemicals (TICs)

A diverse range of toxic industrial materials might also be used to contaminate water supplies. These include agricultural pesticides, poisons (e.g. rodenticides), toxic metals, cyanides, and various chemical intermediates and industrial wastes. TICs may be viewed as posing a greater potential hazard than classical CW agents because they are more readily available and easier to handle. They may also be selected for certain desirable properties such as water solubility or hydrolytic stability. Some of the TICs are detectable by analytical techniques used in routine monitoring of water (e.g. pesticides for regulatory compliance) but others are not. This places extra demand on analytical capabilities[11].

2.5 Waterborne pathogens

Examples of waterborne pathogens are shown in Table 3.

Pathogen	Organism	Disease	Symptoms
Cryptosporidium parvum	Protozoan	Cryptosporidiosis	Nausea, diarrhoea, and stomach cramps.
Vibrio cholerae	Bacterium	Cholera	Diarrhoea, rapid dehydration to state of collapse.
	Bacterium	Gastroenteritis	Diarrhoea, abdominal pain, bloody stools.
Escherichia coli 0157:H7	Bacterium	Typhoid	High fever, headache, malaise and rigors.
Salmonella typhi	Bacterium	Shigellosis (dysentery)	Acute diarrhoea or form of gangrene, rapid collapse.
	Viruses	Polio, Meningitis	Meningitis, encephalitis, paralytic diseases
Shigella dysenteriae	Virus	Hepatitis	Infectious hepatitis, epidemic jaundice
Enteric virus/Enterovirus			
Hepatitis A			

Table 3. *Examples of Waterborne Pathogens of Concern*

BW agents are selected for their stability in air, which make them compatible with aerial dissemination, and impact via the respiratory route. While BW agents such as anthrax spores could cause considerable public concern, waterborne pathogens may be a more logical choice in generating mass casualties. They might also be harder to attribute and could be confused with accidental contamination.

3 WHY TARGET WATER?

Availability of drinking water is key to health and denial of supply can quickly create difficulties. Water supply is part of the critical national infrastructure and as such is recognised as a potential target for attack by terrorists and extremist groups. There is also potential to mount a covert attack at various points in the distribution system. There is the capacity to generate significant publicity for causes and to cause mass panic. Demonstration of the ability to carry out such an attack could undermine public confidence, while denial of use could pose almost as serious a problem as the proven use of a harmful contaminant. There are also potential problems associated with clean up operations, which can be difficult and expensive, with the need to provide public reassurance.

Locations vulnerable to attack include the following:

a. Intake sites. Surface water intakes (e.g. rivers, storage (impoundment) reservoirs) are vulnerable due to relative ease of access. However the effect may be reduced or nullified by dilution and the subsequent water treatment processes.
b. Water Treatment Works (WTW). These are less accessible but may contain hazardous materials in the form of water treatment chemicals. They could also be vulnerable to infiltration by a bogus contractor or disgruntled employee.
c. Service reservoirs. These are downstream of WTW and are often sited in remote areas. They are covered but have access hatches. Volumes are much less than impoundment reservoirs and so dilution effects are less.
d. Hydrants. These are very widespread and difficult to monitor.
e. Water storage tanks. These supply smaller populations in conurbations.
f. Distribution systems. Similar to hydrants, distribution systems are potentially vulnerable to injection of contaminants into pipework. This requires relatively unsophisticated equipment and could be carried out covertly from within a building and could be used to target particular groups and individuals assuming some knowledge of the distribution network.

The risk of causing significant illness is greater the nearer the contaminant is introduced to the consumer due to less dilution, but becoming more limited in its impact.

4 THE "IDEAL" POISON

Assuming the attention is to cause serious risk to health, then the material selected is likely to be highly toxic or (if biological) highly infectious. Use of a highly potent material requires a lower "dose", making it easier to transport and conceal, and is more likely to be present in harmful amounts following dilution. General differences between CW and BW agents are summarised in Table 4.

BW agents	CW agents
Natural in origin (potential sources)	Synthesised (man-made)
Non volatile	Volatile to persistent
Infectious agents replicate (low infectious dose)	Do not replicate
Some contagious	Not contagious
Biotoxins highly toxic	Wide toxicity range
Not dermally active	Many dermally active
Difficult to detect	More likely to be detected by taste or smell
Delayed onset (pathogenic organisms)	Acute symptoms (rapid onset)
Diverse medical effects (complicate treatment & countermeasures)	Water soluble materials not removed by filtration

Table 4. *Differences between BW agents and CW agents*

Volume dilution is an important factor, thus a large service reservoir would require a greater quantity of contaminant to be effective than might be needed for introduction via a fire hydrant. Chemicals will need to be sufficiently water soluble to disperse easily. Biological and chemical agents must also be sufficiently stable in water and resist degradation by chlorine (increased chlorine demand also giving an indication of possible contamination). The material should also exhibit little or no taste or smell if it is to avoid discouraging consumers from drinking the water. Lastly, it will help if the contaminant is fairly easy to obtain or manufacture in sufficient quantity. An awareness of harmful properties (e.g. well known poisons or presence of hazard warning labels) may also influence choice of material. Biological materials can be effective in very small amounts and in the case of pathogenic organisms, there can be a delay during the incubation period before symptoms become apparent, after which medical treatment can be more difficult. This makes them suitable for covert attack, with the potential to be non-attributable. Highly toxic chemical poisons and biotoxins may act very quickly and lack suitable antidotes.

Factors, which will reduce the impact of a poison, include the following:

- dilution factor (all stages)
- water treatment works flocculation/filtration (if water contaminated prior to entering WTW)
- chlorination or other treatment
- interactions with pipework, loss of viability (biologicals)
- distribution pattern and usage dynamics within the network
- amount of unboiled tap water consumed
- warning characteristics such as taint/smell

Highly poisonous chemicals such as pesticides are either being phased out in favour of safer alternatives or subject to tighter controls[12]. Limits are also usually placed on the amount of active ingredient in formulated pesticide products, and in some cases the addition of a colour or taste marker is required. Suppliers of specialist chemicals are obliged to carry out checks before they dispatch listed poisons. CW agents and their precursors are controlled by the Chemical Weapons Convention. This became binding on April 27 1997 and has 147 countries bound by its measures, with a further 27 as signatories. The treaty prohibits the use, manufacturing and stockpiling of all chemical weapons (except for riot gases) and includes a number of verification measures. On site inspection is carried out by the Organisation for the Prohibition of Chemical Weapons (OPCW), based in Netherlands in The Hague. However, these controls may not deter the determined terrorist, who may only have to demonstrate a capability to deliver such a threat to cause considerable disruption and alarm. There is also continuing concern that CW and BW agents held in former Soviet facilities could fall into the hands of terrorists. Terrorist organisations are becoming increasingly sophisticated, as demonstrated by recent events, with the threat of an unorthodox CBRN attack becoming increasingly likely, despite international conventions and controls.

5 THE CHANGING HISTORICAL SCENE

Prior to the early 1980's, the risk of deliberate contamination of food or drinking water was mainly of military concern, related to possible acts of sabotage by enemy nations. However, a case of intentional *Salmonella typhimurium* contamination of food in salad bars took place in 1984, in The Dallas, Oregon USA perpetrated by the Rajneeshi, an Indian religious cult, in an attempt to influence the vote in local elections[13], resulting in 751 cases of food poisoning. A number of large-scale accidental pollution events at around this time also demonstrated the potential impact of a water contamination event on civilian populations. These included Milwaukee (1980), Camelford (1988) and Wem (1994), each of which were considered as case studies at the previous Water Contamination Emergencies Conference[14]. Evidence suggests civilians may be more vulnerable to waterborne terrorism, given the wider range of age groups and health conditions (making it more difficult to use data and threat lists based on military populations).

In 1993 the provisional IRA brought a heightened bombing campaign to mainland Britain when the Bishop's Gate Bomb was detonated in London. Although there had been a few isolated incidents previous to this, the targeting of the capital significantly raised the risk of terrorist activities within the UK. This was also the time of water company privatisation in 1990 and increased concern over the safety of water supplies. A new breed of terrorist also began to emerge at about this time when home made sarin CW agent was used in terrorist attacks in Japan by followers of the Aum Shinrikyo cult. The first attack at Matsumoto in June 1994 killed seven and injured 193. In the second attack, on March 20 1995, which occurred when sarin was released within the Tokyo underground, 12 persons were killed and there were around 5,500 casualties[15]. It was later revealed the same group had been developing the capacity to produce biological agents. The spectre of the D.I.Y. terrorist was born.

IRA bombing of London Docklands in 1996 destroyed the Thames Water Millharbour analytical laboratory and brought home the need for water company contingency plans and backup capacity. The link between terrorist activity and drinking water was made in 1999 following a report that an IRA sympathiser had threatened to contaminate UK water supplies with paraquat via water hydrants, giving sufficient detail to make it a plausible threat. As the millennium approached, the IRA threat began to recede, and the more sinister activities of Osama bin Laden and the Al-Qaeda organisation were becoming known. This culminated in the devastating attack on the World Trade Centre (WTC) in New York on September 11th 2001 closely followed the following month with anthrax attacks delivered via the US postal system (although no connection between the two events has been established).

A heightened state of alert has been in place since the WTC bombing, during which time various Islamic extremist groups loosely affiliated to Al-Qaeda have pursued an international campaign of terror. In February 2002, Italian police arrested four Moroccans involved in a plot to contaminate the water supply of the US Embassy in Rome using potassium ferrocyanide. Other evidence found in Afghanistan revealed Al-Qaeda's interest in CB weapons and poisoning. In the UK in 2003, a small cell of extremists with Al-Qaeda connections were arrested and subsequently convicted of attempting to produce ricin.

6 THE NEW FACE OF TERRORISM

The new breed of terrorist to have emerged in recent years is driven by a fanatical or religious ideology. Individuals undertake detailed planning, are well organized and often go to the trouble of gaining detailed local knowledge. However groups with a common cause may be very loosely connected and difficult to penetrate. They are able to tap into wider resources and capabilities, making use of modern electronic networks to communicate, access information on potential targets and promote their cause. They should be thought of as a global phenomenon, encompassing various autonomous local and regional networks. Their aim is to mount asymmetric, covert attacks, possibly with pursuit of long term aims. Perpetrators are intent on causing serious and indiscriminate harm, possibly seeking symbolic or high profile targets, often without concern for their own safety.

Recent events have drawn utilities' attention to the potential for poisoning of water supplies[16]. Information on poisons is freely available in the public domain and various groups and individuals pose a potential threat. Intentions could be any or all of the following:

- to produce mass casualties
- cause public alarm
- seek publicity
- stretch emergency services
- deny use
- damage economic interests
- promote political or civil unrest

The likelihood of a terrorist attack is often therefore put in terms of not "If" but "When?" While there has been a recent focus, for good reason, on the terrorist threat from Islamists, threats from other sources should not be overlooked. Examples include nationalist, extreme right or left wing groups, religious cults, or criminal gangs. The threat posed by the disgruntled employee who has inside knowledge, access and opportunity should also not be ignored and indeed may be more likely to succeed.

7 THE UK RESPONSE

The IRA Bishop's Gate Bombing in 1993 led to a review of disaster recovery procedures. At the same time water utilities were undergoing privatisation, raising concerns over safety. The Drinking Water Inspectorate reviewed potential risks from deliberate contamination of drinking water and initiated research into rapid methods of analysis. In 1994, a voluntary Mutual Aid Scheme was set up between the water companies to provide support in emergencies. The Security & Emergency Measures Direction (SEMD) of 1998, which followed the London Docklands bombing, formalised Mutual Aid Group activities and established the need for event coordination procedures. The Direction promotes policy on Integrated Emergency Management, in which plans are sufficiently flexible that they can be acted upon, on a collective basis with other relevant bodies, to deal with a variety of emergencies from any cause, whether natural, civil or national security. At the same time an emergency call off contract was also set up to provide rapid analysis capability by specialist UK laboratories if contamination of drinking water is suspected, an arrangement which is now in its seventh year of operation. This turned out to be a prescient move in the light of events which were to follow, coinciding with the emergence of a new style of terrorist, with an interest in chemical and biological weapons.

The increase in the terrorist threat arising from the events of September 11 2001 prompted a corresponding response from UK Government. This resulted in measures to improve upon the UK's ability to respond to civil emergency. The Civil Contingencies Secretariat (CCS) was established in 2001 as a co-ordinating body and centre of expertise to improve the resilience of Central Government and the UK. The formation of the Health Protection Agency in 2003/04 included a remit to co-ordinate health response to CBRN incidents and to integrate activities with the UK's emergency services. Regulations require water companies to notify DWI and HPA of any event that gives rise to, or is likely to give rise to a significant risk to health.

In 2004/05, the Home Office launched a CBRN Science and Technology Programme of research to address gaps in civil defence, with additional support provided by the Cabinet Office. The Department for Trade and Industry is supporting research into new sensing technologies based on nanotechnology applicable to detecting terrorist events. A new Government Decontamination Service was set up by DEFRA in Spring 2005 to advise on remediation activities. Various CBRN coordinating groups now meet on a regular basis, involving key government departments and representatives from the water industry. Regular exercises are also conducted with the emergency services. A continuing assessment of potential risks and a strengthening of protective measures to deal with any potential threat supports the above activity.

8 MEASURES ESTABLISHED TO PROTECT SUPPLIES

Measures introduced to counter the threat of deliberate contamination include the following:

a. Improved Site Security Measures. These include restrictions on site access, fitting of locks to hydrants and hatch covers on service reservoirs, introduction of CCTV etc.
b. Mutual Aid Group Activities. Mutual Aid provision of backup facilities in the form of access to equipment (pumps, bowsers etc) and being able to call on additional expertise and analytical capability. A forum and working groups exist for the exchange of information and experience, as well as promoting best practice within the industry. It also coordinates a voluntary proficiency testing scheme and organises technical workshops.
c. Emergency Planning. Water utilities are required to have suitable emergency procedures in place. Ministerial Guidance published in March 2004 under an amendment to the SEMD stated that "Companies should ensure they have sufficient equipment and resources to deal with emergencies and not to over-rely on the arrangements for borrowing equipment from other water companies in times of emergency. They should also ensure that their installations are adequately protected. Consideration should also be given to physical protection of the pressurised water network from unauthorised access".
d. Call-off contract. Specialist UK laboratories are contracted to provide an emergency analytical service 24/7, 365 days a year. The service is available to utilities in UK and Northern Ireland, with arrangements for Scotland. Procedures are in place for sampling and rapid identification of priority materials using validated analytical protocols, and are subject to a quarterly QC check overseen by DWI and UKWIR.
e. Radiological Response. Updating of the so-called Green Book on radiological hazards.
f. UK Resilience. UK Resilience resides within the UK government Cabinet Office, under the Civil Contingencies Secretariat (CCS). The Civil Resilience Directorate (CRD) was established in June 2003 to co-ordinate resilience programmes under the Office of the Deputy Prime Minister. The CRD contributes to the Home Office programme on CBRN resilience, by producing guidance on countering the effects of CBRN incidents on infrastructure.
g. Access to toxicological databases & information sharing systems.
h. Scenario Exercises. Regular tabletop and large-scale exercises are carried out to train emergency planners and first responders, and to test plans and procedures. A recent example being the Atlantic Blue international counter terrorism exercise[17] involving UK, US and Canada from 4 to 8 April 2005.
i. CBRN Committee & Government Interdepartmental Co-operation.

j. International Collaboration (notably with USA via Department of Homeland Security, Environmental Protection Agency and Department of Defence).

9 REMEDIATION & CONTINGENCY PLANS

The following approach is recommended in the protocol for disposal of contaminated water issued by Water UK[18] and DEFRA[19]:

- CONTAIN - stabilise
- IDENTIFY - survey
- TREAT (& VERIFY) – decide target clearance levels and develop phased recovery options
- DISPOSE – transport and disposal of wastes

Selection of the most appropriate treatment option will depend upon a variety of factors. Common treatment options include the following:

- Raising levels of free chlorine
- Filtration (of particulates)
- Reverse Osmosis
- Ozonolysis (oxidation, disinfection)
- UV light (disinfection, photolysis)
- Activated carbon (absorption)

Under certain circumstances, flushing rather than treatment may also be an option. Treatment for disposal needs to consider the environmental impact of potentially large volumes of decontaminant or concentrated waste. This makes treatments based on ozonolysis or UV irradiation attractive, as they leave no harmful residual. Remediation is likely to be time-consuming, involving verification and public reassurance. In the meantime, contingency plans will need to be in place. Various contingency options may be considered:

- Issuing of a boil water notice while a problem is being investigated
- Supplying alternative supplies (by tanker or issuing of bottled water)
- Switching to alternative water supply resources
- Operation of standby water purification systems
- Use of mobile water purification equipment with independent power supply (military versions of which are to CBRN specification)
- Fitting of Point of Entry (POE) /Point of Use (POU) water purification systems

The last option could be an attractive proposition for a use during heightened alert states and to protect likely targets. However, there is no agreed specification available against which to test performance of civilian POE/POU systems.

10 EMERGING CAPABILITIES

Attention is being given to the following:

a. On-going risk assessment. This includes a continued review of potential threat materials and terrorist capabilities.
b. Supporting developments in on-line and at-site detection to provide early warning capability.

c. Faster methods of sampling and analysis to achieve quicker turnaround time and higher sample throughput. DWI sponsors research into rapid methods for high priority materials not already detected by water industry laboratories. Collaborative efforts by the Mutual Aid Group are seeking to establish new analytical protocols for use by the industry, also drawing upon the useful US EPA Response Protocol Toolbox[20].
d. New rapid screening techniques. This can help to prioritise samples or look for unknowns. They can be based on measurement of a toxic effect, detect a change in one or more physical parameters, show a change in spectral response or generate a chromatographic peak. Rapid radiological tests have also been undergoing evaluation.
e. Development of a more proactive means of assuring water safety based on a system of quality management, including Hazard Analysis Critical Control Point (HACCP) principles. Water safety plans (WSP) are intended to cover all activities under the control of the water supplier, from source, through to treatment, storage and distribution[21]. As well as adopting quality management procedures and HACCP principles, WSP also incorporate other elements such as communication plans. HACCP requires the systematic assessment and prioritisation of risk, leading to methods of risk reduction. It is therefore compatible with addressing water security issues.
f. Availability of database information and software support tools. Construction of an expert knowledge base can assist with emergency planning and training, as well as incident management. Software support tools can provide computer based decision aids and event recording.

11 HOW TO BE PREPARED

Defensive measures should be based on a comprehensive approach, which can be summarised as follows:

PREDICT – conduct risk and vulnerability assessments. Undertake "what if" scenarios (these can be facilitated by the use of distribution modelling software).

PREVENT – use control measures, disrupt activity of terrorist cells.

PROTECT – introduce physical security and surveillance to discourage perpetrators (who may then seek an easier target).

DETECT – develop early warning systems and rapid identification methods.

INTERDICT – frustrate attacks, crisis management

RESPONSE & RECOVERY – emergency plans, contingency procedures, remediation protocols.

EVIDENTIAL – establish means of identifying perpetrator (forensics, chain of custody).

Many of these requirements are already well established within the defence CBRN community, which is able to provide advice and support.

12 WHEN ALL ELSE FAILS

Health surveillance and consumer complaints provide a potential source of information on reported illnesses and symptoms due to contaminated drinking water[22]. This could be the first indication of a terrorist attack, as indeed it can be for non-deliberate contamination events. The UK military have introduced a handheld device called the Prototype Remote Illness and Symptom Monitor (PRISM) which enables information to be entered in the field and relayed by satellite phone for real time analysis and mapping[23]. Various studies have looked at this approach[24] for early identification of patterns of illness and there are plans to introduce a similar capability into the latest NHS computer systems.

In terms of communicating with consumers, there is already the possibility of being able to transmit automatic alerts to selected consumers by telephone or text message, should the need arise.

13 ENHANCING OUR RESPONSE

The following suggestions are put forward as areas for further scientific research that could enhance our future capability:

a) Detailed investigations of real-time monitors as early warning systems.
b) Tapping into available expertise in sensor technology (notably within the defence sector).
c) Establishing test facilities to evaluate on-line early warning systems in detecting harmful contaminants and pipework interactions under controlled conditions.
d) Development of expert systems for responding to unusual events.
e) Modelling of contamination events to predict when, and for how long, source water could be affected and likely concentration levels.
f) Evaluating new field detection capabilities.
g) Increasing our understanding of the fate of contaminants in water under normal conditions.
h) Evaluating the effectiveness of candidate treatments.

In conclusion, it is predicted that the difference between CBRN use by nations in warfare and by terrorist groups in a civilian scenario will continue to narrow. The needs of UK Ministry of Defence and the Other Government Department's defence programmes will coalesce to mutual benefit, helping to support external efforts by the water industry in safeguarding our nation's water supply. Deliberate contamination of drinking water is still considered by many to be a low-likelihood, high-impact event. A combined effort at enhancing our response capabilities can reduce the chances of any contamination event having a serious impact as well as discourage terrorists or others from targeting water supplies.

14 FURTHER READING

L.W. Mays. *Water Supply Systems Security.* McGraw-Hill (2004). ISBN: 0-07-142531-4.
W Dickinson Burrows & Sara E Renner. Biological Warfare Agents as Threats to Potable Water. Environmental Health Perspectives, 107(12), 975-984 (1999).
W.M. Grayman, R.A. Deininger, R.M. Males. *Design of Early Warning and Predictive Source-Water Monitoring Systems.* Awwa Research Foundation and American Water Works Association (2001). ISBN: 1-58321-172-1.

References

1 Precautions against the sabotage of drinking-water, food, and other products. *In: Public health response to biological and chemical weapons – WHO guidance. Second Edition (Annex 5).* World Health Organisation, Geneva (2004).
2 A. Oppenheimer. Terrorists pose threat to water supplies. Jane's Terrorism & Security Monitor - June 01, 2003.
3 D.H. Ellison. *Handbook of Chemical and Biological Warfare Agents.* CRC Press (1999). ISBN: 0849328039.
4 J. Qli, L. Rodrigues, M. Moodie: Jane's Chemical Biological Defence Guidebook, Alexandria (1998).
5 J.A.F. Compton: Military Chemical and Biological Agents. Chemical and Toxicological Properties, Caldwell, New Jersey (1988).
6 Organisation for the Prohibition of Chemical Weapons. Convention on the prohibition of the development, production, stockpiling and use of chemical weapons and on their destruction. See www.un.org.
7 A.S. Kan, S. Morse, and S. Lillibridge. Lancet 356 (9236): 1179-1182 (2000).
8 L.D. Rotz, A.S. Khan, S.R. Lillibridge, S.M. Ostroff and J.M. Hughes. Emerging Infectious Diseases, 8(2), 225-230 (2002).
9 Australian Group List of Biological Agents for Export Control, core and warning lists. Available at URL: http//dosfan.lib.uic.edu/acda/factshee/wmd.auslist.htm.
10 C.D. Ferguson, T. Kazi, J. Perrera. Commercial Radioactive Sources: Surveying the Security Risks. Occasional Paper No.11. Centre for Nonproliferation Studies. Monterey Institute (2003). ISBN 1-885350-06-6.
11 A. Clark. Contamination Monitoring: Screening vs Targeted Analysis. *In:* Water Contamination Emergencies – can we cope? By J. Gray and K.C. Thompson (Eds)., RSC Special Publication No.293, 77-99 (2004). ISBN 0-85404-628-3.
12 See Chemical Hazards Legislation. Chemical Hazards Society www.chcs.org.uk. Also, www.defra.gov.uk on a new mechanism for the control of chemicals called REACH (Registration, Evaluation, and Authorisation of Chemicals).
13 T.J. Torok, R.V. Tauxe, R.P. Wise, J.R. Livengood, R. Sokolow, S. Mauvais et al. JAMA 278:389-395 (1997).
14 J. Gray and K.C. Thompson (Eds). Water Contamination Emergencies – can we cope? RSC Special Publication No.293 (2004). ISBN 0-85404-628-3.
15 K.B. Olsen. Emerging Infectious Diseases, 5: 513-516 (1999).
16 C. Copeland and B. Cody. Terrorism and Security Issues Facing the Water Infrastructure Sector. CRS Report for Congress. Order Code RL32189, updated July 5, 2005.
17 Home Office Press Release. Reference 054/2005 – Date: 17 March 2005.
18 Protocol for disposal of contaminated water. Version 2.1 (September 2003). Issued by Water UK. See http://www.water.org.uk/static/files_archive/02003_Protocol_for_the_disposal.doc.
19 Strategic National Guidance. The decontamination of the open environment exposed to chemical, biological, radiological or nuclear (CBRN) substances or material. DEFRA (2004). Available at www.odpm.gov.uk.
20 Response Protocol Toolbox: Planning for and responding to Drinking Water Contamination Threats and Incidents. Response Guidelines, Interim Final, August 2004. EPA 81 7-D-04-001 (2) http://cfpub.epa.gov/safewater/watersecurity/home.cfm?program_id=8#rptb
21 Strategic National Guidance. The decontamination of the open environment exposed to chemical, biological, radiological or nuclear (CBRN) substances or material. DEFRA (2004). Available at www.odpm.gov.uk.

22 J.W. Buehler, R.L. Berkelman, D.M. Hartley and C.J. Peters. Emerging Infectious Diseases, 9(10), 1197-1204 (2003).
23 Dstl Science Spotlight. See http://www.dstl.gov.uk/pr/science_spot/handheld_health.htm.
24 R. Lazarus, K.P. Keinman, I. Dashevsky, A. DeMaria, R. Platt. Using automated medical records for the rapid identification of illness syndromes (syndromic surveillance): the example of lower respiratory infection. BMC Public Health 1:9 (2001).

SCREENING ANALYSIS OF RIVER SAMPLES FOR UNKNOWN POLLUTANTS

A. Gravell

National Laboratory Service, Environment Agency.

1 INTRODUCTION

Analysis and identification of unknown pollutants from the general background noise using GC/MS in full scan mode can prove challenging. For pollution investigation work the use of mass scan screening allows for the determination of a wide range of semi-volatile organic compounds from a single one litre sample. Identification of compounds at the specified detection limits was not possible using existing methodologies.

2 RETENTION TIME LOCKING

It was therefore decided to use a new fast screening technique known as Retention Time Locking (RTL) which is designed to aid in the identification of unknown compounds. The software calculates the expected retention times for all compounds contained in a database. Identification is achieved by integrating four unique fragment ions from a compound which this is then compared with information contained in a database and an indication of match is specified. Because the software extracts only four ions from the spectral data the sample matrix effect is greatly reduced.

Retention time locking (RTL) is used extensively by many environmental laboratories. RTL works on the principal that the retention time is the fundamental qualitative measurement of chromatography. The process of RTL determines what adjustments in inlet pressure are necessary to achieve the desired retention times. Inlet pressure is determined by developing a retention time versus pressure calibration curve and locking the method on a single pesticide. From this the RTL software can accurately calculate the expected retention time for every compound contained in a searchable database.

The database supplied from Agilent Technologies contained 567 of the most common pesticides and endocrine disruptors known. Sample data is automatically compared with the target ion fragment information in the database as well as with the expected retention time. Compound matches that are confirmed with high confidence are indicated with an “X” and questionable matches are indicated by a “?”, informing the analyst that further review of the data is required.

The RTL database was customized so that it included all of the pesticides, endocrine disruptors and polyaromatic hydrocarbons (PAH’s) analysed for using routine methods. This required an additional 45 compounds to be added to the database, which now contains 612 compounds.

Positive identification is achieved by a match quality of ≥90%, but below this the analyst must use their experience in deciding whether the compound has been correctly identified. The detection limit for compounds identified this way may be higher than 100ng/L, however, any significant compound(s) identified this way would be added to the RTL database.

When a peak is found in the sample chromatogram that cannot be identified, as no reference data exists in the RTL library, the spectral data obtained can be compared with those contained in two additional libraries, the NIST02 (175,214 compounds) and Wiley 7N edition (316,934 compounds). The unknown compound is searched against a possible 440, 000 mass spectra contained in the libraries and the reference compound that has the best match is displayed.

3 CONCLUSIONS

The RTL software and screening library have shown the potential of this technology for environmental analysis, in particular for screening. Additional reference libraries are used to identify compounds that are not contained in the RTL database; this database can be constantly updated. The resultant mass scan method is suitable for the unambiguous identification of organic pollutants in environmental waters.

Since its completion the mass scan method has been successfully used to solve several pollution incidents where low levels of organic pollutants were suspected. Typical compounds identified include Mecoprop, Fluoxypyr, Epoxiconazole, Chlorpyrifos and Cypermethrin.

I. E. Tothill

School of Industrial and Manufacturing Science, Cranfield University, Cranfield, Bedfordshire, MK 43 0AL, UK.

1 INTRODUCTION

The risks from microbial contamination and the presence of pathogens in drinking and recreational waters are of high concern to the public, regulators, and service and support providers. The problem in drinking water can occur due to a range of factors and these include; sudden failure of drinking water treatment plans; accidental contamination of distribution networks; disruption of water supply services; water contamination by natural disasters; incidents of biological agents intentionally introduced in water and re-growth phenomena / biofilm formation. Thus, the rapid detection of pathogenic microorganisms in water is critical regardless of the cause of the problem. There are however, different levels of detection, which may be required and this could be from a simple yes / no results to quantitative data to meet legislative requirements. Water can be analysed for the presence of specific microorganisms, but different organisms have been linked to waterborne disease outbreaks. It is also impossible to analyse the water for every potential pathogen using traditional methods since they are often time consuming and cumbersome. Thus there is an increasing demand for simple, sensitive and rapid analytical tools for decentralised analysis and also on-line/ real time methods for risk assessment and management. Information regarding microbiological methods of analysis is reported in several standard reports which should give validation and comparison of the methods[1, 2, 3, 4]. A range of methods have been developed relying on the biochemical and physical properties of micoorganisms. Conventional methods of microbial detection have a number of drawbacks including being labour intensive, time-consuming and sometimes expensive. Drinking water treatment process is a continuous operation and water is usually consumed within hours after treatment. Therefore, within the water industry corrective actions and safeguard of public health require real-time and on-line analyses. Rapid methods should ideally demonstrate sensitivity and specificity as the standard methods and be able to distinguish between viable and non-viable cells.

Methods developed for microbial detection usually measure total numbers, viable numbers, activity, mass or components. Most methods have been developed for bacterial detection, but many of these can be adopted for single cell algae and yeasts. Methods developed for viruses, fungi and protozoa present their own particular problems. Indicator organisms of faecal contamination are used globally as a warning of possible water contamination and risk to public health. The presences of coliform, *E. coli* and entrococci in water have been used as an index of theoretical risk of water contamination with other pathogenic microorganisms. However, these indicator microorganisms have a short survival time in water and are more susceptible to water disinfection processes than viruses and protozoa which make them poor models. Also, the absence of these microorganisms in drinking water cannot be used as a risk assessment tool and an indicator that the water is free of biological agents introduced intentionally in drinking water. This paper describes the microbiological risks and analysis issues facing the water industry today and will cover the range of conventional, new and novel technologies available and being developed for assessing the microbiological safety of water.

1.1 Conventional Methods and Analysis Issues

Classical microbiological methods are usually based on a multi-step procedure: isolation; identification and then if needed colony forming unit counting (CFU) and rely mainly on specific microbiological and biochemical markers. These methods are time consuming and require a skilled operator to interpret the data, but they are sensitive and inexpensive and able to give qualitative and quantitative information. Since the occurrence of pathogens in drinking water is usually at very low numbers, an enrichment step is needed to increase the number of the contaminating microorganisms to a detectable level. Methods such as the viable count method (motility test) are used to estimate microbial populations. This method relies on the growth of the cells in either liquid culture or on an agar media. Serial dilution of the sample is usually carried out before spreading the sample on agar plates. The method requires the plates or cultures to be incubated at an appropriate growth temperature for between 12 - 72 hours depending on the type of microorganisms being analysed. The number of colonies is counted and this calculated as a colony-forming unit (cfu ml^{-1}) obtained in the original sample. It is possible to use the Most Probable Number (MPN) technique to estimate the original concentration of the target organism in the water sample by carrying out a dilution technique.

Identification of contaminating microorganisms is generally based on biochemical, immunological or genetic characteristics. As an example for microbial identification, specific and selective compounds are incorporated into the growth media which inhibits the growth of the non-target species and promote the growth of target microorganisms. The use of chromogenic media which change colour in response to growth and biochemical events are widely used for microbial identification. Examples include; MacConkey agar for coliforms; Sartorius Nutrient Pad Sets used for filter viable counts; and Chromagar.

The use of highly porous cellulose acetate membrane filters (0.45 µm) has also been implemented in the detection of indicator organisms. Membranes that allow the passage of large volume of liquid sample but prevent the passage of bacteria or fungi have been used. Microorganisms retained on the membrane are incubated on a specific agar media and the appropriate room temperature. The number of colonies is subsequently counted on the filter. The main advantage of this method is the large sample volume that can be applied on these types of filters which is required for drinking water analysis. However, these methods needs incubation time between 16 to 24 hours which is not suitable for real time analysis. Testing water for the presence of specific viruses or protozoan parasites such as *Cryptosporidium* and *Giardia* may take as long as weeks for complete analysis [5].

Conventional methods suffer from the disadvantage of being laboratory based, and require skilled operator. The time required for obtaining confirmed results remain the major limitation since this is hazardous in water testing as the water may have already been distributed to the consumer. Detection of indicator organisms and pathogens in water may underestimate the actual microbial population due to sublethal environmental injury, inability of the target organisms to take up nutrients and other physiological factors which reduce bacterial culturability[6]. To obtain good analytical data appropriate sampling procedure should be undertaken where the sample is representative of the water to be analysed. Other factors such as growth medium, incubation conditions, nature and age of the water sample can all influence the species and count of isolated microorganisms.

Turbidity is also widely used for the estimation of cells in suspensions by using a spectrophotometer. The ability of microbial cells to scatter lights and hence appear turbid in a solution is utilised in this technique to measure the concentration of the cells. The scattered light of a microbial suspension is proportional to the number of cells present. Measurements are usually carried out at 600 nm of bacterial analysis using a spectrophotometer. A standard calibration curve of log I_o/I against either the total count or the dry weight is used[7]. The calibration curve applies only to a particular microorganism grown under a particular set of growth conditions. But this technique is unable to differentiate between viable and non-viable cells. Park and co-workers[8] used spectrofluorometric assay to detect total and faecal Coliforms in water samples. Microscopy is also an important technique in the diagnosis of microorganisms, since it allow the view of the cells under the microscope. The most important stain procedure in microbiology is the Gram stain. Using this method bacterial cell morphology and Gram reaction may be examined via the use of Gram stain microscopy, which provide the information of whether the organism is a gram negative or gram positive based on the differences in bacterial cell walls. Gram staining is a rapid procedure that can be performed in minutes. Further details on the methods are given in Dart [9]. However, this method may not be suitable when the water sample contain low level of microorganisms.

1.2 Direct Detection Techniques

These techniques are based on detecting microbial cells and cover a range of methods which are listed below.

1.2.1 Epifluorescence technique. Fluorescent microscopy is a direct counting method for microbial enumeration. The method is rapid and does not require incubation step. The principle of the method is similar to membrane filtration and takes about 25 min to complete. The method involves filtering the sample through a polycarbonate membrane, retaining the bacteria present in the sample on the filter surface. The bacteria is then stained using a fluorochrome (acridine orange or diamidino-2-phenylindole) for a contact time of a few minutes. The membrane is rinsed with distilled water and the microorganisms are counted using epifluorescence microscopy. Under UV light, acridine orange stains deoxyribonucleic acid (DNA) green and ribonucleic acid (RNA) orange and therefore the method can distinguish between active from inactive microorganisms based on their higher RNA content [10, 11]. A fluorescent redox probe may be used to identify actively respiring bacteria. System such as the Bactoscan Automated Microbiology System (Foss Electric, Hillerød, Denmark) has been developed for the detection of bacteria in food and drinks samples and can be used for water analysis. This device is rapid, with an estimated 60 to 70 samples undertaken every hour. The MicroFoss is a user-friendly instrument and has been used especially in dairy and meat segments and gives rapid microbiological analysis based on microorganisms growth, pH change and dye indication. The products also include ready-to-use vials for enumeration of Total Viable Count, *Enterobacteriaceae*, *Coliform*, generic *E. coli*, and Yeast. The MicroFoss showed the ability to detect counts as low as 0.5 cfu ml^{-1} .

Antibodies conjugated to fluorochrome such as fluorescein iosthiocyanate has been used for microbial detection. By applying tagged antibody for the target microorganism to the sample, the microorganism will fluoresce due to the complex forming with its complimentary tagged antibody. The number of fluorescing cells is then counted using an epifluorescence microscope. This method has had wider applications in the filed of on-line estimation of fermenter biomass[12]. Nakamura *et al*[13] coupled *E. coli* separation using magnetic particles with detection by flourescein isothiocianate.

Fluorescent in-situ hybridization (FISH) combines fluorescent dyes with hybridisation of nucleic acid. This method allows staining of specific types of bacteria. FISH analysis has been implemented in the detection of nitrifying biofilms using targeted oligonucleotide probes. Its use has increased in identifying microorganisms involved in different stages of water treatment plants.

1.2.2 Particle counting. Coulter counters measures changes of resistance as particles move through a small aperture (sensing zone). The pulse generated by each cell is amplified and recorded electronically, giving a count of the number of cells flowing through the aperture. The Coulter method of sizing and counting particles is based on measurable changes in electrical impedance produced by non-conductive particles suspended in an electrolyte.

In this case cells can be counted in the medium in which they are growing. A range of products are commercially available such as the COULTER COUNTER® Z1 Series, (Coulter International Corporation, Miami USA), and Multisizer™ 3 Coulter Counter® (Beckman Coulter, UK). CellFacts I, developed and manufactured by CellFacts Instruments Ltd, UK, also uses electrical flow-impedance determination to count and size particles in a sample[14]. The analysing principle of CellFacts I is it counts and sizes every particle in a sample introduced to the instrument and provides detailed information on the microbiological status of that sample with applications in microbiological research and the food, biotechnology, water, cosmetics, and pharmaceutical industries. Its most powerful applications are in on-line monitoring of fermentation processes and cell cultures.

Devices based on acoustic resonance densitometry have been reported by Clarke *et al*[15], which could provide effective real-time and *in situ* determination of biomass in downstream processes. The technique is based on the change in the acoustic resonance of a fixed volume of water sample due to microbial contamination.

1.2.3 Flow Cytometry Flow cytometry is a technique that allows the user to measure several parameters in a sample and it is one of the most reviewed methods for bacterial detection[16, 17]. Parameters such as physical characteristics as cell size, shape and internal complexity can be examined. The principle behind the technique is that a sample is transported through a laser beam. Biomass is analysed by light scattering methods and fluorescence of each individual cell. Forward light scatter is a function of size and fluorescence measurement will detect auto-fluorescence (e.g. chlorophyll) or a fluorescent marker by staining of chemical components such as DNA. The light energy is converted into an electrical signal by the use of photomultiplier tubes[18]. This method has been expensive and prone to drift, but current systems claimed to be able to detect 10^3 cells ml^{-1} such as Microcyte from Aber Instruments, CyFlow® from Partec GMbH. BactoScan FC ™ is now an industrial standard for counting bacteria in raw milk. Flow cytometry can be combined with Fluorescent in situ hybridisation (FISH) methods which label specific nucleic acid sequences inside intact cells. Flow cytometry application for the detection of indicator bacteria is limited since it requires a microbial density of 10^2-10^3 cells in the sample. The Microcyte (Optoflow, Norway) is a portable, battery operated flow cytometer has the ability to reduce auto-fluorescence of non-target organisms and particles[19]. This instrument has a detection limit of 10^1-10^2 cells ml^{-1}, but for optimal signal a 10^4-10^6 cells ml^{-1} are required. This instrument is used for the analysis of microorganisms in river. This instrument has been evaluated for bio-warfare application.

1.2.4 Electron Microscopy The use of scanning electron microscope for counting bacteria on membrane filters has been reported[20]. However, this technique suffers from the high cost of the instrumentation and operator skills required.

The ChemScan® RDI (Chemunex, France) is a laser-scanning instrument used for the detection of fluorescently labelled micoorganisms captured by membrane filtration. This instrument is able to detect *Cryptosporidium*, *Giardia*, coliforms and *E. coli*. The sample is concentrated through a filter and the microorganisms are then labelled prior to laser scanning of the filter. The assay takes about three and a half-hours to complete. Visual validation can be achieved using epifluorescence microscopy.

1.2.5 Immunoassay Techniques Immunodetection using antibodies has been successfully employed for the detection of microbial cells, viruses and spores[21]. Antibodies can be easily produced for a range of microorganisms. Immunoassay techniques have been developed for microbial detection using different labels to generate the signal. Radioisotopes were the first to be used, but enzymes became more attractive due to cost and environmental issues. Lateral flow immunoassay tests such as the RapidChek™ for *E. coli* O157 marketed by SDI (Strategic Diagnostics Inc. Hampshire, UK) is available for an 8 hour enrichment time and 10 min test time. Clearview™ and REVEAL® are range of products marketed by Oxoid (Basingstoke, UK) and Adgen Ltd. (Ayr, UK) respectively for the detection of *E. coli* O157. Most of these tests are based on isolation and enumeration of the bacteria before detection. This is due to the concentration of the bacteria in the sample is usually too low to be detected directly. All of these methods can be used for water analysis[22]. Similar tests are also available for *Salmonella*, *Listeria* and *Campylobacter*. Routine application of immunoassay to water analysis has been hampered by the non-specific substances and interfering organisms producing false-positive results.

1.3 Cellular Components Detection

Measuring the concentration of a biochemical component of the target microorganism has been used as a method of biomass estimation. The presence of these compounds needs to be determined with the appropriate precision if they are to be used for microbial estimation. A range of compounds such as lipids and their derivatives[7], cell carbon / phosphate[23] and total nitrogen / proteins[24] have all been used for microbial quantitation. Some of these methods are dependent on the physiology of the cells and therefore their validity may be questionable. Chitin, ergosterol, adenine triphosphate, enzymes and DNA probes have all been developed for fungal detection[25, 26]. In this section the more applied methods for water analysis will be covered in details.

1.3.1 Metabolising Enzymes The metabolic activity of specific cellular enzymes has been used as indicators for the presence and detection of specific microorganisms such as *E. coli*, total coliforms and enterococci[27, 28]. For example total coliforms and *E. coli* contain the enzyme β-D-galactosidase which can be used as an indicator for their presence. *E. coli* also contain the enzyme β-D-glucuronidase, which is used to indicate the presence of *E. coli* in the water sample. The enzyme β-D-glucosidase is used for the detection of entrococci. Different companies have implemented the analysis for these enzymes to develop products for microbial detection. The Colilert® Quanti-Tray ™ and Enterolert® Quanti-Tray ™ (IDEXX, Westbrook, Maine) are colorimetric tests based on the detection of these enzymes through their interaction with the substrate. These tests are recommended for drinking water analysis in the United States and for recreational water analysis in Australia and New Zealand[6]. The Colilert® Quanti-Tray ™ is also reported to be approved by USEPA, UBA and also the Agency for the Environment as an alternative method in Germany to the ISO 9308-1 reference method for drinking water analysis. The Colilert test is capable of detecting 1 CFU 100 ml^{-1} in 24 hour of incubation[29]. The Enterolert ® Quanti-Tray ™ is used to detect enterococci through the detection of the enzyme β-D-glucosidase. These methods are very easy to perform and interpret.

The Colifast® Analyser has been developed for the detection of total and thermo-tolerant coliforms, *E. coli*, faecal streptococci, *Pseudomonas aeruginosa* and total viable organisms in water samples, providing a yes /no results[6]. This analyser as been applied as an early warning operational tool. Enzymes have also been used as indicators for the early detection of fungal activity[30, 31]. However, most of the studies concentrate on food deterioration detection.

1.3.2 API tests The API test kits are the best known biochemical tests for microbial identification. API tests marketed by bioMerieux Inc (France) usually contain about 20 miniature biochemical tests, which may detect all bacterial groups and 550 species. The procedure involves the inoculation of each of the 20 mini test tubes with a saline suspension of a pure culture. The samples are then incubated for 18-24 hours before the colour change is read. An API system based on enzyme detection (Rapidec) is also marketed by bioMerieux Inc. The test is followed either by a colour change directly or following the addition of appropriate reagents[32].

1.3.3 Bioluminescence These tests are based on the rapid and sensitive analysis of adenosine 5' triphosphate (ATP) which is the energy molecule for all living cells (animal, vegetable, bacteria, yeast and mould cells). The tests measurement is based on the use of the firefly enzyme luciferase:

$$\text{Luciferin} + \text{ATP} + O_2 \xrightarrow{\text{Luciferas}} \text{Oxyluciferin} + \text{AMP} + CO_2 + \text{PPi} + \text{Photon}$$

PPi = pyrophosphate

Light is produced depending on the concentration of ATP in the sample, which can be interpreted to the sample microbial content. Several instruments are commercially available for the estimation of microbial biomass which often used to demonstrate absence of microorganisms. The Clean-*Trace*™ products such as the Biotrace Uni-*Lite*® and Uni-*Lite*® XCEL instruments (Biotrace Ltd., Bridgend, UK) are instruments use the above principle. Other companies such as Celsis International plc (Suffolk, UK) and Biotest (New Jersey, USA) also market instruments based on this technology. ATP Instruments can usually detect low levels of contamination (10^3 cells ml^{-1}) and tests take between 10-20 minutes. New developments use ATP measurement to give estimates of cell numbers. To implement the test for water analysis, a concentration step will be needed before a sample is taken to be analysed[22].

1.3.4 Molecular techniques Nucleic acids which are the building blocks of DNA and RNA are present in all living cells and can be used as a general indicator of microbial biomass. The principle of the tests is based on the hybridisation of a characterised nucleic acid probe to a specific nucleic acid sequence in a test sample followed by the detection of the paired hybrid. Hybridisation methods have been used for the detection of pathogens in water such as bacteria, viruses and parasites[33, 34, 35]. *In situ* hybridisation has also been implemented where the bacteria can be detected with a microscope or flow cytometer after the hybridisation taken place[36, 37]. The use of oligonucleotide probes directed at the rRNA of the microorganism will ensure only active microorganisms are detected. The use of the polymerase chain reaction (PCR) has been frequently applied for microbial detection to enhance the sensitivity of nucleic acid-based methods[38, 39, 40]. The PCR technique will selectively amplify a gene sequence specific to a group of organisms or a single species aiding in the ability of detecting low level of microorganisms in the sample. The technique has been used for qualitative analysis, but quantitative measurements are important and methods have been developed to make the test quantitative. The use of fluorimetry for PCR product analysis has been implemented for rapid and sensitive tests. Quantitative PCR methods are shown to be very sensitive with detection limit of 10 cells ml^{-1} and analysis time of approximately 3 h is reported[41]. PCR methods have been used to detect viruses[42, 43], bacteria[44], indicator bacteria[45] and protozoan parasites [34, 46, 47, 48] in water samples. However, these methods can be expensive, time consuming, require skilled workers and sometimes complicated.

The GEN-PROBE hybridisation protection assay (HPA) technique uses a specific DNA probe, labelled with an acridinium ester detector molecule that emits a chemiluminescent signal. Two methods which have used this technology successfully are the AccuProbe® and the FlashTrak (Gen-Probe Incorporated, San Diego, USA). In these systems the DNA probe is targeted against the ribosomal RNA of the target organism. The nucleic acids are then hybridised to form a stable molecule. The chemiluminescence is then measured by the gen-probe luminometer (Thomson, 2001). This method has been developed for Fungi and bacterial detection and take 5 hours to achieve the results with 92-100% sensitivity and specificity. Molecular Devices Corporation (Sunnyvale, USA), has developed products for total DNA assay which can be applied for microbial detection. DNA array marketed by Affymetrix (Santa Clara, USA), such as the GeneChip *E. coli* Antisense Genome Array is used for examining expression of all known *E. coli* genes. The GeneChip technology can be used mainly for the broad spectrum of nucleic acid analysis applications including sequence analysis, genotyping and gene expression monitoring. Other companies such as MediGenomix GmbH (Germany) are also expanding into DNA-analysis. Molecular techniques are gaining popularity in the water industry today and methods are being developed for water analysis.

1.4 Electrochemical methods

A range of electrochemical methods have been developed for microbial detection[49]. The most reported techniques for water analysis will only be covered in this section.

1.4.1 Impedimetry and conductivity Changes to the ionic conductivity of the culture media due to microbial growth have been used to measure microbial content[50, 51]. Impedance measurements have been utilised in the development of microbial devices. The Bactometer marketed by bioMérieux (bioMerieux Inc., Marcy-'Etoile, France) is one of these instruments which provides rapid and cost effective system for the detection of microbial contamination. It offers quantitative and qualitative tests including total counts of enterobacteriaceae, coliforms, yeast and mould. The Malthus system (Malthus Instruments Ltd, Bury, UK), is also based on the detection of microorganisms by measuring the changes in the flow of an electric current passing through a medium. This system can detect a range of microorganism such as *Coliforms*, *Salmonella*, yeast and mould. The RABIT system (Don Whitley Scientific Ltd, UK) has also been developed for the detection of microbial groups such as coliforms and enterococci.

1.4.2 Fuel cell technology The fuel cell device uses an anode, a cathode and a supporting electrolyte medium to connect the two electrodes, and an external circuit to utilise the electricity. Micoorganisms can be incorporated into a fuel cell and will generate a current through their activity. The use of microorganisms for the generation of electric currents has been reported to be enhanced by the incorporation of traces of potassium ferricyanide or benzoquinone in the solution[52]. Devices based on the use of mediated systems[53] and non-mediated systems[54] have been used for monitoring microbial growth. The analytical sensitivity of the fuel cell method has been increased by the use of the mediator phenosine ethosulphate for the detection of *E. coli* (detection limit of 4×10^6 cells ml^{-1}) with a 30 minutes assay time.

1.4.3 Amperometry Several devices based on amperometry have been applied for microbial sensing. Redox mediators (such as Potassium hexacyanoferrate (III), benzoquinone and 2,6-dichlorophenolindophenol) have been used which are reduced by the micoorganisms as a consequence of substrate metabolism. Kal'ab and Skl'adal[55] have evaluated the use of different mediators for the development of amperometric microbial bioelectrodes. The reduced mediator diffuse to the working electrode where it is subsequently re-oxidizes. The current flow measured has been shown to be proportional to the reduced mediator concentration and hence the microbial concentration. This device can detect 5×10^4 cells ml^{-1} *E. coli* in 15 minutes[29]. Several devices have been developed based on this principle and commercial instruments were also marketed, but were then withdrawn due to poor reproducibility when tested on a range of microorganisms. Hitchens *et al.*[56] measured bacterial activity using mediated amperometry in a flow injection system. The Medeci analyser (Medeci Developments Ltd, Harpenden, UK) is under development for medical application. The device is based on the use of screen-printed three-electrode configuration incorporated into a novel wall-jet flow-cell design, with electrochemical measurement of bacterial concentration by hydrodynamic coulometry. This process is similar to amperometry in that it measures current, but instead of current at a point in time it measures current over a period of time, hence total charge passed[57]. Devices based on the use of the Clark oxygen electrode have also been developed for microbial detection.

1.4.4 Cyclic and square wave voltammetry A three electrode system using working electrode, reference electrode (SCE or Ag/AgCl) and counter electrode (platinum) has been developed using cyclic voltammetry (CV) for the detection of yeast and bacteria. The application of dyes for bacterial detection using square wave voltammetry (SWV) has also been applied[58]. Dyes such as carbocyanine, 3,3'-dihexyloxacarbocyanine and safranin O have been investigated with concentrations in the range of 10^4 -10^8 cells ml^{-1} being detected.

1.5 Immuno - Sensor Configuration

Microbial detection using immuno- and affinity sensor configuration has been used for a range of microorganisms. Different types of transducers have also been applied in the development of these sensors.

1.5.1 Amperometric sensors Detection of microorganisms using amperometric transducers is widely used and it involves the measurement of the current produced through an oxidation/ reduction mechanism catalysed by microbial enzymes. Amperometric transducers have also been applied in affinity sensor format to detect micoorganisms where the antibody marker produce an electrochemical signal. Devices based on a flow through, immunofiltration and enzyme immunoassay in conjunction with an amperometric sensor was used for the detection of *E. coli*[59, 60]. An amperometric enzyme-channelling immunosensor has been developed[61] and was able to detect *S. aureus* cells in pure culture at concentrations of 1000 cells ml^{-1}. Ivnitski *et al.* [62] developed an amperometric immunosensor based on supporting planar lipid bilayer for the detection of *Campylobacter*. Sensors for *E coli* 0157 using the paramagnetic beads have been developed coupled with electrochemical detection[63]. The system was based on flow injection analysis (FIA) detection of viable bacteria. Antibody derivatized Dynabeads were used to selectively separate *E. coli* from the sample. The immunomagnetic separation was then used in conjunction with electrochemical detection to measure the concentration of viable bacteria[22]. A calibration curve of colony-forming units (cfu) against the electrochemical response was obtained with detection limit of 10^5 cfu ml^{-1} in a 2 h assay time including the bacterial concentration step[64]. Coupling flow injection analysis with immunosensor configuration is very attractive for on-line detection of microorganisms[65]. This type of system can be applied for on-line real time measurements for microbial risk management and assessment in drinking water.

1.5.2 Potentiometric sensors A light addressable potentiometric sensor (LAPS) based on field effect transistor (FET) has been used for the detection of microorganisms[60, 66]. A commercially available LAPS (Threshold® Immunoassay System) marketed by Molecular Devices (USA) uses silicon semiconductor with antibody as the receptor. An enzyme generates a potentiometric signal (urease) as the enzyme marker. This system has been used by several researchers for microbial cells detection[67, 68]. A cell concentration of 2.5 x 10^4 cells ml^{-1} of *E. coli* O157:H7 was detected using this system[69].

1.5.3 Acoustic wave- based sensors Acoustic wave based devices have been applied for microbial detection. The mass sensitive detectors operate on the bases of an oscillating crystal that resonates at a fundamental frequency. Antibodies are usually immobilised on the crystal surface and the sensor is then exposed to the sample containing the microorganism of interest. A change in the resonant frequency of the crystal surface related to the mass change is quantifiable and depends on the microbial concentration in the sample. This type of sensors offers label free and on-line analysis of microorganisms[70, 71]. A piezoelectric crystal immunosensor for the detection of enterobacteria in drinking water has been reported by Plomer *et al.* [72]. A piezoelectric biosensor for the detection of *Salmonella* [73], *Helicobacter pylori* [74], *Listeria monocytogenes* [75], *Legionella* and *E. coli* [76] has also been developed using antibodies as the receptor.

1.5.4 Optical sensors Optical transducers are attractive sensing systems since they allow real-time and label-free detection of microorganisms. Optical sensor based on surface plasmon resonance (SPR) detection and evanescent wave (EW) have shown promise in microbial detection[77, 78]. The BIAcore™ (BIAcore AB, Uppsala, Sweden) has been applied for *E coli* sensing and also for *Salmonella* and *Listeria* detection[78]. In these devices binding events are monitored between two molecules, such as an antibody and its antigen (in this case the microbial cell) using SPR technology. Direct microbial detection using the BIAcore™ achieved a detection limit for *E.coli* O157:H7 of 5 x 10^7 CFU ml^{-1} [79]. The IAsys® systems (Affinity Sensors, Cambridge, UK), which is also an optical biosensor based on resonant mirror has also been used for microbial detection by immobilising the antibody on the chip surface. The above devices use small samples volume and therefore high bacterial content in the water sample is required. Therefore, a pre enrichment (as in

ISO 11290-1[80)] and immunoseparation or concentration step is needed to enhance the detection limit of the sensors for pathogen detection[81, 82]. Watts *et al.* [77] reviewed optical biosensors for microbial cells monitoring. A new range of devices has emerged recently based on resonant mirror configuration and SPR and these are reviewed by Leonard *et al*[66].

1.6 Electronic Nose

Rapid development in sensor technology has enabled the production of senor array formats able to interact with different volatile molecules. The change in resistance or conductivity provide a signal which can be utilised effectively as a fingerprint of the volatiles produced by microorganisms. Metal oxide, conducting polymer and piezoelectric crystals have all been utilised in different array formats to try and qualitatively and semi-quantitatively obtain information on, and differentiate between volatile production patterns produced by different microorganisms. The results are usually combined with multivariate data analyses systems to enable rapid interpretation of volatile patterns. Although most electronic nose systems are qualitative for QA, the level of detail required determines the use of the instrument. If a simple yes/no answer is needed then it can be very appropriate. Recent studies have demonstrated that real-time evaluation can be made in approx. 10 min per sample for discrimination of fungal contamination[83, 84]. Recent studies have suggested that changes in bacterial populations in milk and water can be detected at between 10^3-10^4 CFUs per ml[85, 86]. The take up of this technology had been slow because of problems with consistency and the high price of the technology. However, as the development of sensor arrays becomes cheaper the potential for exploitation of this technology should improve rapidly. The E-nose technology should also be suitable for on –line detection.

1.7 Commercial Products and On-line monitoring

The European water industry produces about 900 000 000 tones of water per hour for domestic and industrial use[6]. It is estimated that 1100 waterworks in Europe have large numbers of samples routinely analysed for microbiological contamination. Therefore, automation and on site measurement is very important not only from public health perspective but also from economical perspectives.

Due to the increased need for detecting micoorganisms in water and food many devices have been developed and marketed. The success of any instrument is based on the level of detection and cost. Table 1 lists some of the commercial products available for microbial detection[22]. Most of the instruments developed for laboratory analysis are large with detection limits of 10^3 -10^5 cells ml^{-1} with analysis time ranging from 10 min to 8 hours. Smaller instruments are in demand especially for on site testing. On-line methods of microbial testing have been developed by many researchers[87, 88, 89]. Sample preparation is important in most cases to reduce interferences.

Physical, chemical and biological parameters associated with drinking water have been used to monitor water quality using grab sampling in source waters, treatment plants and distribution systems, followed by laboratory analysis and manual or computer-assisted data handling[6]. In order to develop an on-line monitoring system, automation of sampling, analysis and reporting functions have all had to be automated. The definition of an on -line system is that the system has to be unattended only require routine maintenance and also allow real-time feedback for water quality characterisation. Such a system could be used as an early warning system, but samples should still be taken for checking water quality. Many European countries today perform on-line monitoring throughout the entire water cycle from source to distribution. Most of these systems only monitor physical-chemical parameters such as turbidity, pH, particles, chlorophyll, dissolved oxygen, temperature,

DETECTION METHOD	DETECTION LIMIT (CELLS ML^{-1})	TIME OF ANALYSIS	COMMERCIAL INSTRUMENT
Bioluminescence	10^3	10-20 min	Clean-Trace™ (Biotrace Ltd., Bridgend, UK)
Electronic particle analysis	10^5	20 min	Ramus 265 (Orbec Ltd., Surrey, UK
Coulter counter	5 x 10^4	30 min	Coulter Counter Inc., Canada)
Enzymes	1 cfu 100ml^{-1}	24 h	Colilert (Palintest Ltd., Gateshead, UK)
Surface plasmon resonance (SPR)	10^5	1-2 h	BIAcore (Pharmacia, Uppsala, Sweden)
Epifluorescence	0.5 cfu	10-20 min	FOSS Electric, Hillerød, Denmark
Impedance	10^5	2.5-8 h	Bactometer (bioMerieux Inc., Marcy-'Etoile, France)
Electronic nose	10^3-10^4	1 h	Alpha MOS, France

Table 1 *Examples of commercial instruments available for microbial detection*[22]

etc. Example of this is the YSI deployment sensor system. However, on-line microbiological testing for water quality remains unrealised.

Monitoring of biological parameters can be carried out by either developing systems capable of detecting pathogens such as *Giardia*, *Cryptosporidium* and indicator organisms or chemical toxins. The literature reports a range of on-line systems for microbial detection, but these systems still need development for a realistic deployment in water testing. Problems such as sensitivity, reproducibility, and on-line capability still need to be resolved. Many of the bio and immunosensors covered in this paper can be developed for on-line monitoring coupled with a microbial concentration step to enhance the sensor sensitivity for drinking water analysis.

Systems based on particle characterisation using multi-angle, multi-wavelength sample analysis have been developed for on-line detection of pathogenic protozoa such as *Giardia* and *Cryptosporidium*. The resulting spectra from these types of systems allow the identification and quantification of particles having size range of 10 nm to 20 μm. This type of measurement approach uses UV-visible light adsorption and scattering at several observation angles. The scattering data are used to measure particle size and molecular weight distribution. While the absorption spectrum is used to estimate particle concentration, density and chemical composition[6]. Therefore *Cryptosporidium* oocysts (4 to 6 μm) and *Giardia* (8 to 15 x 10 μm) can be identified in the sample. However, interfering particles need to be identified accurately for this system to be accepted to eliminate the high incidents of false negative or positive results. Also the growth phase and age of the microbial cells also need to be characterised to increase the specificity of the system[90].

The multiangle light scattering (MALS) technique which is based on simultaneous measurement of various scattered light angles has been used to detect microorganisms such as *E. coli, Cryptosporidium,* phytoplankton and algae. The water flows through the sensing device and particles passing through the laser beam generate a light-scattering fingerprint which can be compared with an existing library of optical fingerprints. Through this analysis particles can be identified instantaneously. The device is still being characterised for different type of water as well as its ability to detect morphological changes of microbial cells such as *Cryptosporidium parvum*[93].

Most of the approaches listed above are still under development and further systems characterisation need to be finalised before they can be implemented for on-line analysis. Ideal methods of analysis should be capable of real time and *in situ* analysis.

1.8 Conclusions and Future trends

There have been considerable efforts to develop new and advanced techniques for the rapid detection of microbial contamination in water. This is to ensure the full protection of drinking water by implementing technologies based on early warning monitoring system. However, most of the methods are complex require expensive equipment and specialised expertise. The range of microorganisms that needs to be detected and the type of water requiring analysis also make it difficult to come up with a single standard procedure. Classical methods usually require long analysis times but give good sensitivity. Recently developed rapid methods may give faster results but usually suffer from low sensitivity. To improve sensitivity in drinking water analysis, attention should be given to concentration and separation methods. Lack of discrimination between viable and non-viable biomass can also result in errors. There are several technologies available today for microbial detection, but their feasibility as early warning systems still need to be investigated. Methods based on enzyme detection and ATP analysis may prove to be good for early warning systems but automation and sample concentration will be required for on-line real time systems. Molecular methods such as the DNA microchip could revolutionise water testing, but they still a long way from on-line robust technology systems. Using techniques based on image analysis, robotics and biosensors, the possibility of rapid, automated monitoring for specific microorganisms is possible in the foreseeable future.

References

1 Drinking Water Inspectorate for England and Wales (2001). Comparison of microbiological methods of analysis, organisation and supervision of performance tests. DWI Contract 70/2/128.

2 Drinking Water Inspectorate for England and Wales (2000). Comparison of methods for drinking water bacteriology- Cultural Techniques. http://www.dwi.gov.uk/regs/infolett/2000/info500.htm (See also reference 3)

3 ISO 17994 (2004). Water Quality – Water quality -- Criteria for establishing equivalence between microbiological methods

4 ISO TR 13843 (2000). Water Quality – Guidance on validation of microbiological methods, Technical report, International Standards Organisation.

5 Fricker, C.R. (2002). Protozoan parasites (*Cryptosporidum, Giardia, Cyclospora*), in World Health Organisation, Guidelines for Drinking-water quality. Addendum Microbiological agents in drinking-water , Geneva, World Health Organisation, 129-143.

6 Bonadonna, L. (2003) Rapid analysis of microbial contamination of water. In Rapid and on-line Instrumentation for Food Quality Assurance, Tothill, I.E. (Editor), CRC and

Woodhead Publishing Limited. pp 161-182.
7 Singh, A., Kuhad, R.C., Sahai, V. and Ghosh, P. (1994). Evaluation of biomass. Adv. Biochem. Eng. Biotechnol., 51, 48-66.
8 Park, S.J., Lee, E-J., Lee, D-H., Lee, S-H. and Kim, S.J. (1995). Spectrofluorometric assay for rapid detection of total and faecal Coliforms from surface water. Appl. Environ. Microbiol. 61, 2027-2029.
9 Dart, R.K. (1996). Microbiology for the analytical chemist. Chapter 5, 7 and 11. The Royal Society of Chemistry, Cambridge, UK.
10 Allen, M.E. (1990). Applications for mediated amperometric biomass sensor technology. M.Phil.Thesis, Cranfield Biotechnology Centre, Cranfield University Bedford, UK.
11 Matsunaga, T., Okochi, M. and Nakasono, S. (1995). Direct Count of bacteria using fluorescent dyes: Application to assessment of electrochemical disinfection. Anal. Chem. 67, 4487-4490.
12 Armiger, N.B., Zabriski, D.W., Meanner, G.F. and Forro, T.F. (1984). Analysis and process control of feed batch production of *E. coli* culture fluorescence. Presented at Biotech. 1984, Washington, DC. On-line publications, Pinner, UK.
13 Nakamura, N., Grant Burgess, J., Yagiuda, K., Kudo, S., Sakaguchi, T. and Matsunaga, T. (1993). Detection and removal of *Escherichia coli* using fluorescein isothiocianate conjugated monoclonal antibody immobilised on bacterial magnetic particles. Anal. Chem. 65, 2036-2039.
14 Gentelet, H., Carricajo, A., Rusch, P., Dow, C. and Aubert, G. (2001). Evaluation of a new rapid urine screening analyser: CellFacts. Pathologie Biologie, 49, 262-264.
15 Clarke, D.J. Calder, M.R., Carr, R.J.G., Blake-Coleman, B.C., Moody, S. C. and Collinge, T.A. (1985). The development and application of biosensing devices for bioreactor monitoring and control. Biosensors, 1, 213-320.
16 Jepras, R.I, Carter, J., Pearson, S.C., Paul, F.E., and Wilkinson, M.J. (1995). Development of a robust flow cytometric assay for determination numbers of viable bacteria. Appl. Env. Microbiol. 61, 2696-2701.
17 Attfield, P.V., Gunasekera, T.S., Boyd, A., Deere, D. and Veal, D.A. (1999). Application of flow cytometry to microbiology of food and beverage industries. Australian Biotechnology, 9,159-166.
18 Okada, H.,Sakai, Y., Miyazaki, S. Arakawa, S., Hamaguchi, Y. and Kamidono, S. (2000). Detection of significant bacteriuria by automated urinalysis using flow cytometry. J. Clinical Microbiology, 38, 2870-2872.
19 Davey, H.M. and Kell, D.B. (2000), A portable flow cytometer for the detection and identification of microorganisms; in Rapid methods for analysis of biological materials in the environment, Stopa, P.J. and Bartoszcze, M.A., (editors) The Netherlands, Kluwer Academic Publisher, 159-167.
20 Borsheim, M., Bratbak, G. and Heldal, M. (1990). Enumeration and biomass estimation of planktonic bacteria and viruses by transmission electron microscopy. J. Appl. Environ. Microbiol., 56, 352-356.
21 Iqbal, S.S., Mayo, M.W., Bruno, J.G., Bronk, B.V., Batt, C.A., Chambers, P. (2000). A review of molecular recognition technologies for detection of biological threat agents. Biosens. Bioelectron., 15, 549-578.
22 Tothill, I.E. and Magan, N. (2003). Rapid detection methods for microbial contamination. 136-160. In: Rapid and on-line instrumentation for food quality assurance, I.E.Tothill (Ed). CRC and Woodhead Publishing. Cambridge, England.

23 Galnous, D. S. and Kapoulos, A. (1966). A rapid method for the determination of organic nitrogen and phosphorus based on a single perchloric acid digestion. Anal. Chim. Acta. 34, 360-366.
24 Garg, S.K. and Neelkantan, S. (1982). Production of SCP and cellulase by *Aspergillus terreus* from Bagasse Substrate. J. Biotechnol. Bioeng., 24, 2407-2417.
25 Magan N (1993). Early detection of fungi in stored grain. Int. Biodet. Biodeg. 32: 145-160.
26 Tothill IE, Harris D and Magan N (1993). The relationship between fungal growth and ergosterol in wheat grain. Mycol Res 11: 965-970.
27 Hernandez, J.F., Guibert, J.M. Delattre, J.M, Oger, C., Charriere, C. Hughes, B., Serceau, R. and Sinegre, F. (1991), Miniaturised fluorogenic assays for enumeration of *Escherichia coli* and enterococci in marine water, Wat. Sci. Tech, 24,137-141.
28 Budnick, G.E., Howard, R.T., and Mayo Dr (1996). Evaluation of Enterolert for enumeration of enterococci in recreational waters, Appl. Environ Microbiol., 26, 332-336.
29 Hobson, N.S. (1996) Development of an amperometric biosensor for the clinical diagnosis of bacteriuria. PhD. Thesis, Cranfield Biotechnology Centre, Cranfield University, Bedford, UK.
30 Keshri, G. and Magan, N. (2000). Detection and differentiation between mycotoxigenic and non-mycotoxigenic strains of *Fusarium* spp. using volatile production profiles and hydrolytic enzymes. *Journal of Applied Microbiology* **89**, 825-833
31 Keshri, G., Vosey, P. and Magan, N. (2002). Early detection of spoilage moulds in bread using volatile production patterns and quantitative enzyme assays. *Journal of Applied Microbiology* **92,** 165-172.
32 Batchelor, B.I.F. (1995). Identification of urinary tract infection: an overview. American J. of Medical Science. 314, 245-249.
33 Knight, I.T., DI Ruggiero, J. and Colwell, R.R. (1991), Direct detection of enteropathogenic bacteria in estuarine water using nucleic acid probes, Wat. Sci. Tech, 24, 262-266.
34 Abbaszadegan, M., Gerba, C.P. and Rose, J.B.(1991). Detection of *Giardia* cysts with a cDNA probe and applications to water samples, Appl. Environ. Microbiol., 57, 927 - 931.
35 Dubrou, S., Kopecka, H. Lopez, Pila, J.M., Marechal, J. and Prevot, J. (1991). Detection of Hepatitis A virus and other enteroviruses in wastewater and surface water samples by gene probe assay, Wat. Sci. Tech, 24, 267-272.
36 Manz W., Szewzyk U., Ericsson, P. Amann R., Schileifer K.H. and Stenstrom T. (1993). In situ identification of bacteria in drinking water and adjoining biofilms by hybridization with 16S and 23S rRNA-directed fluorescent oligonucleotie probes, Microbiol, 59, 2293-2298.
37 Manz W., Amann R., Szewzyk R., Szewzyk U., Stenstrom T.A., Hutzler, P.and Schileifer K-H (1995). In situ identification of *Legionellaceae* using 16S rRNA-targeted oligonucleotide probes and confocal laser scanning microscopy, Microbiol., 141, 29-39.
38 Richardson, S. (2001), Water Analysis, Anal Chem. 73,2719-2734.
39 Baker, G.C., Tow, L.A and Cowan, D.A. (2003). PCR-based detection of non-indigenous microorganisms in pristine environments. J. Microbiol. Methods, 53, 157-164.
40 Cook, N. (2003). The use of NASBA for the detection of microbial pathogens in food and environmental samples. J. Microbiol. Methods, 53, 165-174.

41 Paton, A. W., Paton, J. C., Goldwater, P.N. and Manning, P.A. (1993). Direct detection of *Escherichia coli* shiga-like toxin genes in primary faecal cultures by polymerase chain reaction. J. Clin. Microbiol., 31, 3063- 3067.
42 Graff, J., Ticehurst, J. and Flehmig, B. (1993). Detection of hepatitis A virus in sewage sludge by antigen capture polymerase chain reaction, Appl. Environ. Microbiol., 59, 3165- 3170.
43 Traore, O., Arnal, C., Mignotte, B., Maul, A., Laveran, H., Billaudel, S.(1998). Reverse transcriptase PCR detection of astrovirus, hepatitis A virus and poliovirus in experimentally contaminated mussels: comparison of several extraction and concentration methods. Appl. Environ. Microbiol., 64,31, 18-22.
44 Fach, P. and Popoff, M.R. (1997). Detection of enterotoxigenic *Clostridium perfringens* in food and faecal samples with duplex PCR and slide latex agglutination test. Appl. Environ. Microbiol., 63, 4232-4236.
45 Alvarez, A.J. Hernandez-Delgado E.A. and Toranzos G.A. (1993). Advantages and disadvantages of traditional and molecular techniques applied to the detection of Pathogens in water, Wat, Sci, Tech, 27, 253-256.
46 Stinear, T., Matusan, A. Hines, K. and Sandery, M. (1996). Detection of a single viable *Crypotosporidium parvum* oocyst in environmental water concentrates by reverse transcription-PCR. Appl. Environ. Microbiol., 62, 3385-3390.
47 Mayer, C.L. and Palmer, C.L. (1996). Evaluation of PCR nested PCR and fluorescent antibodies fro detection of Giardia and Cryptosporidium species in wastewater. Appl. Environ. Microbiol. 62, 2081-2085.
48 Kaucner C. and Stineart T. (1998). Sensitive and rapid detection of viable *Giardia* Cysts and *Cryptosporidium Parvum* oocyts in large volume water samples with wound fibreglass cartridge filters and reverse transcription-PCR, Appl. Environ Microbiol, 64, 1743-1749.
49 Paddle, B. (1996). Biosensors for chemical and biological agents of defence interest. Biosensors & Bioelectronics, 11, 1079-1113.
50 Richards, J.C.S., Jason, A.C., Hobbs, G., Gibson, D.M., and Christie, R.H. (1978). Electronic measurement of bacterial growth. J. Phys., 11, 560-568.
51 Colquhoun, K.O.,Timms, S. and Fricker, C.R. (1995). Detection of Escherichia coli in potable water using direct impedance technology. J. Appl. Bacteriol.,79, 635-639.
52 Davis, J.G. (1963). Generation of electricity by microbial action. J. Adv. Appl. Microbiol., 5, 51-64.
53 Benjamin, T.G., Camara, E.H. and Selman, J.R. (1979). Fuel cell efficiency. Proc. 14th Intersoc. Energy Conservation Conference, 1, 579-582.
54 Matsunaga, T., Karube, I. and Suzuki, S. (1980). Electrochemical determination of cell populations. Eur. J. Appl. Microbiol., 10, 125-132.
55 Kal'ab, T. and Skl'adal, P. (1994). Evaluation of different mediators for the development of amperometric microbial bioelectrodes. Electroanalysis, 6, 1004-1008.
56 Hitchens, G.D., Hodko, D., Miller, D.R., Murphy, O.J. and Rogers, T.D. (1993). Bacterial activity measurement by mediated amperometry in a flow injection system. Russ. J. Electrochem. 29, 1527-1532.
57 Thomson, C.R., (2001). The evaluation of an amperometric biosensor for the detection of bacteriuria. MSc Thesis, Institute of Bioscience and Technology, Cranfield University, Bedford, UK.
58 Lafis, S. (1992). Rapid Microbial Detection. Ph.D. Thesis, Cranfield Biotechnology Centre, Cranfield University, Bedford, UK.

59 Abdel-Hamid, I., Ivnitski, D., Atanasov, P. and Wilkins, E. (1999). Flow-through immunofiltration assay system for rapid detection of *E. coli* O157: H7. Biosensors & Bioelectronics, 14, 309-316.
60 Ivnitski, D., Abdel-Hamid, I., Atanasov, P. and Wilkins, E. (1999). Biosensors for detection of pathogenic bacteria, Biosensors & Bioelectronics, 14, 599-624.
61 Rishpon, J. and Ivnitski, D. (1997). An amperometric enzyme-channelling immunosensor. Biosensor & Bioelectronics, 12, 195-204.
62 Ivnitski, D., Wilkins, E., Tien, H.T. and Ottova, A. (2000). Electrochemical biosensor based on supported planar lipid bilayers for fast detection of pathogenic bacteria. ELECOM, 2, 457-460.
63 P'erez, F.G (1998). Immunosensor for the detection of viable microorganisms. Ph.D. Thesis, Cranfield Biotechnology Centre, Cranfield University, Bedford, UK.
64 P'erez, F.G., Mascini, M., Tothill, I.E. and Turner, A.P.F. (1998). Immunomagnetic separation with mediated flow injection analysis amperometric detection of viable *Escherichia coli* O157. Analytical Chemistry, 70, 11, 2380-2386.
65 Bouverette, P. and Luong, J.H.T. (1995). Development of a flow injection analysis (FIA) immunosensor for the detection of *Escherichia coli*. Int. J. Food. Microbiol. 27, 129-137.
66 Leonard, P., Hearty, S., Brennan, J., Dunne, L., Quinn, J., Chakraborty, T. and Kennedy, R.O. (2003). Advances in biosensors for detection of pathogens in food and water. Enzyme and Microbial Technology, 32, 3-13.
67 Dill, K., Song, J.H., Blomdahl, J.A. and Olson, J.D. (1997). Rapid, sensitive and specific detection of whole cells and spores using the light-addressable potentiometric sensor. J. Biochem. Biophys. Methods. 34, 161-166.
68 Dill, K., Stanker, L.H. and Young, C.R. (1999). Detection of *Salmonella* in poultry using a silicon chip-based biosensor. J. Biochem. Biophys. Methods, 41, 61-67.
69 Gehring, A.G. Patterson, D.L., and Tu, S. (1998). Use of a light addressable potentiometric sensor for the detection of *Escherichia coli* O157:H7. Anal. Biochem., 258, 293-298.
70 Bunde, R.L., Jarvi, E.J. and Rosentreter, J.J. (1998). Piezoelectric quartz crystal biosensors. Talanta, 46, 1223-1236.
71 Babacan, S., Pivarnik, P., Letcher, S. and Rand, A.G. (2000). Evaluation of antibody immobilisation methods for piezoelectric biosensor application. Biosens. Bioelectron., 15, 615-621.
72 Plomer, M., Guilbault, G.G. and Hock, B. (1992). Development of piezoelectric immunosensor for the detection of enterobacteria. Enzyme Microb. Technol., 14, 230-235.
73 Pathirana, S.T., Barbaree, J., Chin, B.A. Hartell, M.G., Neely, W.C. and Vodyanoy, V. (2000). Rapid and sensitive biosensors for *Salmonella*. Biosens. Bioelectron., 15, 135-141.
74 Su, X., and Li, S.F.Y. (2001). Serological determination of *Helicobacter pylori* infection using sandwiched and enzymatically amplified piezoelectric biosensor. Anal. Chem. Acta., 429, 27-36.
75 Vaughan, R.D., O'Sullivan, C.K., Guilbault, G.G. (2001). Development of a quartz crystal microbalance (QCM) immunosensor for the detection of *Listeria monocytogenes*. Enzyme Micob. Tech., 29, 635-638.
76 Howe, E. and Harding, G. (2000). A comparison of protocols for the optimisation of detection of bacteria using a surface acoustic wave (SAW) biosensor. Biosens. Bioelectron., 15, 641-649.

77 Watts, H.J., Lowe, C.R. and Pollard_Knight, D.V. (1994). Optical biosensors for monitoring microbial cells. Anal. Chem., 66, 2465-2470.
78 Haines, J., Patel, P.D. and Wahlströn, L. (1995). Detection of the foodborne pathogens, *Salmonella* and *Listeria* using BIAcore biosensor. Fifth European BIAsymposium, Stockholm, Sweden.
79 Fratamico, P.M. (1998) Detection of *Escherichia coli* O157:H7 using a surface plasmon resonance biosensor. Biotechnol. Tech., 12, (7), 571-576.
80 - ISO 111290-1. Microbiology of food and animal feeding stuffs -- Horizontal method for the detection and enumeration of Listeria monocytogenes -- Part 1: Detection method)
81 Kaclikova, E. Kuchta, T. Kay, H. and Gray, D.(2001). Separation of *Listeria* from cheese and enrichment media using antibody-coated microbeads and centrifugation. J. Immunol. Methods, 46, 63-67.
82 Quinn, J.G. and O'Kennedy, R. (2001). Detection of whole cell:antibody interactions using BIAcore's SPR technology. BIA J, 1, 22-24.
83 Magan, N. and Evans, P. (2000). Volatiles in grain as an indicator of fungal spoilage, odour descriptors for classifying spoiled grain and the potential for early detection using electronic nose technology: A review. *Journal of Stored Product Protection* **36**, 319-340.
84 Evans, P., Persaud, K.C., McNeish, A.S., Sneath, R.W., Hobson, N. and Magan, N. (2000). Evaluation of a radial basis function neural network for determination of wheat quality from electronic nose data. *Sensors and Actuators B* 69, 348-358.
85 Magan, N., Pavlou, A. & Chrysanthakis, I. (2001). Milke sense: a volatile sensory system for detection of microbial spoilage by bacteria and yeasts in milk. *Sensors and Actuators B*, **72**, 28-34.
86 Canhoto, O. and Magan, N. (2003). Potential for the detection of microorganisms and heavy metals in potable water using electronic nose technology. *Biosensors and Bioelectronics*. 18, 751-754.
87 Ashley, N. (1991), On-line microbial detection. Dairy Industries International, 56, 39-43.
88 Chunxiang, X., Gang, L. Haobin, C. and Yue, X. (1993). Microbial population sensor for on-line determination of microbial populations in a fermenter. Sensors and Actuators B, 12, 45-48.
89 Silley, P. (1994). Rapid Microbiology -is there a future? Biosensors & Bioelectronics, 9, 15-21.
90 Compagnon, B., Robert, C., Mennecart, V., De Roubin M.R., Cervantes, P. and Joret, J.C. (1997). Improved detection of *Giardia* cycts and *Cryptosporidium* oocysts in water by flow cytometry, in Proceedings of the AWWA Water Quality Technology Conference, Denver, Co, Am Water Work Ass.
91 Gregg, M. (2000). Real- time on-line monitoring for protozoa in drinking water, in Proceedings 2000 of the AWWA Water Quality Technology Conference, Denver, Co, Am Water Work Ass.

REAGENTLESS DETECTION OF CB AGENTS

H. J. Harmon[1], A. Oliver[1], and B. Johnson-White[2]

[1]Department of Physics, Oklahoma State University, Stillwater, OK, 74078 USA
[2]Center for Bio/Molecular Science & Engineering, Naval Research Laboratory, Washington, DC 20375 USA

1. INTRODUCTION

The ability to detect agents of mass destruction in chemical, biological, nuclear, or explosive/energetic form is of great importance. While the military need is apparent, civil authorities such as police and fire departments, medical workers including emergency technicians and emergency room physicians, hazardous response teams, and the citizenry also need detection capabilities. An act of terrorism could be directed against infrastructure capabilities such as transportation centers but has a far more disruptive effect on common services, such as the maintenance of a potable water supply. Water supply integrity can also be compromised or challenged by more conventional chemical species such as toxic industrial chemicals and pesticides/herbicides. These compounds as well as the more conventionally-viewed weapons of mass destruction (WMD) can be used in a terrorist scenario which can disrupt lives and erode confidence in a crucial municipal system.

Monitoring and detection of the water supply is needed to determine chemical use as well as assure that the chemicals have been removed by prudent remediation techniques and water treatment processes. The detection procedures used therefore must be as near real-time as possible and be as sensitive as possible; the detection monitoring system must warn of the presence of harmful agents so that appropriate measures can be taken and then clearly indicate that they have been removed to acceptable healthy levels. Detecting "classes" of compounds as opposed to specific compounds may be desirable since the exact nature of a chemical challenge will be unknown. Which neurotoxic agent will be used? If you have a sensor that is specific for an agent such as Sarin, will you be able to detect a more available compound such as an organophosphate pesticide? If you have a sensor that is specific for an agent such as Sarin, it may not detect VX. These questions suggest the value of a broad spectrum detector that may be used to signal further evaluation by more analyte-specific means. In this communication reagentless real-time sensors will be described that can detect pesticides, cyanide, toxic industrial chemicals (TICs), and chemical warfare agents such as Sarin) in water at or below acceptable and healthy standards for drinking water.

2. MATERIALS AND RESULTS

2.1 Experimental Set-up

The sensor has been described in great detail elsewhere[1] and consists of 3 basic systems: 1) an optical system to deliver light to and from 2) a surface containing a near-monolayer of immobilized enzyme or protein and 3) an optical spectrometer.

The enzyme or protein that is immobilized is one that binds the chemical agent in question at the catalytic active site of that enzyme or protein. For example, acetyl- or butyrylcholinesterase (AChE or BChE) is used to detect nerve agents such as Sarin or VX; organophosphate hydrolase (OPH) is used to detect organophosphates such as diazinon; myoglobin or hemoglobin can be used to detect H_2S, CO, or cyanide. In each of these enzyme/proteins, the analyte to be measured binds at the active site of the enzyme and therefore exhibits great selective specificity for a small number of compounds. Because an enzyme might be able to bind different analytes (AChE can bind VX, Sarin, Tabun, Soman, organophosphates, and drugs such as drofenine and eserine, for example); the sensor is a "broad spectrum" detector that will indicate the presence of any of a family of compounds that inhibit the cholinesterase by binding at the active site. Any compound that binds at a site other than the active site of a cholinesterase will not be detected.

A porphyrin (Figure 1) is then incorporated into the active site of the immobilized (and catalytically-active) enzyme. The porphyrins that we use are themselves inhibitors of the enzymes and bind at the active site[2]. In the presence of the appropriate analyte (Sarin, for example), the analyte binds to the active site of the enzyme and *displaces* the porphyrin from its location as shown in Figure 2. As we have indicated previously[2], only those analytes that bind at the active site can displace the porphyrin. Porphyrins are highly colored compounds (chlorophyll is one) whose absorbance spectrum changes in response to its immediate surroundings[3,4]; it is an effective and sensitive indicator of its microenvironment. Displacement of the porphyrin from the active site into another microenvironment will result in a change in the absorbance intensity as well as wavelength spectrum of the porphyrin; this spectral change is readily measured by a spectrophotometer such as an Ocean Optics (Dunedin, FL) unit.

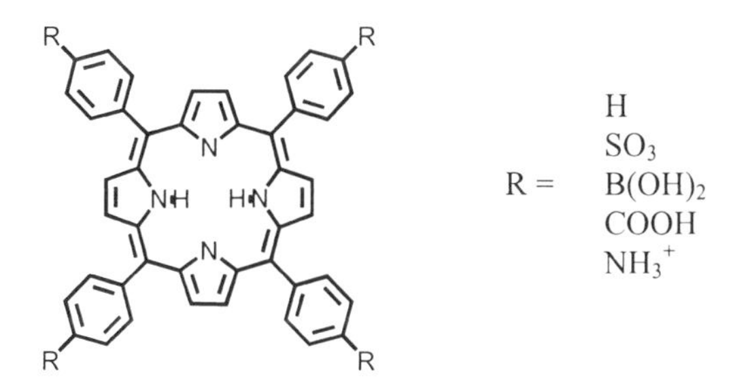

Figure 1 *Porphyrin structure*

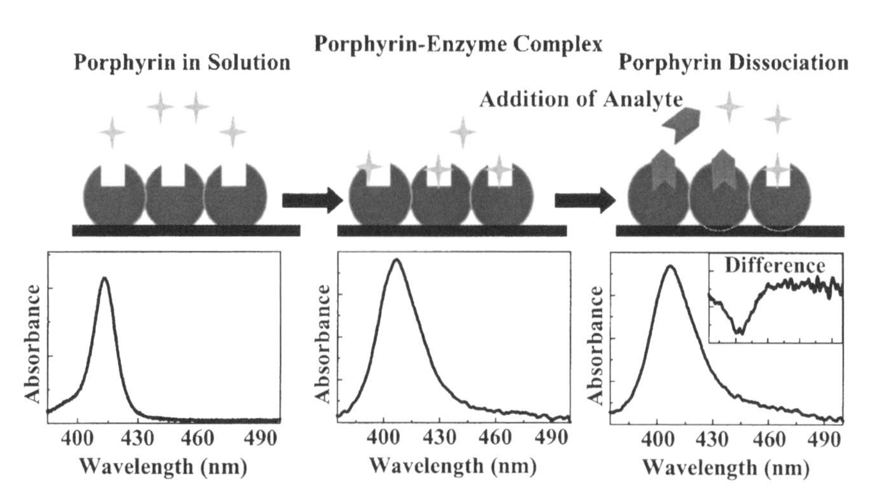

Figure 2 *Diagrammatic representation of 1) the porphyrin and enzyme (blue) the spectrum of the non-complexed porphyrin in solution; 2) binding of porphyrin to yield the enzyme porphyrin complex with the porphyrin bound at the active site and the absolute spectrum of the porphyrin, and 3) displacement of the porphyrin from the active site into the medium by an inhibitor of the enzyme and the absolute spectrum of the porphyrin as well as the difference spectrum of the complex in the presence of inhibitor minus the porphyrin complex.*

Light from a blue-emitting LED illuminates the side of a glass slide or other light transmitting surface such that the slide acts as a planar waveguide. A portion of the light enters the medium to a depth of about 100 nm, illuminating the surface layer of immobilized enzyme-porphyrin complex, and is propagated across the slide to an optical wave guide or directly to the spectrometer itself. The waveguide is positioned to block light transmitted *through* the slide by total internal reflection such than only the evanescent light propagated across the surface is captured and detected.

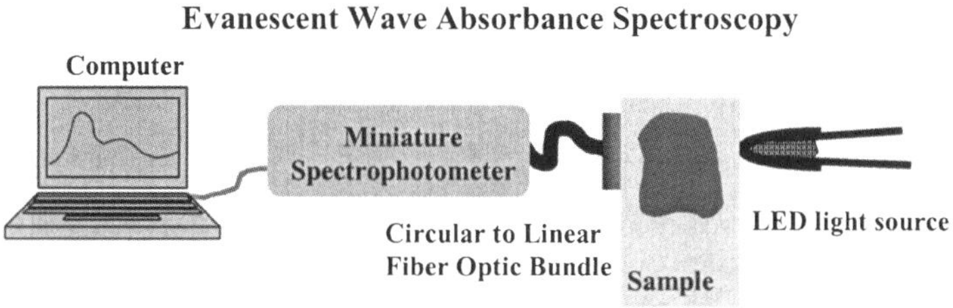

Evanescent Wave Absorbance Spectroscopy

Computer

Miniature Spectrophotometer

Circular to Linear Fiber Optic Bundle

Sample

LED light source

Figure 3 *The experimental setup.*

In all measurements, the absorbance spectrum of the porphyrin bound to the enzyme is measured and stored digitally. The water sample is placed on the slide for a few seconds and removed by capillary action into a tissue. The spectrum of the pre-exposure slide is subtracted from that of the exposed surface to reveal the change in porphyrin spectrum via a difference spectrum.

2.2 Detection of Cholinesterase Inhibitors

AChE[1], BChE[5], or both[6] can be immobilized and retain enzymatic activity as well as the ability to bind inhibitors at the active site. Monosulfonato-tetraphenyl porphyrin ($TPPS_1$) is incorporated as the colorimetric reporter molecule. A typical spectrum of the porphyrin before and after exposure to sarin as well as the $TPPS_1$-AChE + sarin *minus* $TPPS_1$-AChE difference spectrum is shown in Figure 4A. The difference spectrum shows a trough or loss of absorbance at the wavelength absorbed by the porphyrin when in the enzyme and a peak or increased absorbance at a different wavelength; this is due to the change in the microenvironment of the porphyrin as it is displaced which is manifested in a change in absorbance spectrum. As shown in Figure 4B, the absorbance change is proportional to the concentration of analyte present.

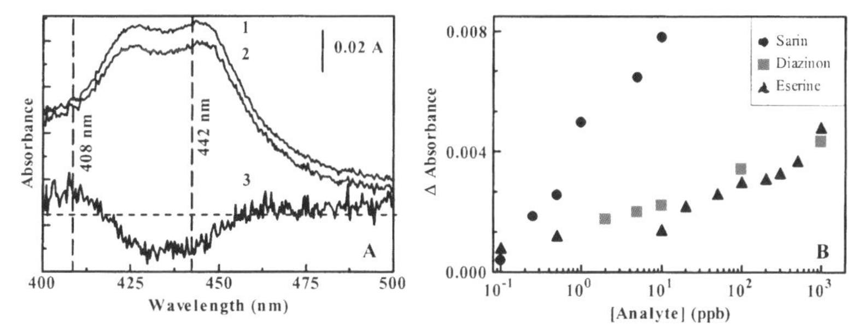

Figure 4 *A. Absorbance spectra of AChE surface before (1) and after (2) exposure to sarin and the difference between the two spectra (3). B. Concentration dependence for detection by the AChE surface shown for sarin, diazinon, and eserine.*

The AChE surfaces have been shown to be effective in detecting the presence of Sarin (GB) in both aqueous and air media. Sarin (GB) in air can be measured above 25 ng levels and in liquid down to 100ppt[7].

2.3 Detection of Organophosphate Compounds

Organophosphate hydrolase breaks down organophosphates and can be used to detect organophosphates such as paraoxon, malathion, diazinon, and coumaphos at 7 ppt, 1 ppb, 800 ppt, and 250 ppt levels, respectively[8]. These compounds can also inhibit the cholinesterase compounds.

For a specific enzyme, binding of its inhibitor or substrate (in the case of OPH) will result in a change in the absorbance at a particular wavelength. If AChE is used, the changes will always be at 446 nm. However, with BChE, the changes are at 421 nm and with OPH the changes are at 412 nm. Thus, any member of the inhibitor family will elicit a response at the same wavelength from the same enzyme. However, different analytes bind to different enzymes. Some compounds will bind to OPH but not to cholinesterases and some will bind to AChE and not to BChE [6]. This allows us to differentiate between inhibitors on a limited scale. Binding of an analyte to AChE results in absorbance changes at wavelengths different than if the analyte were to bind at BChE or OPH. Thus, with 2 or more different enzymes immobilized on the same surface, the response of the different enzymes to the analyte(s) can be determined and quantified for each enzyme as though they were present alone; the enzymes operate independently of each other [6].

2.4 Other Enzyme Systems

This same general procedure has been used to measure glucose in water using glucose oxidase [9] and CO_2 using carbonic anhydrase [10]. Recently, we have been able to utilize the (serine protease-type) enzyme carboxypeptidase to indicate the presence of the nerve agent stimulant and inhibitor of cholinesterases, eserine. The limit of detection, as seen in Figure 5, is about 10 ppt, similar to that determined using AChE. It must be noted that the wavelength changes for AChE and carboxypeptidase are different; their combined use on a single surface gives corroborative measurements of the analytes.

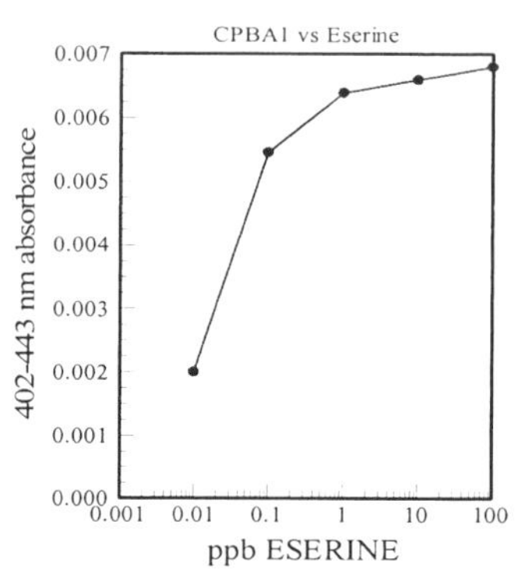

Figure 5 *Peak minus trough absorbance changes at 402 and 443 nm of $TPPS_1$ incorporated into carboxypeptidase as a function of eserine concentration.*

Heme proteins and cytochromes contain Fe-porphyhrins. Myoglobin and hemoglobin are known to not only bind molecular oxygen but also combine with cyanide, CO, azide, F^-, S^-, and other molecules such as NO_x with resultant changes in their absorbance spectra. We have been able to immobilize myoglobin and use it for the detection of cyanide in gaseous and aqueous phases with detection limits down to 1 ppm and 1.5 ppb, respectively [11].

2.5 Advantages of Using Evanescent Measurements

Since the surface contains the indicator(s) as well as the light used for measurement, the measurements are not affected by turbulence, bubbles, turbidity, or color of the solution being measured since the measuring spectroscopic light beam does not cross the medium. The sensor surface can be incorporated into a pipe or other container of any diameter; with wide pipes, the intensity of light will decrease due to the long path and interference by materials contained in the medium (colored material, solids, dirt, etc). In addition, the propagation of light along the surface results in a long pathlength which increases sensitivity and decreases noise (increases S/N ratio).

2.6 Advantages of a Binding Assay

Many commonly used enzyme-based sensors detect toxic or inhibitory agents such as nerve agents by measuring the decrease in enzymatic activity of unaffected enzyme [12-19]. For example, the presence of nerve agents has been detected by their inhibition of acetylcholinesterase activity. Here, the rate of the enzyme in the presence and absence of the inhibitor is measured and compared. Such a system requires that a substrate for the esterase be present as well as a properly buffered solution; the assay is frequently not made in real time since two enzymatic rates must be measured with the time needed to measure the activity dependent on the amount of agent present. Further, the measurements should be made more than once for reliability.

Other enzymatic assay systems utilize two enzymes. For example, glucose oxidase oxidizes glucose and produces hydrogen peroxide which is then the substrate for a peroxidase [20].

Yet another type of assay utilizing enzymes is based, as in the case of OPH, on the rate of reaction involving the analyte (OP) as substrate or co-factor. Here, a single measurement (in duplicate) is made to determine the concentration of analyte based on standard curves of concentration vs. activity, but at lower the levels of analyte concentration, longer measurement time is needed to reach measurable product levels. Again, this precludes real-time measurements, particularly at low analyte concentrations.

Each of the aforementioned techniques, regardless of whether the enzyme is immobilized or not, requires catalysis. The detection protocols described here, however involve only the binding of the analyte to the enzyme and are not dependent on the enzymatic catalytic constant. Since the enzyme is a thin layer, diffusion of the analyte to the active site is limited and the rate of binding is extremely rapid. The extent of binding is a function of the association constants of the porphyrin and the competing analyte. In general, as shown previously [5, 21], the extent of porphyrin displacement from the enzyme is greater for those molecules with a lower IC_{50} value and implied affinity for the enzyme active site. It must be emphasized that inhibitors or molecules that do not bind at the active site do not displace the porphyrin [2].

As with other enzyme-based assays, the specificity of the enzyme for its substrate or inhibitor is the primary determinant of specificity. This specificity rivals that of an antibody for its antigen and is specific enough to distinguish between molecules differing in only one atom as demonstrated with glucose oxidase [8]. Two additional levels of specificity in the measurements described here are afforded by the wavelength changes of the porphyrin. First, the wavelength changes are specific for a particular enzyme porphyrin complex. Second, a simultaneous loss of absorbance at the wavelength corresponding to the enzyme-porphyrin complex is observed at the same time an absorbance increase is seen at the wavelength corresponding to the free porphyrin; this results in a "coincidence" measurement and absorbance changes that are at incorrect wavelengths or at only one wavelength can be disregarded.

2.7 Detection of Other Organic Compounds

Not all compounds have a corresponding enzyme that can be used in the enzyme-based technique. For those molecules, a surface of immobilized porphyrins can be made. The interaction or "docking" of the analyte with the porphyrin distorts the pi-electron rings above and below the porphyrin resulting in a change in the absorbance spectrum [3,4]. Further, the stronger the association between organic molecule and the porphyrin, the greater the shift in absorbance spectrum [22]. As shown previously, the absorbance at the wavelength of the parent porphyrin decreases and the absorbance at a new wavelength of the analyte-porphyrin complex increases as the complex forms. In this system, the interaction of the analyte with the porphyrin is measured; in the enzyme-based system, the interaction of the porphyrin with the enzyme is measured.

This has been shown to be an effective means for the detection of methylphosphonic acid (the breakdown product of nerve agents), the sulfur mustard agent precursor thiodiglycol [23], and the pesticide pentachlorophenol (PCP). PCP can be detected down to 0.5 ppb levels in water in seconds using immobilized porphyrin on glass or cellulose (tissue) fibers [24]. White, et al. [25] were able to show the detection of dipicolinic acid, a major constituent of *Bacillus* and *Chlostridium* spores using amino- and sulfonate-derivative porphyrins. This concept of spore identification by identifying or detecting its components has been advanced to include the detection of the sugar/carbohydrate components of the spore as well using meso-tetraboronic acid porphyrin [26].

While myoglobin can be used to detect cyanide in solution, Cu-porphyrin is also effective in detecting NaCN in water down to 1.6 ppb levels in 6 seconds [11].

Meso-tetraphenylsulfonate porphyrin (TPPS) has also been used to detect the presence of amino acids in solution down to 20 nM levels [27]. Arginine, glycine, histidine, and serine bind with TPPS and give separate unique spectral shifts in the absorbance spectrum of the porphyrin.

In addition, immobilized porphyrins such as TPPS can complex with aromatic compounds such as benzene and naphthalene with resultant analyte specific wavelength shifts; the TPPS-benzene complex appears at 419 nm while the TPPS-naphthalene complex appears at 421 nm. Complexation with formaldehyde results in a peak at 435 nm. Numerous other organics including methanol, ethanol, and other short chain molecules also complex with porphyrins.

2.8 Stability of the Porphyrin Surfaces

Surfaces of immobilized porphyrin are stable and reactive even after 4.5 years of storage in air at room temperature (unpublished results). The AChE-based surfaces stored in light vacuum food storage bags are enzymatically stable for over 322 days and the AChE-$TPPS_1$ (porphyrin incorporated into the active site) surfaces have been shown (unpublished) to detect eserine after 480 days storage in vacuum food bags at room temperature.

3. CONCLUSIONS

Chemicals and chemical warfare agents can be optically detected using a solid-state layer of immobilized porphyrins or enzyme-porphyrin complexes. Addition of reagents or solutions other than the sample is not required and the detection process is complete in less than 6 seconds. The porphyrin surfaces are stable for up to 4.5 years and the enzyme-based surfaces are functional for up to 480 days following storage at room temperature under mild vacuum conditions. The use of evanescent absorbance measurements allows for the sensor surfaces to be integrated into the fluid flow without obstruction of the flow; measurements are unaffected by turbidity, mixing, bubbles, or color of the medium. The detection limits for most analytes is at or below regulatory levels. Enzyme-based sensors are best suited for measurements of classes of compounds (such as nerve agents or organophosphate compounds) and are "broad spectrum" detectors that may not be able to identify a particular compound. Immobilized porphyrin surface, on the other hand, can distinguish between chemicals when the appropriate porphyrin(s) is chosen as the indicator molecule. This represents a technique suitable for in-line real-time reagentless detection of chemical compounds in water treatment and supply systems.

References

1 B.J. White, J.A. Legako and H.J. Harmon, *Biosensors and Bioelectronics*, 2002, **17**, 361.
2 B.J. White and H.J. Harmon, *Biosensors and* Bioelectronics, 2002, **17**, 463.
3 D. Mauzerall, *Biochemistry*, 1965, **4**, 1801.
4 J.A. Shellnut, *J. Phys. Chem.*, 1983, **87**, 605.
5 B.J. White, J.A. Legako and H.J. Harmon. *Biosensors and Bioelectronics*, 2003, **91**, 138.
6 B.J. White, J.A. Legako, and H.J. Harmon, *Sensors and Actuators B*, 2003. **89**, 107.
7 B.J. White and H.J. Harmon, *Sensor Letters*, 2005, in press.
8 B.J. White and H.J. Harmon, *Biosensors and Bioelectronics*, 2005, **20**, 1977.
9 B.J. White and H.J. Harmon, *Biochem. Biophys. Res. Comm.*, 2002, **296**, 1069.
10 B.J. White and H.J. Harmon, *Sensor Letters*, 2005, **3**, 1-7.
11 J.A. Legako, B.J. White, and H.J. Harmon, *Sensors and Actuators B*, 2003, **91**, 128.
12 A.K. Singh, A.W. Flounders, J.V. Volponi, C.S. Ashley, K. Waly, and J.S. Schoeniger, *Biosensors and Bioelectronics*, 1999, **14**, 703.
13 G.A. Evtyugin,, L.V. Ryapisove, E.E. Stoikova, L.B. Kashevarova, S.V. Fridland and V.Z. Latypova, *J. Anal. Chem.*, 1998, **53**, 869.
14 B.D. Leca, A.M. Verdier and L.J. Blum, 2001, *Sensors and Actuators B*, **74**, 190.
15 D. Yao, A.G. Vlessidia, and N.P. Evmiridis, 2002, *Anal. Chim. Acta*, **462**, 199.
16 L. Pognacnik and M. Franko, 2002, *Annali Di Chimica*, **92**, 93.
17 P. Mulchandani, W. Chen and A, Mulchandani, 2001, *Env. Sci. Technol.*, **35**, 2562.
18 H.B. Mao, T.L. Yang and P.S. Cremer, 2002, *Anal. Chem.*, **74**, 229.
19 N. Rupich and J.D. Brennan, 2003, *Anal. Chim. Acta*, **500**, 3.
20 R.D. Richins, A. Muklchandani and W. Chen, 2000, *Biotech. Bioeng.*, **69**, 591.
21 B.J. White, J.A. Legako and H.J. Harmon, *Biosensors and Bioelectronics*, 2003, **18**, 729.
22 H.-J. Schneider and M. Wang, *J. Org. Chem.*, 1994, **59**, 7464.
23 H.J. Harmon, *Proceedings of the First Joint Conference on Point Detection for Chemical and Biological Defense, Williamsburg, VA, 23 October, 2000.*
24 A.M. Awawdeh and H.J. Harmon, *Biosensors and Bioelectronics*, 2005, **20, 1595.**
25 B.J. White, J.A. Legako and H.J. Harmon, *Sensors and Actuators B*, 2004, **97**, 277.
26 B.J. White and H.J. Harmon, *IEEE Sensors Journal*, 2005, in press.
27 M.A. Awawdeh and H.J. Harmon, *Sensors and Actuators B*, 2003, **91**, 227.

BE PREPARED, THE APPROACH IN THE NETHERLANDS

Ben Tangena[1] and Joep van den Broeke[2]

[1]National Institute for Public Health and the Environment, PO Box 1, 3720 BA Bilthoven, The Netherlands
[2]Kiwa Water Research, PO Box 1072, 3430 BB Nieuwegein, The Netherlands

1. INTRODUCTION

Ever since the attacks in the USA (9/11), the security of the drinking-water supply in the Netherlands has been high on the agenda of water companies, the government, politicians and research institutes. The possibility of unlawful and deliberate (terrorist) attacks causing major disturbances to the drinking-water supply has since been perceived as a real threat. The way of dealing with this problem in the Netherlands is to use a so-called 'safety chain', consisting of five stages: pro-action, prevention, preparation, response and recovery & follow-up (see Figure 1).

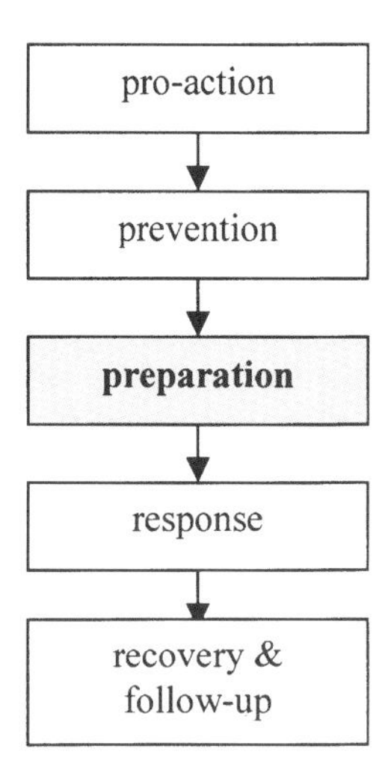

Figure 1 *The safety chain.*

Although attention has been given to all stages in the safety chain, this paper will focus only on the activities performed in the preparation stage. This includes how the Dutch water sector works together with the Ministry of Housing, Spatial Planning and the

Environment (VROM), the General Intelligence Services and Dutch centres of expertise such as the National Institute for Public Health and the Environment (RIVM) and Kiwa Water Research, the Dutch water industry's research centre. Despite all the preventive measures, there is still the possibility that a toxic or infectious substance will be introduced into the system. If we are to be prepared for such an intrusion, knowledge is essential. We have to know what agents are possible, how can they be detected and how harmful are they?

After a short introduction to the Dutch drinking-water sector, we will focus on agent fact sheets, priority agents, early warning systems, laboratory facilities and public health assessment.

2. THE DUTCH DRINKING-WATER SUPPLY: A VARIED PICTURE

In the Netherlands, with its 16 million inhabitants, about 1300 million m^3 drinking- water is produced and distributed by 13 utilities every year. Sixty per cent of this drinking water is produced from groundwater and the remaining 40% comes from surface water, with a major contribution being made by the rivers Rhine and Meuse. Abstraction, treatment and distribution are carried out at 225 locations, which differ significantly in size and layout. Groundwater pumping stations are small and mostly located in woodland or pasture areas. Surface-water stations are large and located near the rivers and lakes. Especially relevant to the Netherlands is the infiltration of pretreated surface water into the dune area, a well-protected nature reserve. The distribution network spans about 112,000 kilometres, with a high density in the cities and a low density in the countryside. Most pumping stations are coupled to ensure reliability. Chlorine is not used as a disinfectant in drinking-water treatment; the reason for this is the formation of carcinogenic trihalomethanes. See Table 1 for some characteristics of the drinking-water supply in the Netherlands.

	Groundwater	**Direct treatment of surface water**	**Dune infiltration of surface water**	**Totals**
Number of stations	210	7	8	225
Total production (million m^3/y)	801	267	250	1318
Average production per station (million m^3/y)	4	38	31	6
Largest station (million m^3/y)	25	100	70	

*Source: VEWIN, RIVM

Table 1 *Characteristics of the Dutch drinking-water supply (2003)**

3. FACTS AND FIGURES ON AGENTS

Facts and figures have been incorporated into an inventory that currently contains 84 fact sheets on chemical, biological and radiological agents that could pose a serious threat if introduced into the drinking-water supply system (see Table 2).

Chemical agents	Warfare agents	17
	Pesticides	21
	Rodenticides	6
	Natural toxins	20
	Other poisons	2
Biological agents	Bacteria	9*
	Viruses	0*
Radiological agents		9

* Six fact sheets for bacteria and 6 for viruses are still in preparation.

Table 2 *Fact sheets on chemical, biological and radiological agents*

A starting point for the selection of threatening agents was formed by lists circulating in the public domain, one of these being the CDC lists in the United States. The substances on these general lists were screened for their significance in the contamination of water. All the chemical agents not suited for use in the contamination of (drinking) water, because of their high reactivity or very low solubility in water, for example, were disregarded in the process of selecting and describing priority threats. Examples here are phosgene, red phosphorus and adamsite. Furthermore, compounds that are relatively harmless were deleted, so that no chemicals with an oral LD_{50} value above 500 mg/kg body mass were considered. Finally, a group of pesticides and rodenticides, along with some natural toxins with a high toxicity (LD_{50} < 500 mg/kg body mass) not included in the initial lists, were added. From the available lists of microbiological agents, those showing instability in water, e.g. *Variola major* (smallpox), were not considered. Radiological agents were chosen on the basis of the following criteria: high radio-toxicity, good water solubility, a minimal half-life of several weeks and the availability of a strong radiological source.

The fact sheets, as presented in Table 3, provide information on general data and properties, availability, analytical techniques, toxicological or infectious properties, fate in the environment and (drinking) water, assessment of (potential) removal in drinking-water treatment and recommended countermeasures.

4. TOWARDS A SHORTLIST

The information in the fact sheets is used to prioritise agents, or in other words, to select the most threatening agents. Four criteria were used for the prioritisation of chemical substances: toxicity, solubility, stability and availability. The value of these criteria for determining the relevance of each threatening substance is represented by three relevance classes: class 1, indicating the highest relevance, class 2, medium relevance and class 3, lowest relevance (see Table 4). To obtain a final score the class number is multiplied by the weighting factor of the criterion and the individual scores are added up. The four criteria were weighted as follows: availability at 50%, toxicity at 30%, solubility at 10% and stability at 10%. Availability was considered to be the most important criterion, because an agent that is difficult to purchase or produce will be less likely to be introduced into drinking water. A brief sensitivity analysis of the influence of the weighting factors and class divisions on the final results showed no significant changes in the shortlist. The selection criteria for biological agents are risk of infection and mortality, both having equal weight in the selection. The stability in water is also basically regarded as a criterion, but the agents in the fact sheets are all stable in water for a period of a few days, which is long enough to bring on an infection. Again, three classes were used for weighting the criteria (see Table 5).

Radio nuclides were divided into three classes according to their radio toxicity. Seven are considered to be the most dangerous ones because of their high to very high toxicity.

Subject	**Chemical**	**Biological**	**Radiological**
General data	Chemical name Chemical formula Synonyms Molecular weight CAS number	Name Synonyms Disease Infectious properties CDC category Biosafety level	Chemical name Chemical formula Atomic weight CAS number
General properties	Melting point Boiling point Water solubility Vapor pressure Volatility Colour Odour Taste	Growth conditions Gram colouring Shape Mobility Natural host	Radio-active decay Half-life Specific activity Water solubility
Availability	Origin Production methods Field of application	Origin Production methods On sale	Origin Production methods Field of application
Analytical techniques	Detection limit Duration Laboratory name Laboratory capacity	Detection limit Duration Laboratory name Laboratory capacity	Detection limit Duration Laboratory capacity
Toxico-logical and infectious properties	Lethal dose No-effect level Acute exposure guidelines	Infectious dose Risk of infection Mortality Incubation period	Dose Conversion Coefficient (acute and long-term exposure)
Fate in environment and drinking water	Environ. distribution Hydrolysis products Half-life	Environ. distribution Stability Growth Disinfection	Environ. distribution Decay products Decontamination methods
Fate in drinking-water treatment	Behaviour in 11 treatment processes	Log removal CT value for disinfection	Behaviour in 11 treatment processes
Counter-measures	Drinking-water ban Shutting off the water Boiling advice Chlorination Draining the network	Drinking-water ban Shutting off the water Boiling advice Disinfecting methods	Drinking-water ban Shutting off the water Draining the network

Table 3 *Fact sheet contents*

Criterion	Weight factor	Property	Class 1	Class 2	Class 3
Toxicity	30%	LD_{50} oral (mg/kg body mass)	<5	5-50	>50
Solubility	10%	mg/l	>10,000	100-10,000	<100
Stability	10%	Half-life (hours)	>100	1-100	<1
Availability	50%	-	large	limited	small

Table 4 *Weighting factors and classes for chemical agents*

Criterion	Class 1	Class 2	Class 3
Risk of infection	$> 7*10^{-3}$	$7*10^{-6}$-$7*10^{-3}$	$< 7*10^{-6}$
Mortality	> 50%	5%-50%	<5%

Table 5 *Classes for biological agents*

The priority list is drawn up according to the way the final scores are ranked. The list consists of 26 chemical agents (e.g. insecticides, rodenticides and natural toxins), and nine pathogenic bacteria. The results of the selection process are presented in Table 6.

		Number of agents considered		
		Available lists	First evaluation	Short list
Chemical agents	Lethal warfare agents	33	14	1
	Incapacitating agents	15	3	0
	Pesticides	18	21	10
	Rodenticides	0	6	6
	Natural toxins	22	20	8
	Other poisons	0	2	1
Biological agents	Bacteria	19	15	9
	Viruses	8	6	0
	Protozoa	3	0	0
Radiological agents		21	9	7
Totals		139	96	42

Table 6 *Results of the selection of agents*

5. EARLY WARNING IS ESSENTIAL

Early detection of harmful agents is necessary for a proper response. Of course, analytical methods are available to detect those agents in drinking water, but rapid response is essential to take measures in the case of an attack. Using existing laboratory techniques, one would be too late to take effective countermeasures to prevent exposure to the public.

An inventory containing the early warning systems (EWS) that are potentially suitable for the timely detection of chemical and biological agents was drawn up using the priority list of threatening agents. Thirteen potential techniques were evaluated using the following criteria:

- Detection of a broad range of substances:
  An ideal EWS will be able to detect as many potential threats as possible.
- Toxicological relevance:
  It is not only beneficial for an EWS to detect a broad range of contaminants, but what is specifically important is that the EWS detects the most critical ones, i.e. the substances in the shortlist.
- Possibility of identification:
  An ideal EWS will not only sound the alarm, indicating there is something wrong with the water quality (as a generic detector does), but will also provide a (rough) identification of the agent.
- Sensitivity:
  Assuming that a person of 70 kg drinks 2 litres of water a day and that the lethal dose is 5 mg/kg body mass, the lethal concentration will be 175 mg/l. Therefore, a EWS must have a detection limit of approximately 100 mg/l or lower.
- Time-to-result:
  Judging from the rate of flow in the network, a response time of less than 30 minutes is deemed desirable.

The evaluation is outlined in Table 7.

Technique	Detection of a broad range [a]	Possibility of identification [b]	Toxicological relevance [c]	Sensitivity [d]	Time-to-result [e]
UV-probe					+
Online HPLC-UV	+	+	+	+	-
Online GC-MS	+	+	+	+	-
Refractive index detector	+	--	-	-	++
ELSD[f]	+	--	-	0	++
Ion selective electrodes	-	+	++	0	++
Fluorescence Detector	-	0	-	++	++
Optical chemical techniques	-	+	++	+	+
Electronic Nose					+
Daphnia and Zebra fish monitor					
CLEAR			+	+	
TOXcontrol					
PCR	-	+	++	++	-

[a] **Table 7** *Evaluation of early warning systems*

+: large variety from different substances; −: limited number of substances (e.g. only cholinesterase inhibitors)

[b] +: identification possible, 0: limited possibility for identification, −: only identification of a broad group of substances, − −: identification not possible

[c] ++: response always indicates toxicological relevant substance, +: toxicological relevance very likely, −: toxicological non-relevant and relevant substances detected.

[d] ++: sub-ng/l, +: sub-μg/l, 0: μg/l level

[e] ++: < 10 minutes, +: < 20 minutes, 0: < 30 minutes, −: > 30 minutes

[f] ELSD: Evaporative light scattering detector

The evaluation has led to the selection of five systems with EWS potential:

- UV-probe
- Electronic Nose
- Combined *Daphnia* and Zebra fish monitor
- Chemiluminescence Enzyme Activity Response (CLEAR)
- Luminescence bacteria (TOXcontrol)

The five systems were selected because they cover a broad spectrum of potential threats, show rapid response, and are either commercially available or in an advanced stage of development. Costs of acquiring the systems were also considered. All these five systems were found potentially appropriate for online detection of chemical agents; at present, there is no feasible technique available for online detection of biological agents.[7]

The first two systems are physical monitors: i.e. they measure physical effects in water, such as absorption of UV light and electrical conductivity. The other three are so-called effect monitors: i.e. they measure impacts of an agent on organism properties, such as cell respiration and the function of an enzyme system. Both sorts of systems are complementary and partly proven in practice. Practical application is preferably carried out by combining physical and effect monitoring. Besides these favourable EWS systems, a 'toolkit' of fast screening methods, should be at one's disposal. Examples are PCR for biological agents, and HPLC and GC equipment for chemical agents. PCR is the only reasonable method for detecting biological agents, making further optimisation desirable.

In order to evaluate the five favourable systems under operational conditions, test-phase research will be carried out in cooperation with drinking-water utilities and equipment suppliers.

6. UV PROBE AND TOXCONTROL ARE PROMISING

Both the UV-probe[1] and the TOXcontrol (in the form of the Microtox assay[2], have been briefly evaluated for suitability as EWS, independent of the efforts described in this paper. The capabilities of a UV-probe for detecting pesticides and surrogates for nerve gases in drinking water were tested (see Table 8 for results).

[7] As chlorine is not used in the Netherlands, measurement of chlorine reduction is not appropriate.

Compound	Detection limit (µg/l)
Aldicarb	100
Mevinphos	100
Oxamyl	100
Azinphos-methyl	10,000
Methamidophos	1000
Isoproturon	50
Linuron	5
Soman and Sarin surrogates	50,000

Table 8 *Test results of a UV-probe*

These test results showed that chemicals displaying strong UV absorption in drinking water could be detected down to the µg/l levels. While it is important to note that a UV-probe will only detect the presence of a contaminant, it will, in most cases, be unable to positively identify the substance. Two pesticides from the list of priority threats were measured as well, and can be detected down to the 100 µg/l level. The sensitivity of the instrument towards a number of nerve agents was evaluated using surrogate, structurally related, compounds. The results revealed a poor detection limit for such agents, since they lack strong UV-absorbing structural features.

The TOXcontrol system itself has not been evaluated as EWS, but the Microtox assay, which is an identical test within a different instrumental setup, has been evaluated in EPA's Environmental Technological Verification Program (USA). In this programme, the luminescent bacteria were exposed to various pesticides, a blood poisoning agent, at least two biotoxins and at least two nerve agents. In these tests the assay proved sensitive to aldicarb and cyanide on mg/l levels. However, as a bacterium does not possess a central nervous system, this assay is not sensitive to chemicals that attack this system.

Further evaluations of both systems are planned in the context of the preparation phase in the safety chain described in section 1.

7. LABORATORIES LINKED

The analysis of chemical and biological agents requires highly specialised laboratory facilities. Along with reference material, specific apparatus and methods will also be necessary. Some of this material is only available upon special governmental permission, as it falls under the Chemical and Biological Weapons Conventions. Moreover, due to the high toxic and infectious properties of most agents, special security measures have to be taken. If we are to be prepared for a real attack, the laboratory needs to be accessible and available on a 24-hour/7-day basis. It is clear that a laboratory can only meet these standards at high costs. RIVM is building a BSL 3/4 laboratory for biological agents.

With this in mind, the Inspectorate of the Ministry of VROM has decided to set up a National Laboratory Network to be coordinated by RIVM. Analytical techniques will focus on the processing of environmental samples, but human samples will also be included. Some nine laboratories, most of them under the government, are working together. A secured website, which is used in crisis as well as everyday situations, plays an important role in the network. A centre of expertise has been established to support information-sharing, including fact sheets on NBCR agents, an (inter)national expert network, research and development of analytical methods and up-to-date procedures for a quick response. In the case of a real attack, analytical results and public health assessment information is exchanged through the website and reported to the national or local authorities.

The laboratory network has been set up to cope with all terrorist attacks, not just those involving drinking-water. In the case of a drinking-water incident the Environmental Inspectorate will order the National Environmental Incident Service to take samples, which are then brought to a high secured BSL-3 facility. There, the samples are screened, treated, split up and stored. Finally, the samples are transported to the designated laboratories. The following techniques are available for screening (Table 9).

Chemical	**Biological**	**Radiological**
XRF	ATP bioluminescence	Liquid scintillation counter
IR	Culture	γ hand monitor
GC-MS	Real time PCR	α/β hand monitor
NMR	Immunoassay	

Table 9 *Screening methods for agents*

8. PUBLIC HEALTH ASSESSMENT

Modeling studies have been performed with a hydraulic network model to get an idea of the consequences of a deliberate attack using an agent in the drinking-water supply system. Here, an agent is introduced into the system of a middle-sized town (70,000 inhabitants) for the duration of one hour. After this the spread of the plume of contaminated water throughout the town's distribution network is followed over a period of several days. The impact of the contaminant varies according to the location where the contaminant was introduced: the closer to the pumping station the contamination takes place, the larger the area that will be affected. When introduced at street level, for example, from a single house, spread of the contaminated water is often restricted to the area directly surrounding this point of introduction.

In assessing the impact of an agent on public health, it is assumed that the action of the agent is only through oral exposure, in which 2 litres per day of total drinking-water consumption and 0.25 litres per day unboiled consumption is taken into account. For chemical agents, the overall impact was assessed using a rough estimate of the number of potentially exposed persons to concentrations of the introduced compound above the LD_{50} level. The number of potential victims was then calculated by summing up all the individuals living at nodes in the network where this LD_{50} concentration was attained during the first 24 hours of the contamination. The impact of a pathogenic bacterium was assessed by calculating the risk of infection using an exponential dose–response model. With the mortality data, the number of potential casualties can be assessed in the same way as for the chemicals. Table 10 presents results for various introduction scenarios and for a pesticide with a $LD_{50,\,oral}$ of 2.2 mg/kg body weight, both a bacterium with an infection risk of 7 x 10^{-6} and a mortality of 25%; all figures hold for oral exposure. As the figures presented in this table refer to the situation without medical treatment, the numbers of casualties in real life will be lower.

Scenario (point of introduction)	Number of potential victims (no medical treatment after exposure to the agent)	
	Pesticide	Bacterium
1 Pumping station	0	725
2 Large supply main	20	550
3 Private home	1000	400
4 Reservoir	0	1

Table 10 *Potential victims after introduction of a chemical or biological agent into the distribution network of a medium-sized town*

The impact of a biological attack is much more difficult to assess as symptoms of infection will often only be expressed after several days. Furthermore, a combination of the risk of exposure and the risk of infection needs to be considered, along with the exposure route when considering pathogens.

The calculations clearly demonstrate the huge impact that can be achieved. Even when a contaminant is introduced at private-home level hundreds or even thousands of consumers will be exposed to it.

Several rough calculations on the impact of introducing radionuclides into the distribution system were carried out for radiological agents. The elements, ^{90}Sr, ^{239}Pu and ^{241}Am were chosen as examples because of their high radio toxicity. Oral drinking-water consumption is assumed to be 2 litres per person per day, and the total flow rate of the distribution system 15,000 m^3 per day (100,000 customers using 150 liters a day). Furthermore, the residence time of the water in the system is assumed to be three days. Results of the calculations are presented in table 11.

Nuclide	Source (Bq)	Source mass equivalent (g)	Dose Conversion Coefficient for ingestion (Sv/Bq)	Resulting dose (Sv)	Impact (after 20-50 years)
^{90}Sr	$1x\ 10^{11}$	0.019	3.4×10^{-9}	0.001	30 extra cancer casualties per 1 million inhabitants
^{239}Pu	1×10^{13}	4300	2.5×10^{-7}	~1	
^{241}Am	1×10^{13}	75	2.0×10^{-7}	~1	

Table 11 *Impact of radiological agents*

The calculations show that the intrusion of an easy manageable100 GBq source, such as 19 mg of ^{90}Sr, will not lead to acute health impacts. Within several tens of years after an incident there will be a slight increase in the risk of cancer. A dose of at least 1 Sv is necessary to bring about acute lethal effects, so for this purpose one would to have a very strong source at one's disposal. If unshielded, such a source is lethal within minutes for any person handling it. For this reason, the threat of radiological agents into the distribution system is not so much the possibility of acute casualties, but rather the higher risk of cancer in the long term.

9. CONCLUSIONS

Preparing for deliberate attacks on the drinking-water supply in the Netherlands is performed in close cooperation with the water sector, the government and centres of expertise. A practical method has been developed to select the most dangerous agents that could be introduced into the supply system. The application of early warning systems is important to be prepared for such an attack. Such systems are evaluated for priority agents, and the most favourable systems need to be tested on laboratory scale and in real life. Rough calculations of the public health impact of chemical and biological agents introduced into the supply system show high numbers of acute casualties. More detailed studies should be done to establish timely countermeasures. And finally, the threat of radiological agents is restricted mainly to increased cancer risks.

References

1 J. van den Broeke and A. Brandt, A short evaluation of the S::can Spectro::lyser, Kiwa Water Research, the Netherlands. Report BTO 2005.002, 2005.
2 R. James, A. Dindal Z. Willenberg and K. Riggs, Environmental Technology Verification Report, Strategic Diagnostics Inc. Microtox® Rapid Toxicity Testing System, Batelle, Columbus, OH, 2003. http://www.epa.gov/etv/pdfs/vrvs/01_vr_microtox.pdf

Overview of the Water Company Challenges

W.T. Sutton

Yorkshire Water Services Ltd, Western House, Western Way, Halifax Road, Bradford, BD6 2LZ

This paper is a personal view and does not seek to represent the views of the industry or Yorkshire Water Services Limited.

1 INTRODUCTION

At the outset I should say that I have been very impressed by this conference, but in particular, I am grateful for the contributions this morning which have put terrorism in perspective within the much larger risk area of contamination risk, detection and incident management. It is the accidental or naturally occurring incidents of contamination that are most prevalent and which we should concentrate on. Business as usual will help prepare us for the risk of deliberate contamination.

Water Companies do not exist in a vacuum and are subject to many if not all of the challenges of any major commercial enterprise and a few more besides in the form of Government, regulatory standards and constraints which control how they do business and how much they can charge their customers for the service and products they provide. Like all commercial ventures they are in the risk business and every day are taking risk based decisions on how to produce and distribute potable water to their customers at lowest cost. They are encouraged to take those risks by our regulatory system which restricts the amount of money companies are allowed to raise through billing, and by the investors in the company who demand a return on their investment which is as good if nor better than other commercial enterprises. The water industry is generally regarded as a safe rather than a speculative investment. Generally the return on investments reflect this – perhaps an indication that the regulatory system has so far served us well as a means of driving forward innovation and efficiency in what was neither an innovative nor efficient industry until it was privatised. The industry and the constituent companies are used to challenges and have consistently responded well to them.

The other major influence is of course our customers, for they take for granted the fact that the water will be there when they turn on the tap, that it will be high quality, that it will be clear bright, and fit to drink. The fact that it hasn't rained for some time in some parts of the country and effected water companies may want to restrict the use of water, by imposing hosepipe bans or advising people not to wash their cars so frequently, means for some customers, and pundits, that the water company is inefficient and gives good cause to complain about how high the bills are.

One could take a cynical view of how water companies are often impacted by environmental and health regulation whether it emanates from Europe, or national Government. That view would be on the lines of government recognising a risk, introduce a new standard or requirement and serve notice of enforcement on the water company to comply. Effectively find the problem and pass it on to someone else (Water Co) to solve within a budget set by our regulator. That way, by government using our regulatory systems, the water company gets criticised for raising bills to pay for Governments environmental policy – this could be construed as yet another stealth tax.

A cynic may also look at the way in which companies behave in order to create efficiencies. The development in new technologies has enabled the automation of many tasks and also on-line monitoring of processing which has released staff from the workforce and enables improved profits. The greater reliance on technology in itself does not increase risk and often reduces it, but the reduction of numbers in staff mean that there are fewer people and resources available to respond when things go wrong. Taking risks to reduce costs calls for strong risk analysis and cost benefit systems.

The cynics view of the customer stance (and I am a customer) is simple – they want it always there, better quality, and cheaper.

It is therefore difficult to place the risk of contamination within that context of competing demands upon water companies, other than to say that without strong guidance from Government and Regulators, convincing water companies to spend large amounts of money on some aspects of this risk would be difficult to achieve.

2 VULNERABILITY TO CONTAMINATION

In general terms the risks from contamination arise from four dimensional areas

Recognition of vulnerabilities

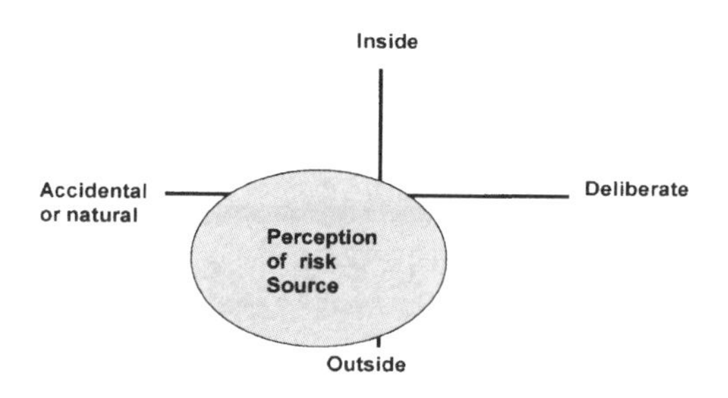

Figure 1. *Recognition of vulnerabilties*

These four areas are:

- Accidental or naturally occurring
- Deliberately caused
- External sources
- Internal sources

The model in Figure 1 shows the four dimensions and my view of where the perception of the risk sources lie.

Figure 2 shows how some of these vulnerabilities might be reduced.

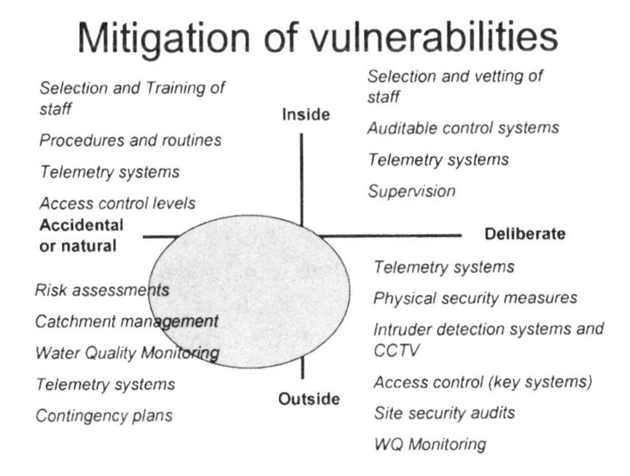

Figure 2. Mitigating vulnerabilities

I have imposed some broad headings as to how water companies may have responded to those risks on the model but in relation to the bottom right hand corner it does not take account of the lack of consistency across the industry in terms of approaches to security risks – matters which are currently being addressed through joint working between Water UK Security and Emergency Planning Group and Government Security Advisers. That is a challenge which is being addressed already, however it may be useful to summarise with an overview of physical security - even if it appears somewhat superficial - it serves as a general indication of the situation.

The physical security of raw water sources is very difficult if not impossible and has not necessarily been assisted by issues such as Right to Roam, nor is the frequency of environmental monitoring by our regulators as frequent as that described in other European countries.

Water Treatment Works are the first real opportunity to provide some physical protection of the water and is where a wide range of protective monitoring equipment may be deployed.

Distribution provides significant physical protection by virtue of the containment of water in underground pressurised pipes. There are some vulnerabilities but the more obvious one have been addressed with improved physical hardening and additional intrusion detection systems.

3 RECOGNISING OF UNUSUAL EVENTS – WATER QUALITY MONITORING

Like the oil, gas, and chemical industry water companies make extensive use of telemetry systems for on-line real-time monitoring of water treatment, quality, quantity including flows, and in some cases control of water supply. In general terms the use of such technology is not evenly distributed

In raw water sources it is partial or non-existent in respect of impounding reservoirs and river sources. In treatment works it is extensive but not always present at the inlet to the works. In the Distribution systems it is not very extensive and tends to be restricted to pumping stations and Service reservoir where it is concerned mainly with flows and stock levels although there is some monitoring of chlorine residual levels. In the distribution mains, up to the point of delivery in the customers premises, on-line water quality monitoring is non existent or negligible at best.

So how do we know the water is safe – we take samples!

Sampling

- 2003 2.9 million (about 8,000 per day)samples tested for regulatory purposes about 80% of them from distribution below Service Reservoirs. Of the total only 3,418 breaches were detected (99.86% compliant).
- There are 53 million consumers in England and Wales which means that about every 18 years one sample is taken and analysed on my behalf.

How many of those sample analysis results helped us prevent an incident impacting on our customers? My view is not many – for most if not all of them the results of the analysis did not become known until many hours after the water had been used – it is a retrospective view of a point in time of a point on a particular system – but it is not to be dismissed lightly for it has served to prove and drive consistently improving standards in water quality.

- In the same year (2003) the DWI reported there had been 99 events which they classified as Incidents affecting a total of 2.5 million customers. The majority of incidents were in distribution.

So how do we recognise those events in our distribution systems? What is our best shot, most frequently used and accurate source of information during the critical early stages of an incident? OUR CUSTOMERS. They are paying us to monitor our systems for us – are we supposed to be a customer service orientated industry – shouldn't we be telling them first?

One day soon a water company will have to answer questions from the media as to why despite 9/11 and all the dire warnings in the newspapers, and more measured advice from other sources, we have still not got on-line water quality monitoring devices in our distribution systems capable of giving warning of a contamination event and providing a trigger for a response.

- What has been done to meet the challenge so far?
  - DEFRA has invested in research
  - Other Govt Dept. have invested in research
  - We have seen at this conference many fine examples of what is becoming available but is not yet through the hoop of being proven in the field and in production at a price which passes the cost benefit test in relation to the risk
  - How much money individual water companies and UKWIR have invested in this technology since 9/11 is not known to me. Nor do I know the proportion of spend on resolving this issue as opposed to the many others which will be making demands on limited R and D budgets

The Challenge

- Develop an on-line system to operate in distribution which will provide real-time contaminant monitoring and warning with the following characteristics:-
  - Low rate of false +/-
  - Rapid response time
  - Highly automated
  - High sampling frequency
  - Able to capture samples for confirmation analysis
  - inexpensive

4 ABILITY TO RESPOND – LABORATORY RAPID ANALYSIS

Knowing or suspecting that there is a contaminant in the water is the trigger for a whole series of responses from within the water companies and industry as a whole. Well rehearsed plans and procedures, with notifications, resources and action plans are in place and waiting to go, like greyhounds in the slips. But we are missing something – information i.e. is there something in the water or not. If there is, what is it? How soon will we know what it is? And what will its effect be? Whilst measures to protect our consumers can be moving rapidly forward the void of information about what if anything is in the water will create huge pressure upon the water company, and inevitably lead to speculation and loss of confidence. It is essential to fill that void rapidly. To do so we need high grade scientists with experience.

At the previous event 2 years ago 'Can we Cope' papers were presented relating to rapid analysis techniques. Speakers also raised the issue of the loss of experienced scientists from the laboratories and commented upon the potential impact this would have on the laboratories ability to provide an efficient rapid analysis service. The problem has been further looked at and I have summed it up as follows:-

" If age does not weary them, boredom and cost efficiencies are likely to."

Our more experienced scientists are getting older and seeking retirement – they have seen their job change over the years and these days see there work in the laboratory more akin to that of a machine operator on the production-line. Targeted analysis is the automated revenue producer for our laboratories. Only rarely are laboratories called upon to undertake more technically challenging work. There is also a problem with recruiting and retaining young scientists to become our future greybeards in such circumstances – keeping their enthusiasm and interest high is a near impossible task.

At the same time, equipping a laboratory to be capable of rapid analysis for a wide range of microbiological organisms, chemical compounds, or radiological contamination is a very expensive undertaking. Especially when there is little demand for regular operational use and training and familiarity with the equipment comes at an additional cost. All this at a time when water companies are seeking to find further efficiencies and reductions in the analysis bills.

The laboratories for the water industry in the UK have for several years met regularly in what has been called a mutual aid meeting, for the purpose of sharing knowledge and best practice. They have developed capability statements which they have shared with the industry and published guidance notes on rapid analysis techniques. This is a group of water company owned and independent commercial laboratories which, to their credit, have co-operated with each other and the industry for the greater good of all. They have recognised these issues of loss of experienced personnel and limited funding with which to purchase sophisticated equipment and brought it to the attention of the water companies. Likewise the DWI has identified the problem and written to each water company seeking re-assurances as to the service levels and capabilities of their laboratories.

To me it seems that small laboratories can no longer sustain the level of service and sophistication required for the wide range of rapid analysis capability needed in the 21st century. There is a need for companies through Water UK to review the future needs of the industry. Issues such as capacity for targeted analysis; technology requirements; and recruitment, training and retention of staff will be essential elements. When the review is complete it may identify the need for regional or national laboratory facilities for the non-targeted work, and the development of centres of excellence in particular types of analysis. The challenge is:

To ensure water company laboratory services meet the requirements of:

- high volume targeted analysis, and
- high technical competence in rapid analysis for unknown chemical and biological samples
- adequate independent radiological analysis capability
- Cost efficiency

5 COMMUNICATIONS

Modern communications and information technology has greatly improved the opportunities and ability of companies to communicate with its customers and consumers quickly and effectively. The same technology enables us to reach many other audiences e.g. regulators, investors, Water Voice, local and regional resilience forums. All have different needs but all require timely, accurate, easily understood information which hopefully will also be re-assuring. More advanced technology is used by the media on a daily basis to get their messages across to their viewers, listeners and readers. We know that the media can and do act responsibly in times of emergency but they are also in competition with each other and for that reason have a need to differentiate between their news and that of the competitors. One of the difficulties the media faces is when there is a void in information. They need to be telling people what is happening, have the latest information, and when they haven't got that information then they will pose questions. Unfortunately those questions will appear as headlines and some will see the headlines as scaremongering or sensationalising the situation. The media will fill the gap we or others leave. If our information to the media is incomplete or misleading then we risk our messages of re-assurance being lost amongst the unanswered questions.

The media can, and do, make or break the public perception of how we handle incidents.

I am not confident that as individual companies, or as an industry, or as government departments, we understand the lines the media are going to pursue when a major contamination of a water supply occurs. We can speculate, but have we ever sat down with them and played the 'what if game'. Surely this is something we should be doing now in co-operation with other agencies, to ensure we do have a clear understanding of their expectations and as a consequence are better prepared to meet them. If we have explained our concerns and strategies to the media in advance, and made them aware of the potential consequences of speculation rather than factual reporting, in terms of undermining public confidence and creating panic, we will have gone some way to ensuring our communications and theirs are following the same lines. In my view this is a challenge for all those who would be involved in responding to such an incident, but it is one for which

Government should be leading by bringing the parties together in a safe environment, where these matters can be talked about openly without the risk of them being the subject of the next TV docu-drama.

The Challenge is:

- For DEFRA, with Water UK and others, to get closer to the media in a safe environment so they and we may be better prepared in responding to water contamination emergencies and providing accurate and re-assuring information to the public.

6 LEADERSHIP

There is one more challenge and it is a challenge for all at this conference as well as the UK water industry as a whole. Where is the leadership coming from within the UK on contamination and related science issues? Is it being provided by Water UK? In my view the leadership is coming from the group of people who organised this conference and the previous one, but take a look at them – they are all getting old and probably within 2 to 5 years will have retired. We will have lost their collective experience in a very short period of time. So who from this audience is going step forward and take on the leadership responsibilities which will take water contamination and other related issues forward for the next 10 years? Whoever you are, you need to start moving now before it is too late. Or are you going to rely on Water UK?

The Challenge is to identify, involve and develop the people who will be the water industry scientific leaders for the next 10 years.

7 PERSPECTIVE

Finally, as the penultimate speaker, and after 3 days of concern about relatively rare events in our lives, we should check out our sense of perspective by having regard to a quotation from a speaker at the 2003 event.

"It is interesting to speculate the finance that would be available if terrorists were killing off millions of children under the age of 5 every year.

This of course is the number of children killed by dirty drinking water and poor sanitation annually."

Dave Clapham
Can We Cope 2004

CLOSING REMARKS

J. GRAY

Drinking Water Inspectorate, Ashdown House, 123 Victoria Street, London, SW1E 6DE

1. INTRODUCTION

As I said in my introduction the events in New York in 2001 prompted the water industry to look a little harder, and perhaps with a little more urgency, at the systems it had in place to protect the public water supply. There are many organisations potentially involved during an emergency affecting drinking water supplies. They are of many disciplines and their different requirements and needs will have to be satisfied. The last few days have put some of these potentially conflicting requirements into perspective. Water treatment is multi-layered, as is security, which itself is part of an overall water safety plan approach requiring risk assessment and review.

We have heard of many different research projects some of which are developing new and exciting technologies and others are applying old and familiar technology in a new way. A number of key issues have been identified which require continuing effort and investment including sampling, preparation for radiological events, emergency response and the need for training and end to end exercising of systems and responses. We should not forget that our job is public health and that whatever happens; we should strive to maintain "business as usual".

However, it is apparent that detection of contamination of water in supply indicates failure since the incident will have obviously occurred. Communications with consumers is key and there is a need to establish prior to any incident a trusted and trusting relationship between water supplier and consumer. There is also the need to maintain communications with each other in the future, continuing the dialogue which we may have begun here (if not previously).

I confirm that the organising committee will feed back your responses and comments to our parent organisations, including water companies, Water UK and Government. The CBRN subgroup of the National Security and Emergency Working Group will also be informed and advised by the outcome of this conference. If you as individuals take back one piece of previously unknown information, or one idea, or have made one new contact, the past few days will have been worthwhile and a success.

On behalf of the committee, thank you for your enthusiastic participation. It's been an honour and a pleasure to be here with you.

Posters

MONITORING OF ORGANIC MICRO CONTAMINANTS IN DRINKING WATER USING A SUBMERSIBLE UV/VIS SPECTROPHOTOMETER

J. van den Broeke,[1] A. Brandt,[1] A. Weingartner[2] and F. Hofstädter[2]

[1] Kiwa Water Research, P.O. Box 1072, 3430 BB Nieuwegein, The Netherlands (Email: Joep.van.den.Broeke@kiwa.nl)
[2] s::can Messtechnik GmbH, Herminengasse 10, A-1020 Vienna, Austria (Email: fhofstaedter@s-can.at)

1 INTRODUCTION

An uninterrupted supply of drinking water of impeccable quality is the primary objective of the Dutch drinking water companies. Up-to-date knowledge on the composition of both raw- and drinking water is essential for achieving this goal, and requires the employment of on-line analytical devices. One such a device is the portable, submersible diode array, UV/Vis spectrophotometer. In the study described here, this spectrophotometer was evaluated for monitoring the presence of organic micro contaminants in drinking water.

2 MATERIALS AND METHODS

A S::can Spectro::lyser UV-probe (Figure 1) was evaluated both under laboratory as well as field conditions. The instrument was tested in off-line and on-line modes in the laboratory at Kiwa Water Research in Nieuwegein, and at the water treatment plant WRK in Nieuwegein.

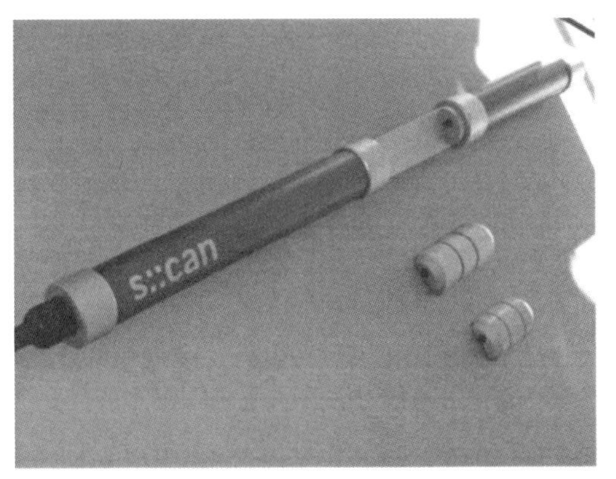

Figure 1: S::can Spectro::lyser, 100mm pathlength and inserts.

This study focused on the ability of the spectrophotometer to detect toxic compounds, such as pesticides and (simulants for) chemical warfare agents (Figure 2). Drinking water samples spiked with a selection of compounds, which are considered as most likely for use in the event of an intentional contamination of a drinking water supply, were analysed (Table 1).

It was demonstrated that substances with a high UV-extinction coefficient can be detected down to the µg/L level in drinking water. However, due to the similarity of the spectra of all compounds present in the water, it was not possible to identify single compounds at these concentration levels by their fingerprint (Figure 3). Compounds with a lower UV-extinction coefficient could be detected at higher concentrations only, e.g. mg/L levels.

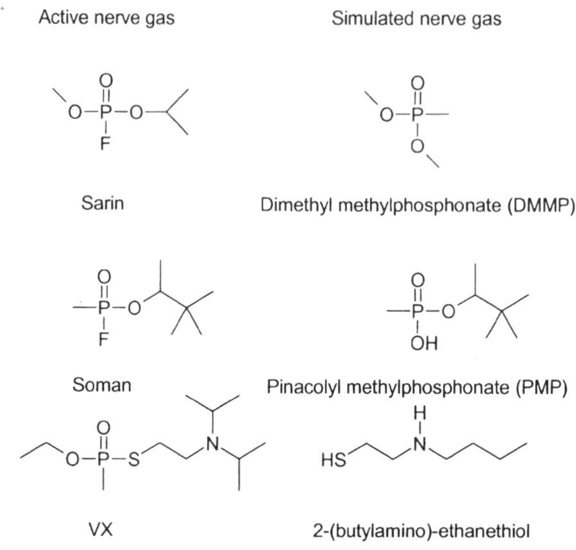

Figure 2: Nerve gas simulants measured using the Spectro::lyser.

Compound	Detection limit
PMP	50 mg/L
DMMP	50 mg/L
Aldicarb	0.1 mg/L
Mevinphos	0.1 mg/L
Oxamyl	0.1 mg/L
Azinphos-methyl	10 mg/L
Methamidophos	1 mg/L
Isoproturon	0.05 mg/L
Linuron	0.01 mg/L

*: 2-(Butylamino)-ethanethiol was also measured but no detection limit was established as the compound contained an impurity with very high UV-extinction coefficient.

Table 1: *Lower detection limit of the Spectro::lyser towards a selection of pesticides and simulant agents.*

The use of the alarm module in the software allows automated detection of anomalies in water composition (Figure 4).

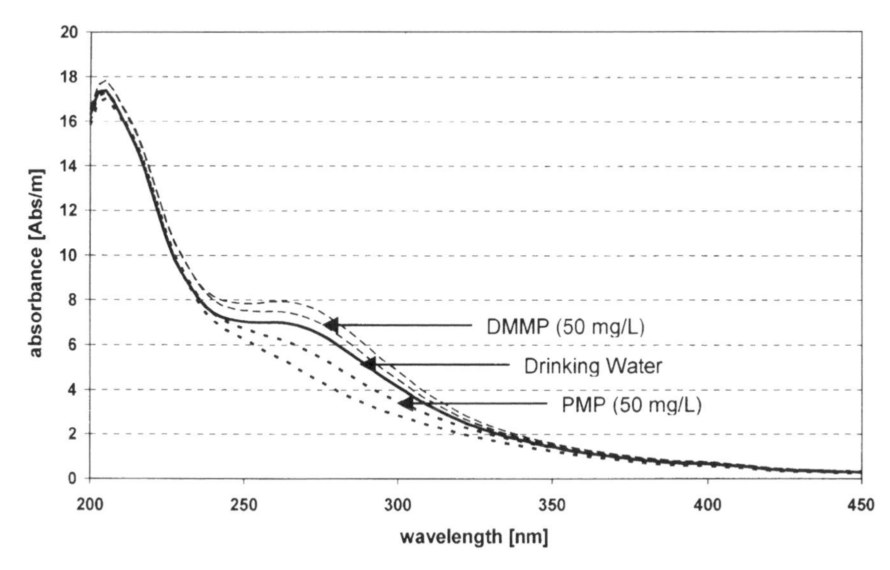

Figure 3: *Fingerprints of drinking water contaminated with simulants DMMP and PMP. Spectra are shown for 50 and 100 mg/L concentrations.*

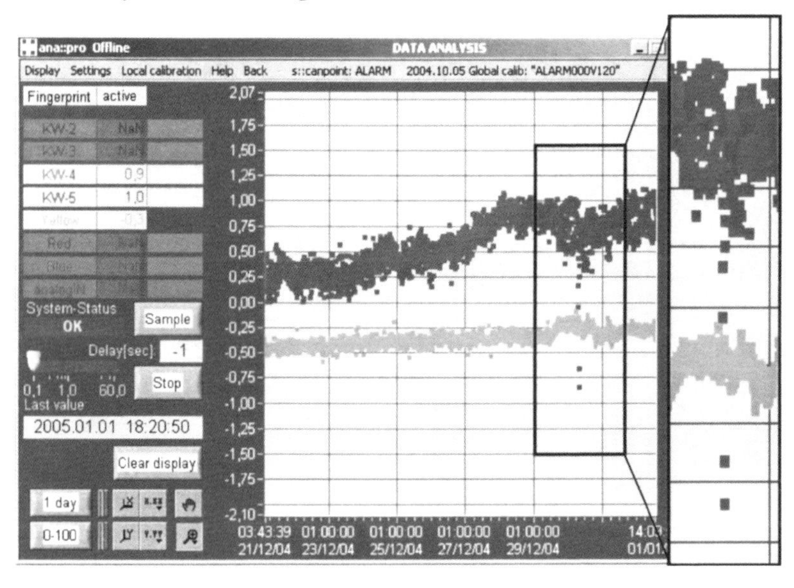

Figure 4: Response of alarm parameters to changing water quality.

3 CONCLUSIONS

The presence of most compounds could be detected down to mg/L or even µg/L levelswithout any pre-treatment of the sample

- it is very difficult to identify contaminants in whole water samples using UV-spectroscopy only
- small anomalies in water composition can be detected. The implies that UV/Vis spectroscopy is a valuable tool for detection of (intentional) contamination of a drinking water supply system.

Acknowledgements
This work was funded by and executed within the joint research programme of the Dutch drinking water companies (BTO, project 111508.030).

[1]: An assessment of priority agents was performed in a project commissioned by the Netherlands Ministry of Housing, Spatial Planning and the Evironment, and executed by the Dutch National Institute for Public Health and the Environment (RIVM) and Kiwa Water Research.

REMOVAL OF HUMIC SUBSTANCES FROM WATER BY MEANS OF CA^{2+}-ENRICHED NATURAL ZEOLITES

S. Capasso, P. Iovino and S. Salvestrini

Department di Environmental Science, University of Napoly II, via Vivaldi 43, 81100 Caserta, Italy . Sante.capasso@unina2.it

1 AIMS OF THE STUDY: to investigate the application of Ca^{2+}-enriched tuff for the removal of humic acid (HA) from water

2 PREVIOUS-KNOWLEDGE: Previous studies have shown that the enrichment of zeolitic tuffs by divalent cations markedly increases their capacity to bind HA.

3 MATERIALS: The tuff sample was from Marano (Napoli, Italy). CEC was 1.90 meq g^{-1} and the zeolite content 54%, with phillipsite = 37% and chabazite = 17%.

Enriching tuff sample by Ca^{2+}. Natural samples, ground to a fineness < 170 mesh, were mixed three times with 1 M $CaCl_2$ solution, washed by bi-distilled water and finally dried at 60 °C in an oven for two weeks. The exchangeable calcium ion located on the NYT surface was determined by eluting the sample up to exhaustion with 1M tetrabutylammonium hydrogen sulfate (0.18 meq surface Ca / g tuff).

Humic Acids (HA) were purified by acid/base precipitation and solubilitation to reduce the content of ash, < 0.1 %.

4 METHODS: The ability of Ca^{2+}-enriched tuff to adsorb HA was evaluated by equilibrium sorption experiments (sorption hysother) and by dynamic experiments carried out by percolating HA solutions through a small NYT column (breakthrough curves).

5 RESULTS: The sorption isotherms are shown in Figs. 1 and 2. Figure 3 shows the breakthrough curve for the sorption of HA onto Ca^{2+}-enriched NYT, obtained by eluting a small column with a HA solution.

6 DISCUSSION AND CONCLUSION: The results strongly support the possible utilisation of Ca^{2+}-enriched Neapolitan Yellow Tuff for removing humic acids from water. Measurements carried out in equilibrium closed systems and by flowing HA solutions through a small bed clearly showed that this relatively cheap material had a very high capacity for adsorbing humic substances from water. Eluting the tuff with water not

containing salts (Fig. 3) markedly decreases salts and the absorption of the HA, thus promoting the release of HA. This provides an easy and cheap method to regenerate the tuff columns after its use.

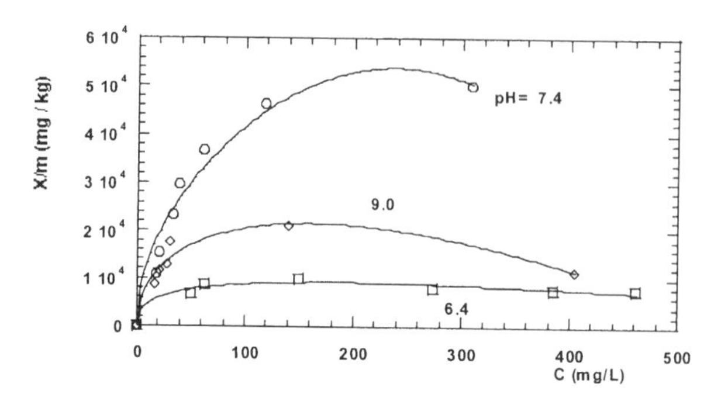

Figure 1. *Dependence of sorption capacity on the pH*

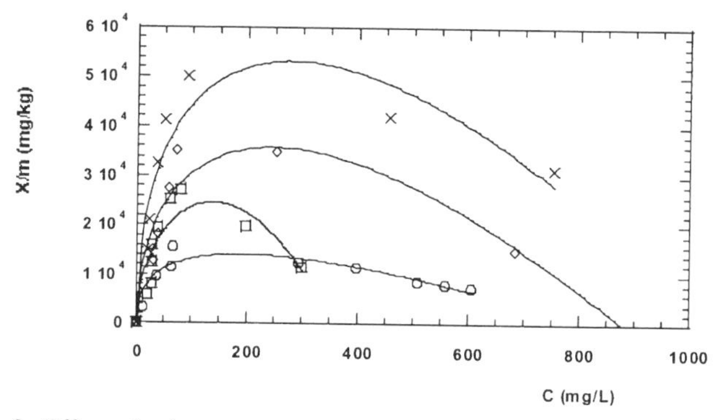

Figure 2. *Effect of salt concentration on the sorption isotherm*

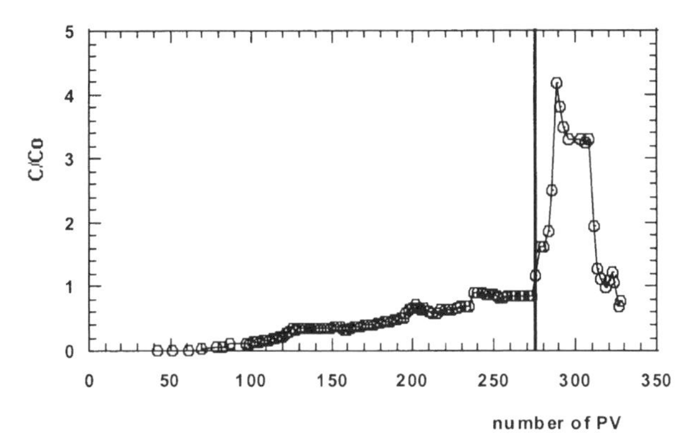

Figure 3. *HA breakthrough plot obtained eluting Ca^{2+}-enriched NYT bed by HA solution, pH 7.4, 0.01 M Tris buffer. C and Co are the HA concentrations in effluent and influent, respectively.*

PV means pore volume. The curve after the vertical line has been obtained eluting with a solution containing only the Tris buffer

PROTECTIVE EFFECTS OF CATHODIC ELECTROLYZED WATER ON THE DAMAGES OF DNA, RNA AND PROTEIN

Mi-Young Lee[1], Kun-Kul Ryoo[2], Yoon-Kyoung Kim[1],Yoon-Bae Lee[2], Jong-Kwon Lee[2]

[1]Division of Life Science, Soonchunhyang University, Asan, Chungnam, 336-745, Korea
[2]Division of Material and Chemical Engineering, Soonchunhyang University, Asan, Chungnam, 336-745, Korea.

1 INTRODUCTION

Electrolysis of water produces oxidized water and reduced water near the anode and cathode, respectively. Recently, the usefulness of electrolyzed water in the medical and agricultural fields is being examined.[1] Cathodic electrolyzed water(CEW) was reported to scavenge active oxygen species such as superoxide anion ($O_2^{-\bullet}$) and hydrogen peroxide.[2] CEW was also reported to have superoxide dismutase-like activity or antioxidant activity, and CEW could protect DNA from oxidative damage. It has been suggested that the protective effect of cathodic electrolyzed water against DNA damage is due to the dissolved molecular hydrogen in CEW.

The CEW used in this study has an oxidation-reduction potential of – 850 mV, and used to investigate the protective effect of cathodic electrolyzed water on the damages of DNA, RNA and protein.

2 METHOD AND RESULTS

2.1 Electrolyzed Water Preparation

The electrolyzed water generation apparatus used in this study consists of three chambers: anode, cathode, and middle. While ultrapure water (UPW) was supplied to each chamber, the diluted electrolyte was supplied to the middle chamber. Electrolyzed water was generated by electrolysis of UPW and diluted electrolyte NH_4Cl with electrolyzing current of 9 A and voltage of 10.5 V. See Fig1 in previous poster

2.2 Protective Effect of CEW on the DNA Degradation

Active oxygen species or free radicals are thought to cause extensive oxidative damage to biological macromolecules. Human lymphocytes were exposed to hydrogen peroxide in the absence and presence of CEW in a total volume of 1 ml. The incubation was performed at 37 °C for 30 min. The reaction mixture was centrifuged at 4000 rpm after the incubation and the supernatant was discarded, and then the pelleted lymphocytes were resuspended in 100 μl of phosphate buffered saline (PBS) and processed further for Comet assay (Single cell gel electrophoresis) . Comet assay was performed under alkaline conditions essentially according to the procedure of Singh[3] with slight modifications. Images from 50 cells (25 from each replicate slide) were analyzed. The parameter taken to assess lymphocyte DNA damage was tail length (migration of DNA from the nucleus, μm) and was automatically generated by Komet 5.5 image analysis system. The effect of CEW on hydrogen peroxide-induced oxidative DNA damage in human lymphocytes was evaluated as % fluorescence in tail. As shown in Figure 1, hydrogen peroxide-induced DNA damages in human lymphocytes were almost completely protected by CEW. Moreover, oxidative damage of plasmid DNA was also prevented by CEW as analyzed on the agarose gel electrophoretic gel (Figure 2).

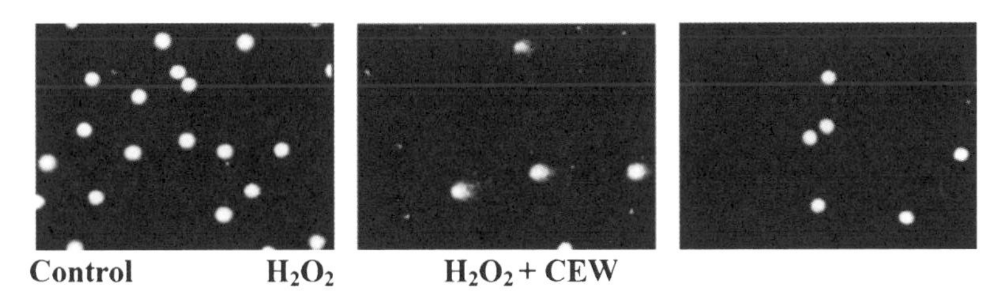

Figure 1 *The protective effect of cathodic electrolyzed water on the hydrogen peroxide - induced oxidative damage of human lymphocyte DNA.*

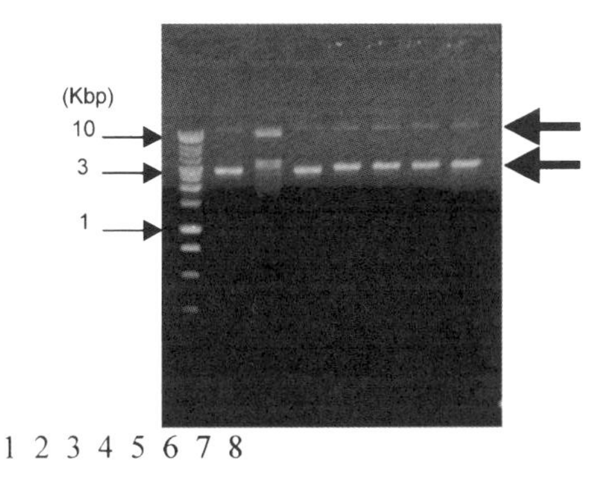

Figure 2 *The protective effect of cathodic electrolyzed water on paraquat-induced plasmid DNA damage.*
Lane 1 : 1 Kb ladder, 2 : DNA in TE buffer, 3 : DNA in TE buffer, 4 : DNA in CEW, 5 : DNA in CEW (1 ul)+paraquat, 6 : DNA in CEW (10 ul)+paraquat, 7 : DNA in CEW (14 ul)+paraquat, 8 : DNA in CEW (18 ul)+paraquat

2. 3 Protective Effect of CEW on the RNA Degradation

Total RNA was extracted serially with phenol and chloroform, precipitated with ethanol according to the standard protocol. Total RNA was incubated with diethylpyrocarbonate (DEPC)-treated H_2O or CEW at 25□, for indicated days. As shown in Figure 3, CEW showed much stronger protective effect on RNA degradation than DEPC-treated H_2O_2 at 25□.

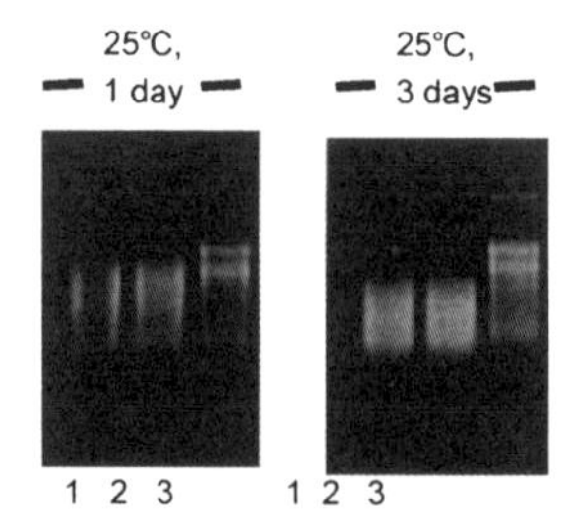

Figure 3 *The protective effect of cathodic electrolyzed water on the RNA degradation Lane 1 : RNA in deionized water, lane 2 : RNA in DEPC-treated H_2O, lane 3 : RNA in CEW*

2. 3 Protective Effect of CEW on the Protein Degradation.

SDS-polyacrylamide gel electrophoresis using horseradish peroxidase was carried out as described by Laemmli.[4] The peroxidase activity with guaiacol as a substrate was assayed. The assay mixture contains 40 mM phosphate buffer, 15 mM guaiacol, 5 mM H_2O_2 and 50 ul of enzyme in a total volume of 1 ml. The reaction was initiated by the addition of H_2O_2, and the increase in absorbance at 470 nm was measured using a UV/VIS spectrophotometer. Several cleaved protein fragments were found in the horseradish peroxidase incubated in the deionized water and AEW for 10 days. However, any cleaved protein fragments could not be seen in CEW-treated sample. This result suggests that CEW could protect protein against oxidative degradation (Figure 4).

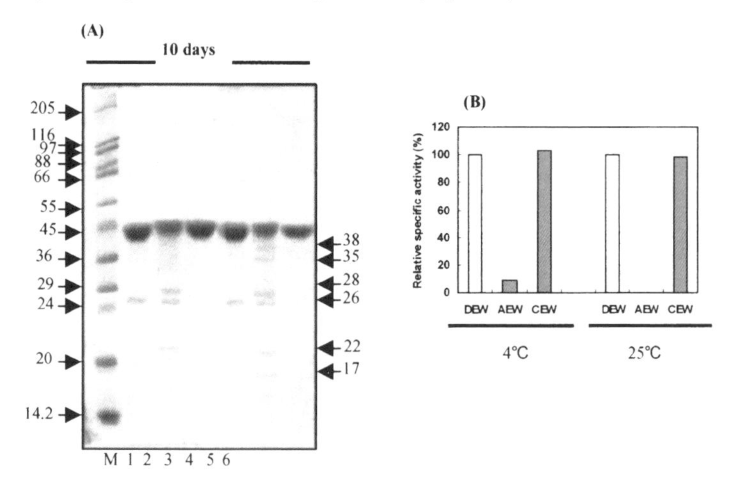

Figure 4 *Effects of cathodic electrolyzed water (CEW) on the horseradish peroxidase degradation (A) SDS-polyacrylamide gel electrophoregram of horseradish peroxidase Lane M : Marker , lane 1 : deionized water at 4°C, lane 2 : AEW at 4°C, lane 3 : CEW at 4°C, lane 4 : deionized water at 25°C, lane 5 : AEW at 25°C, lane 6 : CEW at 25°C (B) Residual enzyme activity.*

3 CONCLUSION

Cathodic electrolyzed water could effectively protect DNA, RNA and protein against oxidative degradations.

References

1 S. Nakagawara, T. Goto, M. Nara, Y. Ozawa, K. Hotta and Y. Arata, *Anal. Sci.*, 1998, **14**, 691.
2 S. Shirahata, S. Kabayama, M. Nakano, T. Miura, K. Kusumoto, M. Gotoh, H. Hayashi, K. Otsubo, S. Morisawa and Y. Katakura, *Biochem. Biophys. Res. Commun.*, 1997, **234**, 269.
3 P. N. Singh, M. T. McCoy, R. R. Tice and E. L. Andchneider, *Exp. Cell. Res.*, 1988, **175**, 184.
4 U. K. Laemmli, *Nature*, 1970, **227**, 680.

Detection of 88 Pesticides on the Finnigan TSQ® Quantum Discovery using a Novel LC-MS/MS Method

W. Gebhardt[1], E. Genin[1], D. Ghosh[1], M. Churchill[1], J. Klein[2] and L. Alder[2]

[1] Boundary Park, Hemel Hempstead, UK;
[2] Federal Institute for Health Protection of Consumers & Veterinary Medicine, PO 330013, D-14191 Berlin, Germany

1 INTRODUCTION

Pesticide residues in food are strictly regulated according to the provisions of US Environmental Protection Agency CFR Title 40. Allowable pesticide residue tolerances may range over numerous orders of magnitude. For example, the tolerance for captan in cattle fat is 0.05 mg/kg, while the tolerance in lettuce and spinach is 100 mg/kg. The lower limit of residue measurement in the EPA's determination of a specific pesticide is usually well below tolerance levels. Pesticide analysis requires a technology platform capable of simultaneous detection of many pesticides over a wide dynamic range.

LC-MS/MS has emerged as the technique of choice for identifying and quantifying pesticides. Following a simple solvent extraction and fast cleanup, extracts are analyzed using matrix-matched standards. To analyze the large numbers of samples whose pesticide treatment history is usually unknown, the EPA uses analytical methods capable of simultaneously determining a number of pesticide residues. These cost-effective multi-residue methods can determine about half of the approximately 400 pesticides and their metabolites with EPA tolerances. Conventional MS/MS methods generally require extensive instrument parameter optimization
for each target analyte or compounds belonging to the same chemical class.

The goal of this work was to develop an automated, generic high throughput LC/MS-MS screening method to detect multi-residue analytes (88) in a short chromatographic timescale following minimal separation by HPLC; demonstrate the utility of using different time segments and scan events and demonstrate evidence of absence of cross-talk between co-eluting components.

2 METHOD

Chemicals and Reagents : Water, methanol and acetic acid were HPLC grade and purchased from J T Baker Chemicals, France.
Samples: Pesticides listed in Table 1 were purchased from Sigma unless otherwise noted. Standards solutions of 0.1, 0.5, 1, 5, 10 and 50 pg/ul were prepared in methanol

3,4,5-Trimethacarb	Clethodim-sulfon	Imazalil	Propoxur
3-Hydroxy-carbofuran	Clethodim-sulfoxid	Imidacloprid	Prosulfuron
5-Hydroxy-clethodim-sulfon	Cyprodinil	Indoxacarb	Pymetrozin
Acephate	Daminozid	Iprovalicarb	Pyridate
Aldicarb	Demethon-S-methyl	Isoproturon	PyridateXX
Aldicarb-sulfoxid	Demethon-S-methyl-sulfon	Isoxaflutole	Pyrimethanil
Aldoxycarb	Desmedipham	Linuron	Quinmerac
Amidosulfuron	Diflubenzuron	Metalaxyl	Quizalofop-ethyl
Atrazin	Dimethoat	Metamitron	Rimsulfuron
Azoxystrobin	Diuron	Methamidophos	Spiroxamine
Bendiocarb	Ethiofencarb	Methiocarb	Tebuconazol
Bensulfuron-methyl	Ethiofencarb-sulfon	Methomyl	Tebufenozid
Butocarboxim	Ethiofencarb-sulfoxid	Metolachlor	Thiabendazol
Butocarboxim-sulfoxid	Fenhexamid	Metsulfuron-methyl	Thiacloprid
Butoxycarboxim	Fenoxycarb	Monocrotophos	Thiodicarb
Carbaryl	Fenpropimorph	Nicosulfuron	Thiofanox
Carbendazim	Fentin-hydroxide	Omethoat	Thiophanat-methyl
Carbofuran	Flazasulfuron	Oxamyl	Triasulfuron
Chlorsulfuron	Fluazifop-P-butyl	Oxidemeton-methyl	Triflumuron
Cinosulfuron	Flufenoxuron	Phenmedipham	Triflusulfuron-methyl
Clethodim	Furathiocarb	Pirimicarb	Vamidothion
Clethodim-imin-sulfon	Haloxyfop-ethoxyethyl	Promecarb	Propoxur

Table 1 *List of pesticide standards used in the assay*

HPLC analysis was performed on the Finnigan Surveyor® HPLC system, using an Aquasil C18 50 x 2.1 mm column. The mobile phase was water/methanol 80/20 V/V (A), and methanol/water 90/10 V/V , both containing 0.05% m/m acetic acid. The program was run at 200 uL/min with a linear gradient of 100% solvent A to 100% solvent B over 11 minutes, hold for 12 minutes, and return to 100% A in 2 minutes.

The MS conditions were as follows :

Source : ESI
Ion polarity : Positive
Spray voltage : 4 kV
Sheath/iAuxiliary gas : Nitrogen
Sheath gas pressure : 40 arbitrary units
Auxiliary gas pressure : 5 arbitrary units
Ion transfer tube temperature : 300° C
Scan type : SRM
CID conditions : Ar at 1.5 mTorr

3 RESULTS

Figure 1a shows the simultaneous detection of 88 pesticides at 50 pg/ul, all within a chromatographic timescale of 16 minutes. An expanded version of this chromatogram is shown in Figure 2b. The complexity of the chromatogram can be seen by expanding the area from 8 to 11 minutes. All peaks are at the same concentration but those with low responses are still detected under other analytes. A summary of the results for these pesticides at 50 pg/ul are tabulated in Table 2. As is clearly evident, excellent linearity was observed for all components with the coefficient of correlation varying from 0.9990 to 0.9998. This is also indicative of the fact that the scan rates used for this assay were sufficient to give enough data points for good quantitation and linearity.

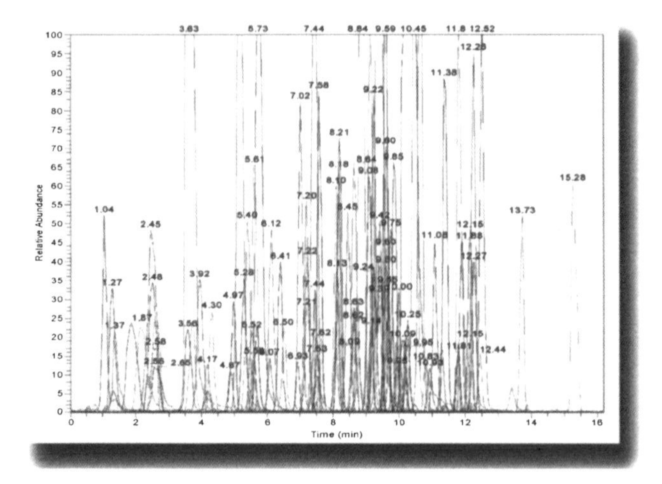

Figure 1a *Ion chromatograms showing the simultaneous detection of 88 pesticides at 50 pg/ul*

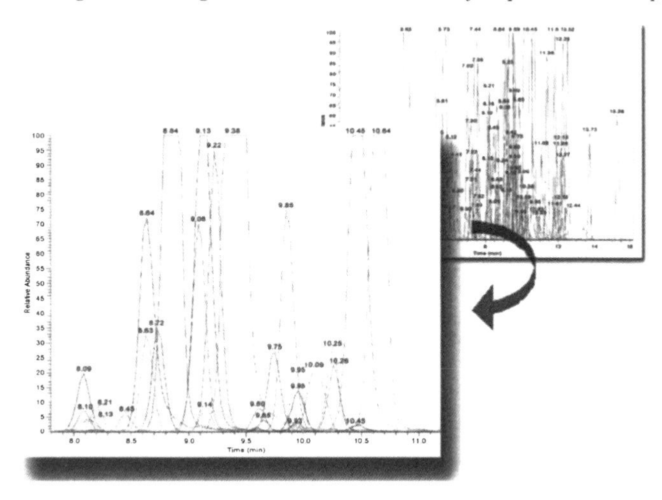

Figure 1b *Expanded ion chromatogram showing detection of minor components under large peaks*

An example demonstrating the type of linearity obtained during the assay is shown in Figure 2, with the associated values (in red) along with a number of other examples are given in Table 2. It should be noted that no internal standards were used during the assay, and that the quantification was carried out by area response only. This figure shows the correlation coefficient varying from 0.9990 to 0.9998 for the five components metamitron, isoxaflutole, iprovalicarb, methiocarb and clethodimin-sulfoxide.

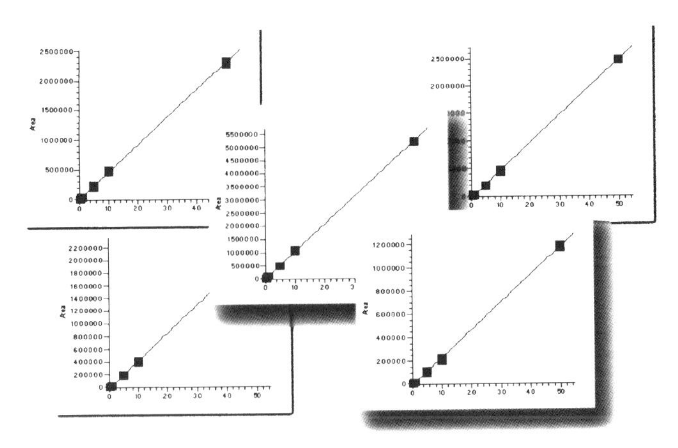

Figure 2 *Demonstration of linearity for five components (metamitron, isoxaflutole, iprovalicarb, methiocarb and clethodimin)*

Component	Retention time (minutes)	SRM	Correlation (r^2)
Metamitron	5.52	203>175	0.9993
Clethodim-imin-sulfon	.67	309>208	0.9992
Isoxaflutole	8.63	360>251	0.9996
Methiocarb	9.95	226>169	0.9990
Iprovalicarb	10.09	321>119	0.9994
Butocarboxim-sulfoxide	2.45	207>75	0.9372
Pymetrozin	2.48	218>105	0.9995
Aldoxycarb	2.55	223>86	0.9977
Carbendazim	3.63	192>160	0.9999
Demeton-S-methylsulfon	3.71	263>169	0.9995
Oxidemeton-methyl	3.92	247>169	0.9989
Monocrotophos	4.17	224>127	0.9964
Ethiofencarb-sulfon	4.30	258>107	0.9983
3-hydroxy-carbofuran	4.97	238>163	0.9871
Ethiofencarb-sulfoxide	4.87	242>107	0.9964

Table 2 *Coefficient correlation values of selected number of pesticides*

4 CONCLUSIONS

A LC/MS/MS screening assay to monitor 100 pesticides using minimal LC separation was developed using the Finnigan TSQ Quantum Discovery. It was possible to analyse all components within a chromatographic timescale of 16 minutes by carrying out SRM transitions with the use of 2 time segments. Using dwell times of 20 ms, it was shown that no cross-talk was present. This type of method will be applicable to both the environmental monitoring and agrochemical industries working under EPA guidelines.

Reference

US Environmental Protection Agency Code of Federal Regulations 40.

Notes: -
Finnigan, TSQ and Surveyor are trademarks of Thermo Electron
AquaSil Siliconizing Fluid for treating glass surfaces is sold by Pierce Chemical Co., Rockford IL

WATER SAFETY PLANS: PREVENTION AND MANAGEMENT OF TECHNICAL AND OPERATIVE RISKS IN THE WATER INDUSTRY

A. Hein, A. Maelzer and U. Borchers[1]

[1] IWW Water Research Centre – Department Management Consulting, Moritzstrasse 26, D-45470 Muelheim an der Ruhr, GERMANY

1 WHY DO WE NEED WATER SAFETY PLANS?

One of the most important requirements to ensure a safe and reliable Drinking Water supply is that:
"Drinking Water Plants should produce drinking water in a quality that will never cause diseases 24 hours per day and 365 days per year"
Water Safety Plans (WSP) [1] can help to achieve this. They are very helpful tools for a state of the art risk management. Their main goals are:

- The identification of hazards from the catchment area to consumers' tap and control of hazards according to individual supply systems and specific demands of water suppliers and consumers.
- A preventive risk management on process level by using effective control measures following HACCP principles.

2 PRINCIPLES AND STRATEGIES OF WATER SAFETY PLANS

The principles and strategies of WSP can be defined in a summarised form as:

- Application of the multi-barriers principle
- Application of HACCP-principles
- Preventive quality management via continuous process and asset monitoring
- Enforcement of own independent surveillance activities of the water supplier

3 STEPS TO IMPLEMENT TECHNICAL RISK MANAGEMENT

The integration of Technical Risk Management into Quality Management schemes is an important step to improve the performance of Water Supply Companies.
Figure 1 demonstrates the continuous cycle for the implementation and adaptation of Risk Management on process level for water supply companies.

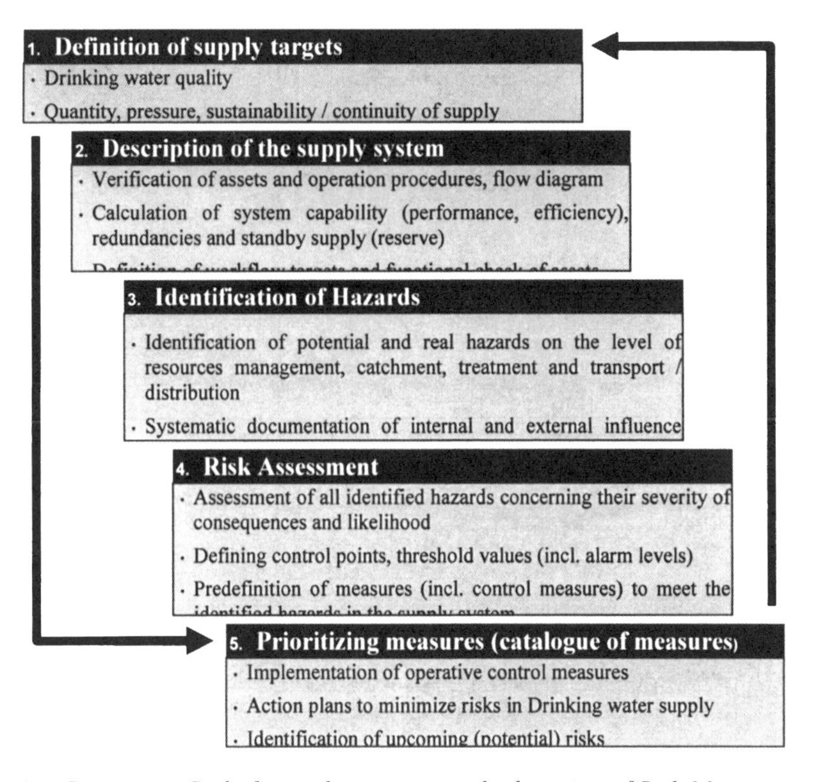

Figure 1 *Continuous Cycle for implementation and adaptation of Risk Management on process level for water suppliers*

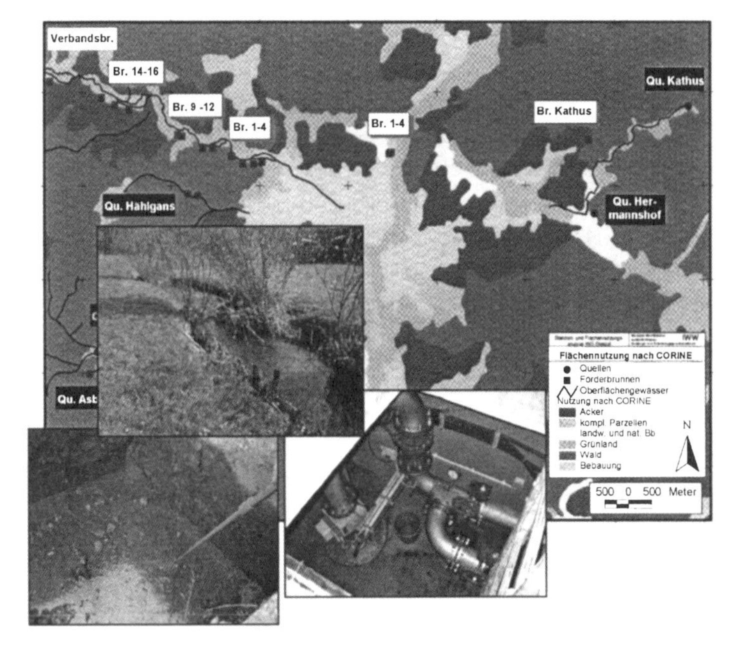

Figure 2 *Examples for hazards and potential risks in water supply*

4 RESULTS AND EXPERIENCES OF TECHNICAL RISK MANAGEMENT PROJECTS IN GERMANY

4.1 Results and Experiences

IWW has been involved in the implementation of the first WSP in Germany. The projects carried out in co-operation with some large water supply companies in Northrhine-Westfalia led to the following results and experiences:

1. The systematic identification and documentation of hazards enables the identification of necessary activities
2. A consequent hazard control increases the quality of supply in the water industry (operating safety, optimisation)
3. A clear and detailed documentation increases the transparency of technical core processes (organisational and technical safety)
4. The established and binding catalogue of measures with priorities defines clear tasks, responsibilities, project planning and budgets
5. The validation of existing structures and workflows plus securing the existing operating expertise and experiences of the staff with addition of external expertise (to avoid blind spots)

4.2 Examples for Implemented Measures

- Installation of smoke detectors (deacidification) with online surveillance
- Renewal of a distribution pipe (risk of breakage)
- Creation of a GIS-based map incl. potential sources of raw water contamination
- Start of additional and more sophisticated analyis
- Evaluation of limit values for well sanitation
- Compilation of written instruction manuals
- Completion of the operation manual

5 CONCLUSIONS

- Water Safety Plans are an effective instrument for risk assessment of technical process competence for resource protection, catchment, treatment, transport and distribution of drinking water
- Water Safety Plan concepts with additional demands (pressure, quantity of supply etc.) reflecting European supply conditions have been successfully adapted to German water supply standards
- Technical Risk Management is a new instrument of systematic and continuous process monitoring for European water suppliers

References

[1] World Health Organisation, *Guidelines for Drinking Water Quality*, Vol. 1, 3rd Edition, WHO, Geneva 2004

Analysis of Aquifer Response to coupled flow and transport on NAOL remediation with well fields

I. David and S. B. Anim-Addo

Institute for Numerical Methods and Informatics in Civil Engineering, University of Technology Darmstadt (TUD), Germany. anim-addo@iib.tu-darmstadt.de

ABSTRACT

The application of optimization methods provides more efficient engineering solutions for remediation systems.

To improve the mathematical and simulation models for the optimization of NAPL Remediation Systems, solutions to non-linear coupled partial differential equations with complex contact, transitional, initial and boundary conditions that consider different types of inner singularities must be addressed.

Optimizing solutions of coupled flow and transport processes lead to non-linear and non-convex objective functions with multiple local minima and both nonlinear models and nonlinear constraints. A thorough analysis of simulation models and scenarios with multi-objective approaches that have economic as well as management constraints have to be studied. The development of these algorithms, however, may have to be preceded by descriptive site-related parameter sensitivity assessments for various remediation scenarios.

This procedure for the estimation of the aquifer response for the most important factors of influence, leads to a more suitable definition of the objective functions, scope and constraints required for the optimization of the remediation system.

Prof. Dr.-Ing. Ioan David, Samuel B. Anim-Addo, MSc.
Institute for Numerical Methods and Informatics in Civil Engineering
University of Technology Darmstadt (TUD)
Germany
anim-addo@iib.tu-darmstadt.de ; david@iib.tu-darmstadt.de

Abstract

The application of optimization methods provide more efficient engineering solutions for remediation systems.

To improve the mathematical and simulation models for the optimization of NAPL Remediation Systems, solutions to non-linear coupled partial differential equations with complex contact, transitional, initial and boundary conditions that consider different types of inner singularities must be addressed.

Optimizing solutions of coupled flow and transport processes lead to non-linear and non-convex objective functions with multiple local minima and both nonlinear models and nonlinear constraints.

A thorough analysis of simulation models and scenarios with multi-objective approaches that have economic as well as management constraints have to be studied.

The development of these algorithms, however, may have to be preceded by descriptive site-related parameter sensitivity assessments for various remediation scenarios.
This procedure for the estimation of the aquifer response for the most important factors of influence, leads to a more suitable definition of the objective functions, scope and constraints required for the optimization of the remediation system.

Governing Equations

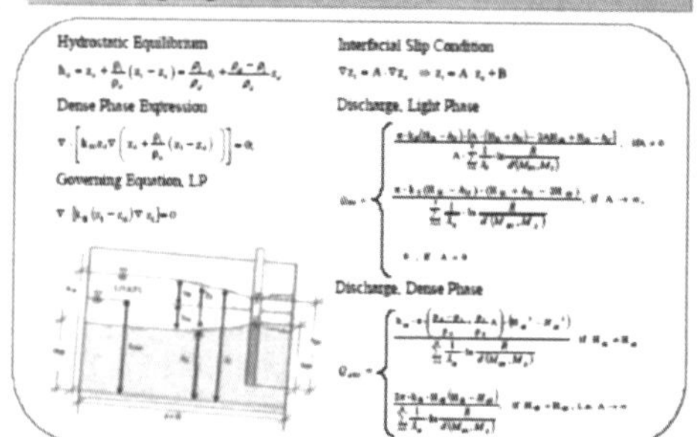

Remediation Scenarios

Contamination and Spread of free-phase LNAPL

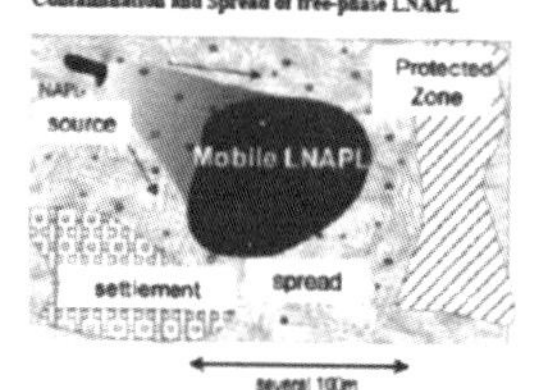

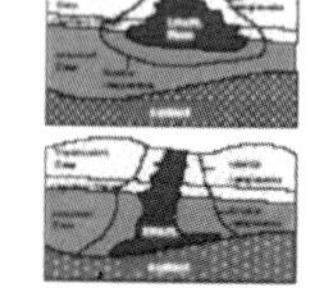

Assumptions

- Sharp interface model
- Dupuit-Forchheimer Approximations
- Fluids are in vertical equilibrium
- Steady state conditions during recovery
- aquifer with sufficient hydraulic conductivity to neglect effects of capillary pressure

Remediation Regimes

Case	Remediation Phase	Constraints
1	LIGHT PHASE ONLY	Skimmer wells that take into account: • *coning* at GW-LNAPL-Interface • constant horizontal GWL (avoidance of *smearing*)
2	DENSE PHASE ONLY	Avoidance of upward coning
3	BOTH PHASES (LNAPL/GW or GW/DNAPL)	Dual pumps under hydraulic and management constraints

Operational and hydraulic Constraints

Concentration constraints

Concentration constrained GW management scenarios have to contend with nonlinearities in the aquifer response and system response to pumping stresses, and the nonlinear form of the objective and the constraints functions.

Hydro-physical Constraints

For a large free-phase NAPL flow domain with known Radius of Influence R, the following hydro-phyiscal constraints are valid

For any point S in the flow domain, $h_l(x,y) = h_{ls}$ and $h_d(x,y) = h_{ds}$

If $A = \frac{\rho_l - \rho_d}{\rho_l}$ then

- Only the LIGHT PHASE can be removed (Case 1)
- The DENSE PHASE shows the characteristic upward CONING in the vertical plane
- Solution for SKIMMER WELL Recovery Systems can be obtained

If $A = 0$, *i.e.* $z_d = H_{1s}$ and $Q_{1i} = 0$ then

- Only the DENSE PHASE can be removed (Case 2)
- Solution for ANY Recovery System can be obtained

If $A = 0$ or $A = \frac{\rho_l - \rho_d}{\rho_l}$ then

- Both LIGHT and DENSE PHASES can be removed.
- Solution for ALL Recovery Systems can be obtained

For large values of A (*i.e.* $A \to \infty$ or $H_{ds} = H_{ls}$)

- interface elevation constant (horizontal), i.e. $z_d = H_{ds}$

Preliminary Results on Analysis of Inter-Parameter Relationships

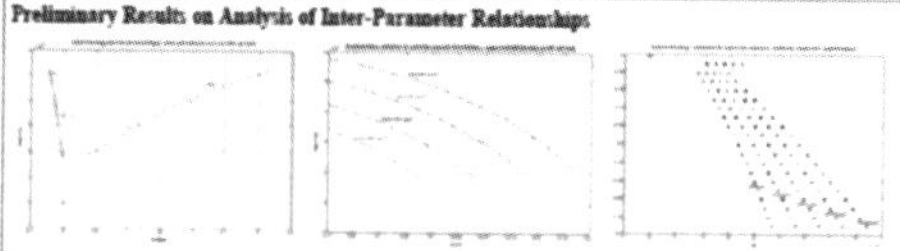

Quantitative Parameter Analyses

Objectives

By Case 1 or 2 (removal of LNAPL / DNAPL only):

- Maximize LNAPL / DNAPL extraction rates
- Minimize Extraction Time at constant LNAPL pumping rates

By Case 3 (Simultaneous Removal):

- Maximize LNAPL/DNAPL extraction rate at minimum / constant GW removal rates

By Case 1-3 (All Cases of Recovery):

- Optimal well placement

Data

H_{ls}=9.2[m]; H_{ds}=10.2[m]; r=0.1[m]; L=100[m]; B=100[m]; k_l=2.62E-5[m/s]; k_w=9.81E-5[m/s]; ρ_l = 800[kg/m^3]; ρ_d = 1000[kg/m^3];

Assumption: Steady-state flow at each time step

Parameter Sensitivity Analysis

- Parameter interdependence
- Time variants
- Functional variations and singularities

Comparative Parameter relationships:

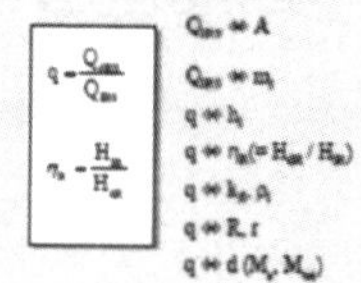

Simulation for $h_l(x,y)$ at constant $h_d(x,y)$

Simulation and Function Derivation

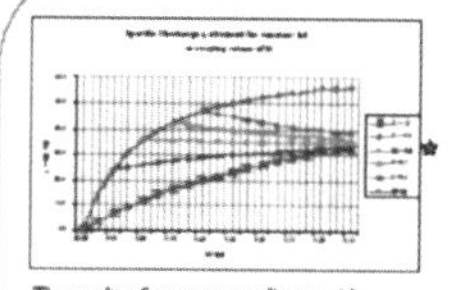

The results of parameter studies provide characteristic curves from which linear and nonlinear relationships within "solution windows" may be generally formulated as

$$Y_s = \alpha X_s^{\,s} + \beta$$

Objective functions are formulated on the basis of these simplified relationships.

The aquifer response to selected parameter relationships lead, under given constraints, to objective formulations

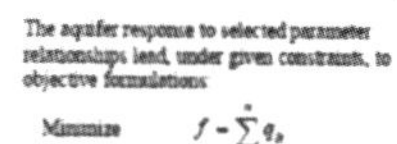

Minimize $f = \sum_i^n q_b$

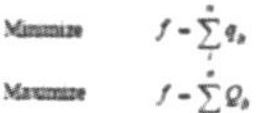

Maximize $f = \sum_i^n Q_b$

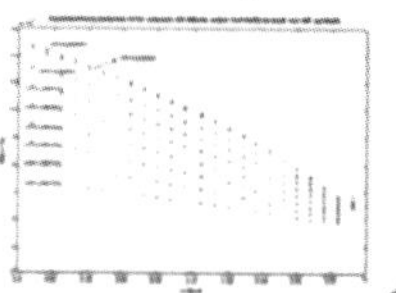

SAFE DRINKING WATER: LESSONS FROM RECENT OUTBREAKS

S. E. Hrudey[1] and Elizabeth J. Hrudey[1]

[2]Department of Public Health Sciences, University of Alberta, Edmonton, Alberta, Canada
steve.hrudey@ualberta.ca

1 INTRODUCTION

The public health implications of the serious drinking water contamination that occurred in May 2000 in Walkerton, Ontario, a community of 4800 residents, hold important lessons for the water industry. More than 2300 individuals experienced gastroenteritis, 65 were hospitalized, 27 developed haemolytic uremic syndrome (HUS), a serious and potentially fatal kidney ailment and seven died. The pathogens identified as being primarily responsible were *Escherichia coli* O157:H7 and *Campylobacter jejuni*. The Government of Ontario established a public inquiry to determine the causes and responsibility for this tragedy and to examine broader questions relating to the safety of drinking water in Ontario.

One of us (SEH) served on the Research Advisory Panel to the Commissioner of the public inquiry. Part of those duties involved advising the Inquiry about outbreaks other than Walkerton that had occurred in recent years. This contribution was developed into a book: *Safe Drinking Water – Lessons from Recent Outbreaks in Affluent Nations* (2004), IWA Publishing, ISBN: 1843390426.

Waterborne disease outbreaks continue to occur in developed countries despite wealthy economies and access to proven drinking water treatment technologies. The obvious question is:

Why do serious outbreak failures continue to occur?

2 METHODS

We undertook an analysis of the major factors contributing to drinking water disease outbreaks by searching the published English language literature over the past 30 years. We screened papers and outbreak investigation reports that discussed specific disease outbreaks and that described some of the failure modes contributing to the outbreak. We did not attempt to review all outbreaks. There was substantial variation in the quality and detail of describing failure mechanisms among these papers. The multiple barrier approach was adopted as a reference framework for analyzing the outbreak literature with a view to converting hindsight into foresight.

The factors reported to have contributed to these selected outbreaks were considered in relation to the documented failures contributing to the Walkerton outbreak. The Results summary is limited by space to considering the Walkerton outbreak alone. Other fatal outbreaks are summarized in Table 1.

Date	Location	Major Failure	Pathogens	Cases Confirmed	Total Cases Estimated	Hospitalizations	Deaths
1983 Feb	Drumheller, AB, Canada	sanitary sewage spill upstream	not identified	1326	3,000	not reported	2
1989 Dec 1990 Jan	Cabool, MO, U.S.A.	sewage infiltration believed to have contaminated a distribution system	*Escherichia coli* O157:H7	243	no estimate	32	4
1990 Oct	Saitama, Japan	contamination of well water supply to a nursery school	*E. coli* O157:H7	42	186	>20	2
1993 Mar-Apr	Milwaukee, WI, U.S.A.	sanitary sewage contaminated drinking water intake	*Cryptosporidium parvum* genotype 1	285	~400,000	4,400	50-70 extra over 2 years
1993 Nov-Dec	Gideon, MO, U.S.A.	bird faeces contaminated insecure storage tanks	*Salmonella typhimirium*	31	650	15	7
1999 Sep	Washington County Fair, NY, U.S.A.	concessions used unchlorinated water from a shallow well located ~11 m from a dormitory septic tank	*E. coli* O157:H7 *Campylobacter jejuni*	171	2,800-5,000	71	2
2000 May	Walkerton, ON, Canada	inadequate chlorination to cope with influx of manure contaminated groundwater, lack of monitoring	*E. coli* O157:H7 *Campylobacter* spp.	163 (E) 105 (C) 12 both	2,300	65	7

Table 1 *Summary of Recent Fatal Waterborne Disease Outbreaks in Developed Countries*

3 RESULTS

3.1 Multiple Barrier Failures

Justice O'Connor adopted the expert evidence that a multiple barrier approach is necessary for providing safe drinking water:

Source: the best possible raw water quality should be maintained

Treatment: effective treatment designed, operated and maintained

Distribution: secure storage and distribution of treated water

Monitoring: appropriate and effective monitoring must be performed

Response: appropriate and effective responses to adverse monitoring or adverse circumstances

The review of other outbreaks reveals that they commonly involve failures in more than one barrier, a finding that reinforces the need for the multiplicity of barriers and the need to assure the effectiveness of each barrier.

3.2 Source

The pathogens (*E. coli* O157:H7 and *C. jejuni*) causing the Walkerton outbreak were attributed to contamination arising from cattle manure from a local farm. These pathogens contaminated a shallow production zone (only 5 to 8m depth) of Walkerton Well 5. The contamination was believed to have occurred following a period of exceptionally heavy spring rainfall, estimated to be a 1 in 60 year occurrence.

3.3 Treatment

Chlorine dosage practice at Well 5 was insufficient to achieve a 0.5 $mg{\cdot}L^{-1}$ residual and it is reasonable to assume that the contamination causing this outbreak was accompanied by a chlorine demand sufficient to consume entirely or almost entirely the low chlorine dose, thereby allowing inadequately disinfected water into the distribution system.

3.4 Distribution

Several scenarios whereby the distribution system at Walkerton could have been contaminated were identified and investigated but the source of contamination of the Walkerton water system was cattle manure from a farm near Well 5.

3.5 Monitoring

Well 5, which was known to be under the direct influence of surface water since 1978, was not required to implement a 1994 policy to provide continuous chlorine residual monitoring. If the Walkerton system had been required to do so the fatal outbreak would have been prevented.

3.6 Response

The operators at Walkerton clearly lacked any substantive understanding of the need for disinfection to inactivate pathogens in drinking water and the serious health consequences that could arise from failing to maintain adequate disinfection of the Walkerton water supply. As a result they failed to provide adequate disinfection and they failed to monitor chlorine residual which would have provided a real time warning of the contamination.

Contaminated water was delivered for 8 days because chlorine residual monitoring was not carried out and appropriate actions were not taken before a boil water advisory was issued. This was only issued after widespread disease with no other likely cause being readily evident in the community.

4 DISCUSSION

A short list of priorities must be understood by anyone responsible for drinking water safety. These include:

1. Pathogenic microorganisms still pose the greatest recurring risk to the health of drinking water consumers. Treatment and disinfection of pathogens that may be present in raw water must be maintained effective, by a substantial margin, to prevent waterborne disease outbreaks.
2. Any sudden or extreme change in water quality, flow or environmental conditions (extreme precipitation, snowmelt, runoff or flooding) should arouse suspicion of adverse conditions that might cause drinking water contamination.
3. The drinking water system must have and continuously maintain robust and resilient, multiple barriers appropriate to the level of contamination challenge to the raw water supply.
4. System operators must have an effective, continuing capacity to learn from past problems, failures and near failures so as to respond quickly and effectively to adverse monitoring signals.
5. System operators must be personally dedicated to continuously providing consumers with safe water. Concerns or complaints about water quality from consumers or other responsible parties must always be investigated.

5 CONCLUSIONS

A multiplicity of failures occurred in Walkerton despite the readily accessible experience from elsewhere warning of many similar failure modes. The challenge for improving drinking water system safety is to reform the pervasive culture of complacency that has been evident among so many key players in drinking water systems. Such complacency must be replaced with a culture of personal responsibility and vigilance.

Finally, a sense of personal accountability, based on an understanding of a manageable set of principles, must be communicated to all those holding responsibility for protecting public health by providing safe drinking water.

6 ACKNOWLEDGEMENTS

The authors are indebted to Justice Dennis O'Connor for documenting so thoroughly what happened in Walkerton and Dr. Harry Swain for guiding the Research Advisory Panel. The authors also acknowledge the role of the Canadian Water Network – Réseau Canadian de l'eau in facilitating this work and the Cooperative Research Centre for Water Quality and Treatment of Australia for providing the base for developing this approach, Alberta Health & Wellness and the Natural Sciences and Engineering Research Council of Canada for financial support.

PREVENTION AND SECURITY MEASURES AGAINST POTENTIAL TERRORIST ATTACKS TO DRINKING WATER SYSTEMS IN ITALY

M. Ottaviani[1], R. Drusiani[2], E. Mauro[2], L. Lucentini[1], and E. Ferretti[1]

[1] Istituto Superiore di Sanità, Department of Environment and Primary Prevention, Section of Inland Water Hygiene - Viale Regina Elena, 299 - 00161 Rome, Italy; e-mail: ottavian@iss.it
[2] Federgasacqua, Via Cavour, 179/A - 00184 Rome, Italy.

1 INTRODUCTION

The dramatic changes in the entire world scenario since September 11, 2001 have required improvement security of many critical infrastructures, and to cope with unprecedented forms of attack.

Water supply systems have been identified as one of the key elements within the government-wide efforts to ensure homeland security both in the US[1] and the EU[2].

Water infrastructure communities in Italy - as in other European countries – rely on long-established measures and procedures to cope with emergencies, prevention involving both natural disasters (storms, blizzards, earthquakes) and simple well known threats (vandalism, theft). However, the recent inclusion of broader malevolent threats by domestic or foreign terrorists called for a new integrated approach to enhance security of water supply systems and moved the focus to preventing and responding to a wider range of human-provoked emergencies. These phenomena and related consequences - *e.g.* disruption of water supply - could involve physical destruction of drinking water infrastructure - *e.g.* use of explosive devices - or water contamination and may exert a tremendous impact on public health, as well as on social and economic activities.

As a first approach to pursue an effective global management of security of water supply systems in Italy, a multidisciplinary group was set up in 2001. This combined expertise in water quality, drinking-water supply chain, up to the point of consumption, water contamination and risk assessment and knowledge of system-specifics in different local and territorial situations. Experts of the group belonged to the Italian Ministry of Health, the National Institute of Health (Istituto Superiore di Sanità, ISS) and the Italian association of water utility companies (Federgasacqua) (Table 1).

The activities have been focusing on the development and sharing of management strategies, operational tools and physical measures able to prevent and handle intentional attacks to water systems in Italy. The workgroup's results are currently being collated in order to set guidelines addressed to Public Health, Civil Protection and water protection agencies to prevent and manage crises related to water contamination incidents.

Ministry of Health: N. Sarti
National Institute of Health: L. Bonadonna, P. Bottoni, G. Citti, G. Donati, E. Ferretti, S. Frullani, E. Funari, L. Lucentini, A. Nusca, M. Ottaviani, E. Veschetti, G. Zapponi
Federgasacqua (Water Service): R. Drusiani, M. Gatta, E. Mauro
Water utility companies: P. Carlino, C. Alaimo, A. Felici (ACEA ATO2 – Roma), D. Davoli, N. Fontani (AGAC S.P.A. - Reggio Emilia), M. Colombino, O. Conio, F. Palumbo (AMGA S.P.A. – Genova), R. Di Marino, F. Donadio, G. Magistrale (AQP S.P.A. – Bari), E. Pafumi (ASM S.P.A. – Brescia), F. Alava (BAS S.P.A. – Bergamo), G. Peterlongo, A. Vitale (CAP Gestione S.P.A. – Milano), L. Minelli, G. Paderni, D. Nasci (HERA S.P.A. – Bologna), C. Pacchiarotti, M. Pieroni, T. Serenelli (MULTISERVIZI S.P.A. – Ancona), D. Burrini, G. Ciatti, D. Santianni (PUBLIACQUA S.P.A. – Firenze), L. Agnoletti, G. Graziani, F. Farina (ROMAGNA ACQUE S.P.A. – Forlì), L. Meucci, L. Cappuccio (SMAT S.P.A. – Torino), R. Brinis, M. Fabris, P. Miana, F. Perissinotto (VESTA S.P.A. – Venezia)

Table 1 *Components of the workgroup on security of water supply systems in Italy (2001-)*

2 METHOD AND RESULTS

2.1 The Italian framework

When we approached the subject of prevention and management of water supply system security against human-provoked emergencies, we were faced with a very complex scenario.

A first order of problems deals with the broad range of very different malevolent acts to be considered. These can be addressed to the infrastructure assets, to the water safety and/or against the workers and consumers and can include physical destruction actions, attacks to the supervisor control and data acquisition (SCADA) or connected cyber systems, biological, radiological and chemical intentional contamination of water systems exerting a risk for public health or impact on acceptability of water.

The second critical element concerns the multifaceted Italian water infrastructure sector. As apparent from data of Tables 2 and 3, this is a large and diverse sector consisting of many thousands of separate infrastructures across the country, varying from very large urban systems serving millions of consumers to small community systems providing water to very limited populations.

Component	Figure	Unit	Source
Catchments	9,137,651,000	m^3	ISTAT 1999
Aquifer	48,6	%	ISTAT 1999
Spring	37,9	%	ISTAT 1999
Surface	13,3	%	ISTAT 1999
seawater	0,3	%	ISTAT 1999
Water distributed	7,855,894,000	m^3	ISTAT 1999
Water supplied	5,692,010,000	m^3	ISTAT 1999
Water use			
domestic	74,79	%	ISTAT 1999
industrial	8,82	%	ISTAT 1999
agriculture	147	%	ISTAT 1999
Other	14,92	%	ISTAT 1999
Water suppliers			
aqueducts	4,635		ISTAT 1999
distribution systems	5,360		ISTAT 1999

Table 2. *Some data on public drinking water systems in Italy*

System size (m^3)	Number
1-100	7,132
101-500	2,826
501-2000	1,195
2,001-5,000	315
5,001-1,.000	145
10,001-20,000	61
20,001-100,000	47
>100,000	7

Table 3. *Size of Public Water Systems*

Ownership and responsibility are both public and private and diverse systems may significantly differ in age, experience and needs. Water supply systems catch ground and/or surface waters, which are transported to treatment facilities by a system of reservoirs, aqueducts and pipes which may be open or closed, running on the surface or below. Simple or multiple plants, in which waters are treated, usually through flocculation, filtration and disinfection, can be used depending on the extension of the water system. Treated waters are finally delivered to the consumers by relatively complex service reservoirs and distribution systems, operating at different hydraulic conditions (*e.g.* pressures, flows) and equipped with many diverse water pipelines, valves and other fittings.

A third element to be considered when approaching security-related issues for water systems is the low intrinsic economic value of water, as it is essential to human society and recognized as a common property good, with the state responsible for regulating access and use. As a result, the economic resources available for government and private sectors for investments, also including water security, are usually very limited.

2.2 Approach

In order to cope with unprecedented forms of potential attacks to drinking water systems, an intense sharing of knowledge is necessary within the water infrastructure communities, involving central and local Authorities in charge of assuring and monitoring water quality, decision makers in emergency situations, water supply managers, bodies for the protection of public health and the detection and management of crises.

The following criteria were established by the Italian workgroup for developing management strategies, operational tools and physical measures able to prevent and handle intentional attacks to drinking water systems in Italy.

- Multiple and different elements involved in the task require a multidisciplinary approach based on expertise in the drinking-water supply chain, up to the point of consumption, knowledge on water contamination and risk assessment issues and a good awareness of system-specifics in different territorial and social contexts.
- On account of dissimilarities among water infrastructures, guidance to drinking water utilities for water security related issues should be diversified (*i.e.* from basic to advanced) and highly flexible, so that they can be tailored to individual water systems.
- In order to optimize the allocation of the scarce economic resources available for the water sector, priority should be given to preventive actions, including basic physical defence and simple monitoring measures, rather than complex management technologies and advanced analytical systems. This is particularly important for small water systems which are less protected and highly vulnerable to even unsophisticated attacks.

- Large drinking water supply systems, mainly located in urban areas, could represent an important target for attacks due to the potential dramatic influence on public opinion. This could prompt hostile actions by skilled terrorists able to get around conventional security measures. Thus, although large water systems have generally already implemented physical security systems against some "traditional" malicious threats (*e.g.* vandalism, theft), additional measures should be applied against more serious unprecedented threats. Such measures should be based on an integrated innovative approach and on advanced technical solutions, involving early crisis detection and decision support to prevent and counter threats and mitigate the effects of attacks towards water .

The key features to be considered are:

1. Background:
- comprehensive information on bioterrorism, chemical and biological agent attacks, past events of intentional water contamination;
- identification of potential objectives of hostile actions and chemical, physical and biological risk agents that could be used.
2. Scenarios:
- vulnerability assessment of different water supply utilities;
- identification of scenarios for possible attacks to water systems.
3. Prevention and response measures:
- protection and monitoring measures for different water systems;
- decision making and emergency plans.
4. Costs.
5. Research and support activities.

2.3 Background

Prevention and management of human-provoked emergencies in drinking water supply utilities should be enhanced through an accurate identification of hostile actions which the specific water infrastructure may have to face. The assessment of potential hazardous events that can have major impact on water quality and on the performance of drinking-water systems is the core element in order to design and implement effective measures to prevent and mitigate risks, assuring an optimal resource allocation.

As a first step, the different workgroup members' expertise allowed a better knowledge of events of toxic/infecting and warfare agents used against human beings and phenomena related to bioterrorism before and after September 11, 2001.

Specific attention was then focused on the international and European counter-terrorism measures and security actions related to drinking water infrastructures, including legislative acts, recommendations (*e.g.* ref.[4]) and guidelines (*e.g.* ref.[5]). Particularly measures taken within the EU to strengthen defences against deliberate release of biological and chemical agents, in terms of preparedness and response were discussed.

Another important issue concerned a review of biological, physical and chemical agents potentially employable for water contamination, their general characteristics, dissemination methods, routes of exposure, health effects, toxic doses, access limitations. Some databases of toxic agents, belonging to relevant international bodies, such as CDC, EPA, EMEA, ISLI, WaterISAC, WHO, were identified and shared within the group. Based on criteria of risk assessment and likelihood of hostile actions, a list of agents was prepared by the workgroup and made available for the water infrastructure community.

2.4 Vulnerability assessment and scenarios

Due to the many differences among Italian drinking water systems (*e.g.* size, population served, water source, treatment complexity, system infrastructure), the identification of a unique water infrastructure security standard is not feasible. The approach was therefore focused on designing a common general scheme aiming at identifying the major critical points of the water chain related to possible adversarial actions. A key element is vulnerability assessment, helping the water systems to evaluate their susceptibility to different human-provoked emergency scenarios. Through the identification of the most likely threats and the critical facilities and assets, appropriate actions for reducing or mitigating risks can be determined. The final objective is the safeguard of public health and the reduction of the potential for disruption of water system infrastructures.

2.4.1. Vulnerability assessment of water supply utilities Vulnerability assessment represents a process of varying complexity, depending on local/territorial evaluation, design and operation of the water system. Many methodologies are reported for vulnerability assessment in US drinking water systems[6]. Although vulnerability assessment methods should be determined by individual water systems, the following criteria should be considered, taking into account the two main risks of hostile actions towards water systems, *i.e.* physical destruction of infrastructure and water contamination with biological, chemical or radiological agents.

- *Site analysis* - The feasibility of the attack is deeply related to the accessibility of the water supply system, thus an important feature of vulnerability assessment deals with location of the systems, *e.g.* urban areas, mountain or countryside areas with easy or difficult access.
- *Water system characterization* - Different ways of attack could be tailored on the system specific features. Physical attacks (*i.e.* use of explosives), aiming at provoking a large damage to users, with an extended and prolonged interruption of water supply, may be reasonably addressed to components involving large water volumes, such as reservoirs, treatment plants, high pressure distribution systems. Water contamination actions could virtually occur at any location, the most accessible being the source waters, many tracts of distribution systems, water pipelines, valves, and terminal appendages such as fire hydrants, backflow check valves and user taps; the feasibility of intentional contamination of pressurised water distribution systems and of system components involving large water volumes was considered relatively improbable, due to the difficulties in obtaining and managing destructive quantities of contaminants and to the existence of multiple barriers to the introduction of contaminants within the water supply units. Important distinctions should therefore be made between elements in which water is or is not under pressure and on the basis of water volumes stored, treated or distributed. The water source is another key element in vulnerability assessment and the possibility of contaminant introduction at the catchment stage should be carefully evaluated (*e.g.* existence of navigable channels in surface water used for drinking water production) since this action, though very difficult to carry out, can exert the gravest effects on user communities. Distribution systems are generally regarded as the most vulnerable tracts of the water supply chain: contamination acts are usually considered less feasible if the contaminant insertion point is far from the final user, on account of existing of physical barriers and of the large amount of agent necessary, even if the impact of the action on consumers' health would be massive. Additional elements to be considered in order to assess vulnerability may include system-specific features such as utility facilities, management and operating procedures, chemical use and storage.

- *Service area* - A critical element of vulnerability assessment is related to the mission and objective of the water supply system, including the identification of sensitive populations and critical customers, which could be possible targets, within the service area (*e.g.*, hospitals, schools, governments, military structures *etc.*). This is an important element not only in the prevention steps (*i.e.* higher likelihood and severity of an adverse attack against the water systems), but also for setting up response measures (*e.g.*, notification to the concerned structures in the event of a specific threat against the utility).
- *Dependence on other infrastructures* - Many assets could be essential for drinking water supply systems. They could include electricity, transportation, connections with other water utilities, transmission systems, supervisor control and data acquisition (SCADA) systems, satellite communication systems, geographic information systems (GIS), and may represent critical points for the processes of the drinking water utilities and for water safety.
- *Existing countermeasures* - Water supply systems currently employ countermeasures in order to protect critical assets from conventional malicious threats or natural disasters. They can include physical security equipment, surveillance systems, online monitoring systems. Within the vulnerability assessment process it is important to consider all the existing countermeasures by determining their performance characteristics and risk reduction capability.

2.4.2. Scenarios of possible attacks to water systems An important preventive action closely related to and often included in the vulnerability assessment concerns the qualitative probability (use of simple categories of probability) estimate of malevolent acts against drinking water supply systems. The possible modes of attack that may reasonably take place should be considered and assessed within the individual drinking water system.

The working group examined potential attacks to drinking water systems taking into account the threats – including the kind of adversary and the mode of attack - which are reasonably predictable in Italy, the possible impact on human health, on service performance and costs. The results, which relied to a significant extent on experts' opinions, are summarized in Table 4. Different risks have been described by identifying the likelihood of occurrence, evaluating the consequences if the hazard occurred and pointing out the most sensitive components of the water systems.

Risk assessment carried out on the Italian water systems assigned a low priority to intentional radiochemical contamination on account of the nature, availability and amounts of materials needed for adverse effects and of the great skills required to perpetrate the attack[3].

An important element to be considered when designing scenarios following human-provoked emergencies is the time elapsing between the hostile action and the effects on the system or population.

In case of intentional contamination the time will depend on the distance between the insertion point and the terminal appendages. Also the system component (*e.g.* pipeline, storage tank) in which the contamination is perpetrated determines the residence time of the agent. The resulting time and possible mixing or dilution phenomena following contaminant insertion are important factors to be considered in designing response measures and mitigating the effects of the hostile action.

Risk scoring	Action	Impact	Severity	Likelihood	Sensitive component
Risk 1	Physical demolitions (explosives) of dams or large reservoirs	H S	C-E A	E	C, M, T
	Massive contamination of large volumes of water with chemical or biological agents	H S	A-B A	E	C, M, T
Risk 2	Contamination of limited volumes of water with chemical or biological agents	H S	B-D C-D	D-E	S, D
	Physical demolitions (explosives or tampering) of secondary reservoirs and water storage facilities or booster pumping stations	H S	D-E C-D	B-C	S, D
Risk 3	Attacks to drain, flushing devices Attacks to hydrants or terminal appendages	H S	D-E C-D	B	S, D

Table 4 *Some risks related to potential attacks to drinking water supply systems in Italy*

Keys
Impact: H: impact on human health; S: impact on the service (interruption of supply) and economic impact
Severity: A: catastrophic; B: major; C: moderate; D: minor; E: insignificant
Likelihood: A: almost certain; B: likely; C: moderately likely; D: unlikely; E: rare
Components of drinking water supply system: C: catchment (intake/well), M: Transmission main; T: treatment; S: Service reservoirs; D: distribution

2.5 Preventive and response measures

Many water utilities in Italy have developed strategies and incorporated defence measures to prevent malevolent actions by vandals, criminals or saboteurs, effects of natural disasters on water infrastructures or accidental water contamination events. Changes in world security prompted water infrastructure communities to reconsider their strategies against human-provoked emergencies. Vulnerability assessment, as above reported, is the key tool for identifying appropriate additional countermeasures against terrorist actions.

2.5.1 Protection and monitoring measures Protective measures are finalised to enhance water system security to protect employees and consumers and to minimize disruption of service from malevolent actions. The measures should be determined on the basis of the vulnerability assessment results and should be tailored to individual water systems. Basic protection measures are usually adopted by small water systems, having less protected and highly vulnerable infrastructures, whereas advanced protecting measures and monitoring techniques are implemented by large urban water systems for which attacks by skilled terrorists are more likely.

The following fundamental protection measures are identified.

- *Training activities* – Training courses and seminars should be addressed to the personnel of the water system to improve awareness of risks related to potential attacks to drinking water supply systems, as identified by vulnerability assessment.
- *Physical security equipments* – A large variety of security devices and equipments is commercially available. Vulnerability assessment, characteristics and requirements of

the areas to be protected, as well as cost-benefit analyses, will determine the most appropriate security equipments for individual water supply systems. Security enhancements may include: fencing, gating, lighting, security doors and locks, equipments for access control and detection of intruders (*i.e.* by security cards, PIN, biometrics), sensors, security cameras, tamper-proofing of manhole covers, fire hydrants and valve boxes. Innovative data fusion modules, able to merge data coming from a number of different sensors (video cameras, sensors for water quality tests, biometric sensor for authentication, etc) can be adopted by high technology large water systems.

- *Monitoring systems* – Measures concerning the quality of drinking water, established in Italy according to EU standards (Dir 98/83/EC, transposed in Italian Law by D.Lgs. 31/2001) are not sufficient to meet requirements in relation to early detection of infective agents and toxicants, potentially inserted in the water supply systems within terrorism events. Routine methods, capable of detecting the relatively restricted list of specific analytes commonly applied to water quality assessment, may fail to monitor many potential threatening agents and routine analyses times are not consistent with the restricted time available during emergency episodes. Moreover, intentional water contamination incidents are relatively brief, but severe phenomena are difficult to identify by the conventional sampling frequency adopted for water quality monitoring.

  Different analytical approaches have been proposed for potentially contaminated water samples allowing specific detection of chemical and biological agents[7]. A wide-range of advanced analytical techniques (*e.g.* liquid chromatography coupled with mass spectrometry and tandem mass spectrometry, LC-MS, LC-MS/MS, inductively coupled plasma mass spectrometry, ICP-MS polymerase chain reaction, PCR, etc.) are applied to the detection of different classes of analytes to perform basic or expanded screen for target and unknown contaminants. Though this is an effective, sophisticated and integrated strategy for identification of intentional water contamination it is a high-cost and labour consuming approach, also requiring advanced skilled technicians. Nevertheless such approach would be the only eligible means to gain unambiguous data on water contamination during emergency episodes and it should be implemented in appropriate centres of excellence to give analytical support to water infrastructure community.

  Although rapid means of detecting the occurrence of biological or chemical attacks on water systems are necessary as a core security measure to protect public health, their development and availability is hampered by several factors. Firstly, the wide range of potential threat agents, significantly differing in chemical and biological properties, require highly flexible analytical systems and diverse detection capabilities. Secondly, the potential introduction of the contaminant anywhere along the systems require relatively inexpensive in-field analytical systems to be used for rapid screening tests.

  On the other hand, following the introduction of contaminants, water quality baseline relating to simple parameters such as pH and conductivity can be changed from "normal" conditions to "anomalous" values. Thus, contaminant detection can be approached relying upon the measure of "water quality surrogate monitors" instead of contaminant specific methodologies.

  "Early warning systems" (EWS) aim at identifying a low probability/high impact contamination incident in a water system allowing sufficient time for an appropriate response that mitigates or eliminates any adverse impact resulting from the incident[8]. An ideal EWS would be fully automated, have a rapid response time and high sampling rate, provide a specific and sensitive screen for a range of contaminants, have a low occurrence of false positives and negatives, be reliable and robust, easy to use, affordable to install and operate[8].

The suite of instruments and water quality surrogate monitors as reported in Table 5 can be proposed as early warning system. The system allows on-line application, even in different sites of water systems - so that the migration of contaminants within the system can be monitored – easy interfacing with automated systems and Information Technologies Systems, with possibility of auto-sampling in case of alarm. With the exception of TOC, measurements are very simple, reagent free and systems are relatively inexpensive. Significant changes in the "usual" water quality as recorded by the system should trigger the agent identification. The core element of the system is therefore the "unusual" water quality that should be assessed against a well defined water quality baseline, taking into account "normal" changes in significant parameters in time and space.
Selecting sampling sites to detect contaminants within the water supply system should be an iterative process based on local conditions and on experimental data to be assessed by individual systems.

Parameter	Chemical surrogate	Microbiological surrogate	Toxin surrogate
pH	+		
Turbidity	+	+/-	
Total organic carbon	+		+/-
Chlorine residual	+	+	
Conductivity	+		
Dissolved oxygen	+		
Nitrate, nitrite	+	+/-	
Phosphate	+	+/-	
Biomonitors	+	+	+

Table 5. *Early warning systems*

Second step (confirmatory) analysis is initiated by early warning system positive results. It is based on advanced and specific analytical methodologies and targeted at avoiding false alarms and unambiguously identify/quantify risk factors to perform correct risk assessment procedures.

- *Information security* – A basic information security policy should be implemented by the water systems in order to assure that no sensitive information on water utility functioning and security measures is available to external unauthorized persons and may be used for adverse actions. Adequate personnel training and implementation of specific procedures/recommendations should be addressed to the staff managing sensitive information on drinking water systems.

2.5.2 Response measures Response actions involve the strategies and measures prepared by the water supply system to respond and recover from malevolent acts with the aim of minimizing the health impact or the disruption of service. They should be implemented following any kind of threat warning, *i.e.* any unusual event, observation, or data that indicates a potential contamination incident

- *Decision making* - The first element to be considered concerns the roles and responsabilities within the drinking water systems during the phase of response to an emergency, particularly during the initial stages of the response. Responsibilities and authorities should be defined by the utility management and may vary depending on the specific conditions of the utility.

A topic highlighted by water utility companies concerned the management of threat warnings notified by different means and particularly by phone by the potential perpetrator. Such warnings are frequently received by drinking water utilities. A crucial step obviously is assessing the credibility of the notification by evaluating the real possibility of acts having consequences on public health, (*e.g.* contamination incidents), infrastructure damages, adverse impacts on the aesthetic qualities of the drinking water, and reduced consumer confidence.

A comprehensive process of evaluation of threat credibility is reported elsewhere together with details on the methodology[6]. The Italian workgroup agreed on the necessity of designing a flexible decision process, tailored on the specific structure utility and possibly defined in a standard procedure, taking into account the following key elements.

- Decision group: a restricted group should be constituted, possibly involving personnel with different expertises, with well defined roles and responsibilities, in charge of the threat management process, taking into account the very limited decision time (*ca.* 1 hour); the contact info for each component should be available, including substitute for key roles, and the contact list should be easy accessible for the utility personnel.
- Input: all available information relevant to the contamination threat should be collected in order to deeply analyse the credibility event.
- Evaluation: collective information should be methodically evaluated to determine the credibility of the contamination threat. Although this is an iterative, flexible process tailored to the individual event/system, the following fundamental stages should be considered during evaluation[6, 9].
  - "possible event": initial evaluation aiming at discriminating between threats requiring further investigation and those that can be dismissed as implausible. If the threat is regarded possible, immediate operational responses should be implemented to contain the suspect water while the investigation continues. Otherwise the investigation is closed, the threat filed and routine activities are restored in the system.
  - "credible event": this is when sufficient information corroborates the evidence that the water quality has changed. If the threat is regarded as 'credible,' adequate response actions should limit the potential for human exposure to the suspect water. Health and Police Authorities should be notified promptly. The investigation will continue together with such response actions in an effort to confirm the contamination incident.
  - "confirmed event": the event is confirmed by definitive information unambiguously establishing that the incident has occurred.

As more information about the threat becomes available and the situation evolves, different organizations may play a major role or take command in crisis management. Particularly Public Health Authorities and Civil Protection may assume responsibility in situations in which there is a potential threat to public health. Public Security Authorities may assume responsibility for incident command in situations where criminal activity is suspected.

Whereas a State/Local Authority assumes responsibility for incident command, the utility should play a supporting role during the threat management process and maintain responsibility for the system.

- Output: conclusions of the threat evaluation; either the event is considered a hoax and normal operative conditions are restored in the water utility or alternatively successive emergency actions are implemented according to emergency procedures as designed within the "Emergency Response Plan".

- *Emergency Response Plan* - Emergency Response Plans are fundamental tools prepared by drinking water supply systems in order to provide standard response and recover actions and procedures to prevent, minimize and mitigate adverse effect resulting from emergencies, either of natural origin or human-provoked.
  Planning should be based on the system risk assessment as determined during the vulnerability assessment, as well as on information gathered from international and European recommendations and guidelines and also during meetings or workshops with local emergency management personnel and first responders.
  A central element within the plan deals with communication and coordination among the water system groups in charge of decision making within crisis management and the State/local emergency response authorities, such as Police, Civil Protection, Public Health Authorities.
  According to EPA Guidelines[9], a typical Emergency Response Plan would contain eight fundamental issues:
  - System Specific Information;
  - System Roles and Responsibilities;
  - Communication Procedures;
  - Personnel Safety;
  - Identification of Alternate Water Sources;
  - Replacement Equipment and Chemical Supplies;
  - Property Protection;
  - Water Sampling and Monitoring.

2.6 Costs

Considerable economic resources are necessary to enhance security level of drinking water supply systems. A preliminary estimate indicates costs for an Italian utility to be about 1% of the operation management costs[10]. Costs are related to the individual utility security upgrades and therefore highly dependent on local factors. It is however to be noticed that many costs could be referred not only to security issues but also to other emergency situations (*e.g.* protection and response to natural disasters).

As in other countries, water policy in Italy is aimed at assuring access to safe drinking water and adequate sanitation to the whole population; even if the price of drinking water in Italy is lower than in many other European countries, costs related to security issues are not easy to pass on the customers. Consistently with this policy an efficient enhancement of prevention and security measures against potential terrorist attacks to drinking water systems should depend in large part on public funding.

In the USA the government has provided very substantial economic resources for security at water infrastructure facilities since 2001 (over $72 million and over $66 million in 2004 and 2005 respectively)[11].

2.7 Support activities and side research

An effective response capable of reducing the risk of human-provoked emergencies and mitigating their adverse effects can be obtained through an integrated approach. This would involve the implementation in drinking water systems of specific strategies, management processes and operative procedures, as suggested by International and Italian guidelines[3].

Support to drinking water infrastructure communities should be provided by a national centre of technical-scientific excellence for setting out both preventive measures - such as vulnerability assessment - and response actions for the identification of possible toxic agents intentionally introduced in water systems. Support activities should include adequate training aimed at improving awareness of drinking water infrastructure communities on human-provoked emergencies and defence measures against unprecedented manmade threats.

Research activities are also important to improve capability of preventing and responding to human-provoked emergencies. Case studies simulating events of intentional contamination with different chemicals and biological agents of water systems are currently carried out at the National institute of Health (ISS). Studies are carried out in hydrodynamic conditions by using a pilot plant consisting of a metallic structure (3.5 m long, 2.5 m wide and 2.2 m high) subdivided lengthwise in two parallel tanks (Fig. 1). Monitoring systems are based on multiple simultaneous quality surrogate measures, as screening methods, and advanced specific techniques (*e.g.* GC-MS, LC-MS) employed as confirmatory methods.

3 CONCLUSION

An integrated approach involving both prevention and security measures has been recently proposed in Italy against potential intentional contamination of drinking water supply utilities.

Specific strategies, management processes and operative procedures – as reported in this paper - have been set out by a cooperative task force involving the Ministry of Health, the ISS and the Italian association of water utility companies (Federgasacqua).

Nevertheless, a real strengthening of security in small and large drinking water systems - and therefore the improvement of consumer's health protection - is related to the availability of adequate economic resources for water utilities and support and research activities through the establishment of a national reference laboratory.

Figure 1 *Pilot plant for case studies*

References

1. Presidential Decision Directive 63" May 22, 1998. "The Clinton Administration's Policy on Critical Infrastructure Protection": See: http://www.fas.org/irp/offdocs/paper598.htm
2. Brussels, 2.6.2003. COM(2003) 320 Final. Communication from the Commission to the Council and the European Parliament on cooperation in the European Union on preparedness and response to biological and chemical agent attacks (Health Security). See: http://europa.eu.int/eur-lex/en/com/cnc/2003/ com2003_0320en01.pdf)].
3. M., Ottaviani, R., Drusiani, L., Lucentini, E., Ferretti, L. andBonadonna. Rapporto ISTISAN 05/04. Misure di prevenzione e di sicurezza dei sistemi acquedottistici nei confronti di possibili atti terroristici 2005, Rapporti ISTISAN 05/04, ISS, Rome.
4. Commission o f t he E uropean C ommunities, B russels, 2 .6.2003, C OM(2003) 3 20 Final Communication from the Commission to the Council and the European Parliament on cooperation in the European Union on preparedness and response to biological and chemical agent attacks (health security
5. Water and Wastewater Security Product Guide. http://cfpub.epa.gov/safewater/watersecurity/ index.cfm - Environmental Protection Agency. United States.
6. Environmental Protection Agency. United States. Response Protocol Toolbox: Planning for and Responding to Drinking Water Contamination Threats and Incidents Module 2: Contamination Threat Management Guide Interim Final - December 2003. Internet address: http://www.epa.gov/safewater/ watersecurity/pubs/guide_response_module2.pdf
7. Environmental Protection Agency. United States. Response Protocol Toolbox: Planning for and Responding to Drinking Water Contamination Threats and Incidents. Module 4: Analytical Guide. Interim Final - December 2003. Internet address: http://www.epa.gov/safewater/watersecurity/pubs/ guide_response_module4.pdf
8. International Life Sciences Institute Risk Science Institute. (ISLI). 1999. Early Warning Monitoring to Detect H azardous E vents i n Water S upplies. I LSI P RESS, W ashington, D C. http://www.ilsi.org/file/ EWM.pdf
9. Emergency Response Plan Guidance for Small and Medium Community Water Systems to Comply with the Public Health Security and Bioterrorism Preparedness and Response Act of 2002. EPA US Environmental Protection Agency. April 2004. http://www.epa.gov/safewater/watersecurity/pubs/ small_medium_ERP_guidance040704.pdf
10 R. Drusiani. Sicurezza delle forniture idriche e terrorismo: un inquadramento generale. *L'Acqua*, **4**, 2002. p. 68.
11. C.,Copeland and B., Copy. Terrorism and Security Issues Facing the Water Infrastructure Sector. CRS Report for Congress Received through the CRS Web. Order Code RL32189. 2004. http://www.ncseonline.org/NLE/CRSreports/04Mar/RL32189.pdf

IMPROVED UNDERSTANDING OF WATER QUALITY MONITORING: EVIDENCE FOR RISK MANAGEMENT DECISION-MAKING

S. Rizak[1] and S. E. Hrudey[2]

[1]Department of Epidemiology and Preventive Medicine, Monash University, Melbourne, Australia
[2]Department of Public Health Sciences, University of Alberta, Edmonton, Alberta, Canada
steve.hrudey@ualberta.ca

1 INTRODUCTION

Drinking water quality monitoring has developed as a means for assessing compliance, verifying system performance, triggering response actions, and conducting process research. The purpose of this monitoring is to limit or reduce the uncertainty and then use the data to make reasoned, appropriate decisions about the water supply system performance. There is some misunderstanding about the meaning of monitoring data in the water industry, public health practice and regulatory sectors. Some individuals interpret water quality monitoring evidence as absolute measures of truth and show little appreciation of the fundamental limitations to monitoring evidence. These limitations include not only the recognized shortcomings of sampling and analytical techniques, but also important sources of error in the use and interpretation of data from monitoring.

Interpretation of monitoring evidence for the purpose of making risk management decisions is not a straight-forward task. Because of the limited resources available to us to fully understand problems, we will often have to judge monitoring evidence despite incomplete information and uncertainty. Drawing an analogy with statistical hypothesis testing, this requires maintaining a balance between making False Positive (Type I) and False Negative (Type II) errors in risk management.[1]

Water quality illustrations of these error types could be:

- False Positive Error: the Sydney Water Crisis of 1998 involved ~4 million consumers in Sydney, Australia experiencing on again – off again boil water advisories over a period of 3 months based on what might have been erroneous monitoring results. In any case, despite reports of high numbers of *Giardia* cysts and *Cryptosporidium* oocysts, no elevated illness was detectable in the community, even with ~33% non-compliance with the boil water advisory.
- False Negative Error: the Walkerton outbreak of 2000 resulted in 7 deaths, 65 hospitalizations, 27 cases of haemolytic uremic syndrome (median age 4), 2,300 cases of gastroenteritis caused by *Escherichia coli* O157:H7 and *Cambylobacter* spp. This outbreak occurred despite many warnings of adverse water quality and failure to act on available warnings and to collect required data on chlorine residual that would have provided "real time" warning of contamination.

These two events illustrate the difficulty in interpreting and understanding monitoring evidence and responding to it appropriately to manage public health risks. Obviously both types of error have negative consequences and while it is generally favoured to avoid false negative errors over false positive errors for public health risk management, it must be recognized that chronically committing false positive errors will serve to undermine consumer confidence in water safety, waste resources and risk complacency developing in the long term, leading to failure to respond when action truly is needed.[1]

The interpretation of data and appropriate follow-up are important features of the monitoring process that tend to be overlooked. Because the risk management decisions that result can have serious consequences and significant social and economic impacts, it is important that the many limitations are clearly understood. Objectives of our research are to examine where misinterpretation and misunderstanding lie in the use of monitoring data and provide greater appreciation and understanding of monitoring data in order to make better use of evidence and more informed risk management decisions.

2 PRINCIPLES OF DIAGNOSTIC TESTING

2.1 Characteristics and Performance of Diagnostic Tests

In medical screening, the performance of a test is indicated by two concepts that reflect the results expected in patients with without the disease. These are the diagnostic sensitivity (***DSe***) and diagnostic specificity (***DSp***) of the test and are determined from the apparent capability of the screening test method. These can be impressive for some tests (>95%), but may often be less than 90% in water quality applications.

Diagnostic sensitivity, ***DSe,*** is the conditional probability that the test will correctly identify something as a hazard (disease), given that it truly is a hazard (disease). In medical diagnostics, the ***DSe*** of a test is determined by identifying the proportion of people with the disease who will show a positive test for that disease when screened.

Diagnostic Specificity, ***DSp***, on the other hand, represents the conditional probability that the test will correctly identify a non-hazard (no disease), given that it is truly a non-hazard (no disease). Similarly, the ***DSp*** of the test is determined by identifying the proportion of people without disease who will show a negative test for that disease when screened.

In medical practice, these terms are used without the modifier "diagnostic", but this modifier is used here to avoid confusion in terminology. For environmental scientists **sensitivity** generally means the lowest level of a hazard (e.g. a pathogen in drinking water) that can be accurately detected, and is typically expressed as a concentration. Likewise, to an environmental scientist, **specificity** refers to the ability of a method to discriminate a particular hazard (say a particular pathogen) from other factors (e.g. other microbes) that may be present. These meanings differ subtly from the medical diagnostic terms, which are both expressed as conditional probabilities (no units).

To determine the diagnostic sensitivity and specificity for a screening test, the relationship between the diagnostic test results and the actual presence of some hazard or disease, i.e. the truth or reality, must be evaluated. These relationships between a test's results and the truth can be explained as shown in Table 1, whereby the rows of the table relate to the evidence obtained from a testing method considered to be either positive or negative, and the columns relate to the true condition of whether some hazard or disease is present or absent.

Table 1 (a, b, c) *Diagnostic Evidence*

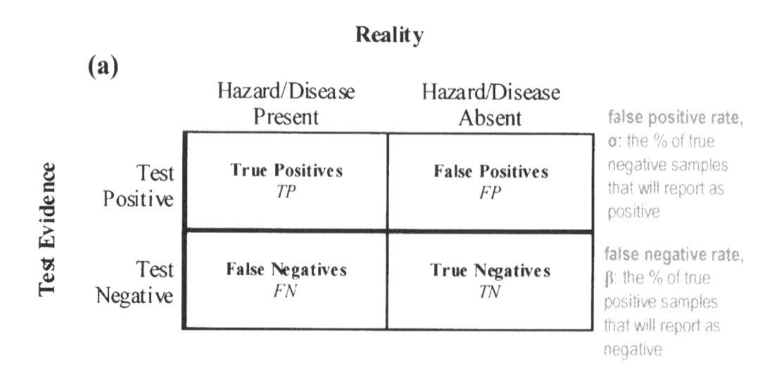

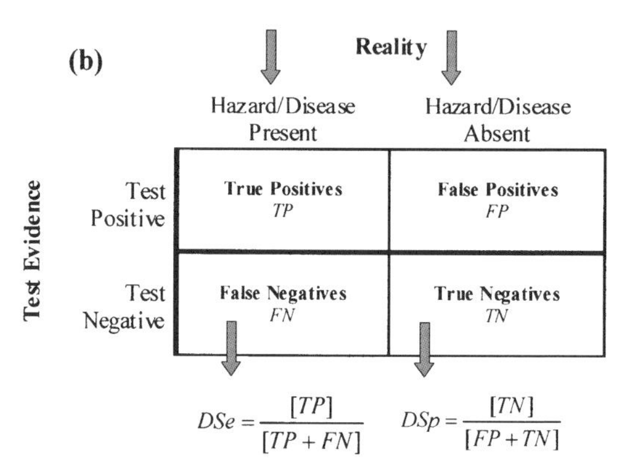

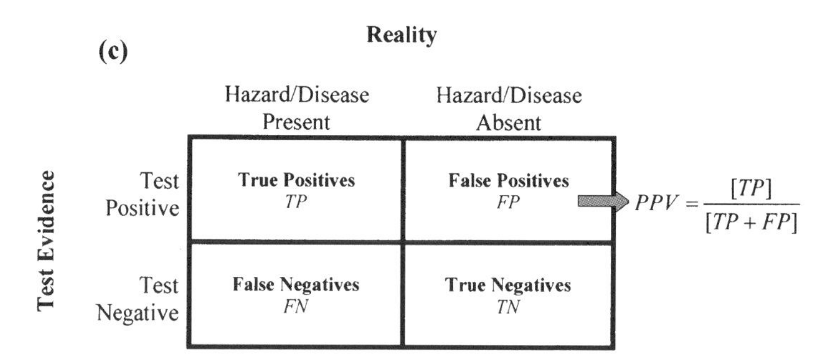

2.2 Applications of Principles to Water Quality Monitoring

States et al. described the importance of false positive rates noting that one hazard screening assay manufacturer claimed a false-positive rate (**a**) of only 3% for the pathogen *Bacillus anthracis* contrasted to competing products with false positive rates of 37 to 83%.[4]

Intuitively, the 3% false positive rate appears good for a rapid screening test. However, suppose that we apply such screening technology to circumstances of rare contamination where, for the sake of illustration, only 1 out of 300 samples is truly hazardous. We can ask: *If we get a positive result (contaminant is detectable or exceeds the standard) from an analytical test, how likely is that "positive" result to be correct?*.[2,3] We will need to screen 299 samples free of detectable levels of the contaminant to find 1 sample that contains detectable levels. With a false positive rate of 3%, we will detect approximately 9 false positives ($299 \cong .03$) in our search for the 1 true positive. Consequently, only 10% of positive results from this analytical test will correctly reflect the presence of a hazard (i.e., the ***PPV*** $\cong$ 10%). This finding is an inescapable reality (as a function of the false positive rate and frequency of true hazards) for any analytical screening test. As the hazard we are searching for becomes more rare, we can expect false positives to exceed true positives unless a test offers a false positive rate approaching the frequency of the hazard (Figure 1).

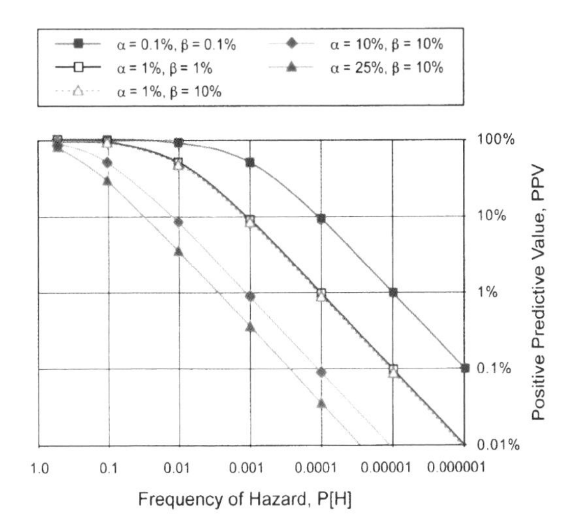

Figure 1 *Positive Predictive Value, PPV*

3 CONCLUSIONS

If the hazard that monitoring evidence is screening for is rare (as we would expect in monitoring treated water quality), the chance of any positive detected being a true positive, the positive predictive value (***PPV***) will be small unless the false positive rate (**a**) is as small the frequency of that hazard.

This quantitative reality depends on the limited capabilities of methods and the frequency of hazards. This reality must be recognized so that monitoring programs can be planned effectively. Responses to positive monitoring results must be appropriate to their likely meaning to ensure that responses will do more good than harm.

References

1 S. E. Hrudey and W. Leiss, *Environ. Health Perspect.*, 2003, **111**, 1577.

2 S. E. Hrudey and S. Rizak. *J. Am. Water Works Assoc.*, 2004, **96**, 110.
3 S. Rizak and S. E. Hrudey, in Proceedings of the 11th Canadian National Conference and 2nd Policy Forum on Drinking Water, Calgary, Alberta, Canada, April 3–6, 2004.
4 S. States, J. Newberry, J. Wichterman, J. Kuchita, M. Scheuring and L. Casson, *J. Am. Water Works Assoc.*, 2004, **96**, 52.

TOOLS FOR THE RAPID DETECTION OF PATHOGENS IN MAINS DRINKING WATER SUPPLIES

S.A. Wilks[1], N. Azevedo[2], T. Juhna[3], M. Lehtola[4] and C.W. Keevil[1]

[1]School of Biological Sciences, University of Southampton, Bassett Crescent East, Southampton SO16 7PX, UK
[2]Centro de Engenharia Biológica, Universidade do Minho, 4700-057 Braga, Portugal
[3]Riga Technical University, Institute of Gaz, Heat and Water Technologies, 16 Azenes Street, 1048 Riga, Latvia
[4]Department of Environmental Sciences, University of Kuopio, Savilahdentie 9, 70211 Kuopio, Finland

1 INTRODUCTION

Contaminated water supplies continue to be a problem, and while there is a better understanding of how to ensure the safety of the bulk water, little is known about the survival of potential pathogens in biofilms. It has been suggested that biofilms can offer a refuge hence protecting cells against disinfection and physical disruption.

The challenge is how to detect individual bacteria within the biofilm, particularly on highly corroded surfaces. Current quantitative methods rely on culturing which requires the mechanical removal of the cells. Such methods take time, and provide no information on the location of cells within the biofilm structure. Molecular methods including denaturing gradient gel electrophoresis (DGGE) fingerprinting can also be used but again rely on the mechanical removal of the biofilm. In order to visualise individual bacteria within the biofilm the cells must be labelled in some way. Fluorescence *in situ* hybridisation (FISH) can be used by designing specific molecular probes, usually targeting sites on the 16S rRNA, which are labelled with a fluorophore. Traditionally DNA-based probes have been used but these have a number of disadvantages which make them unsuitable for use on complex environmental samples.

In this study we have examined the alternative use of peptide nucleic acid (PNA) probes with a FISH-based assay to detect specific pathogens within potable water biofilms. PNAs are synthetic molecules where the phosphodiester backbone is replaced by a 2-aminoethyl-glycine linkage (Fig. 1). The molecule will bind to both DNA and RNA following Watson-Crick hydrogen bonding rules. The unique chemistry conferred to PNAs provides them with a number of advantages over DNA probes for use in FISH assays. They exhibit greater thermal stability and are resistant to enzymatic and ionic stress. These characteristics make them particularly suitable for use in multiplex assays. DNA probes require stringent hybridisation conditions that are generally specific to each individual probe. This causes problems if more than one probe is to be used at any one time, the flexibility of PNA probes means that they do not have the same limitations.

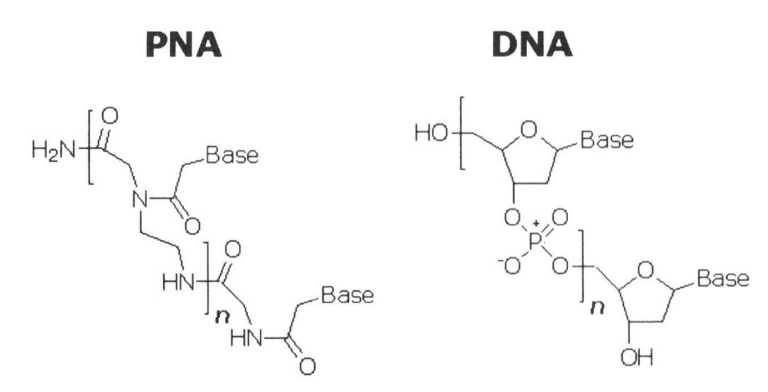

Figure 1. *Structure of PNA and DNA molecules, showing the differences in chemical structure.*

PNA probes specific to the following bacterial pathogens and indicator species have been designed; *Legionella spp.*, *L. pneumophila*, *Helicobacter pylori*, *Mycobacterium avium* and *Escherichia coli* K12 and O157 NCTC 12900. The suitability of these probes for use in complex environmental samples and in multiplex assays has been evaluated. The main aim of this study is the application of PNA probes in a FISH-based study to assess the distribution of these selected species in and on the biofilm associated with mains drinking water supply pipes.

2 METHOD AND RESULTS

PNA probes were designed and tested for target specificity by screening on various ribosomal databases. All PNA probes were designed to have a length of 15-mer, which is optimal for their successful hybridisation. Each probe was synthesised with a fluorophore attached (Eurogentec, Belgium). The specificity of each probe was tested using slide hybridisations on a range of bacterial species. Comparisons were also made with existing DNA probes. Following on from single species slide hybridisations, spiked reactor coupons were used for validation before the use of actual mains drinking water supply pipe samples. Samples were received from partner organisations in the UK, France and Latvia. Pipes had been removed from the ground, wrapped in polythene to retain moisture and transported by courier to each lab. On arrival, they were cut into small coupons (approx. 4 cm x 4 cm) and analysed immediately. Samples were also used for standard culture detection methods.

A standardised protocol was used for PNA hybridisation.[1] Samples were fixed by flaming gently and then dehydrating in 90% ethanol for 10 minutes. Probes were added at a final concentration of 200 nM in hybridisation buffer (30% (v/v) formamide, 0.1 mM NaCl, 10% (w/v) dextran sulphate, 0.1% (w/v) sodium pyrophosphate, 0.2% (w/v) polyvinylpyrrolidone, 0.2% Ficoll, 5 mM disodium EDTA, 0.1% (v/v) Triton X-100, pH 7.6) and incubated at 57°C for 90 minutes. They were then covered with washing buffer (5 mM Tris-HCl, 15 mM NaCl, 0.1% (v/v) Triton X-100, pH 10) and left at the same temperature for a further 30 minutes. After this time, samples were rinsed carefully with filter-sterilised distilled water and left to dry before examination.

Figure 2. *Example of mains drinking water supply pipe supplied for analysis, before and after cutting into small coupons.*

Slides were examined under oil at x1000 magnification. The water pipe sample were examined directly using the episcopic differential interference contrast/epifluorescence (EDIC/EF) microscope.[2] This microscope system has long working distance objectives and episcopic illumination to permit the examination of large opaque samples.

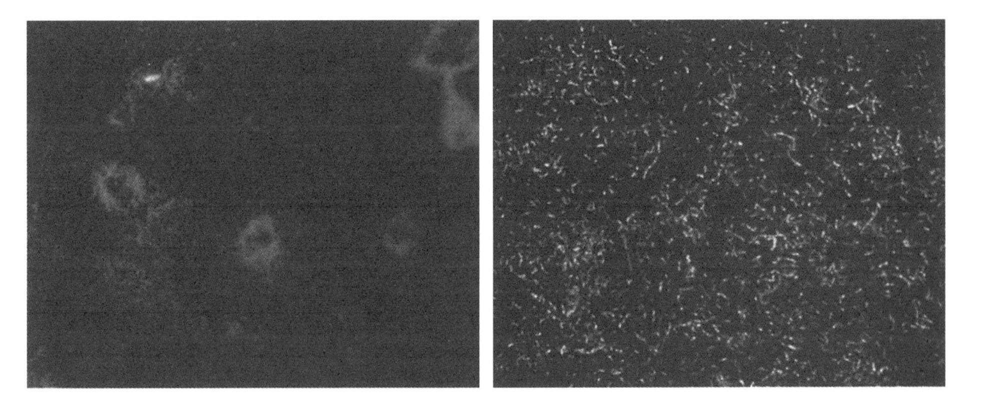

Figure 3. *Comparison of slide hybridisations using DNA and PNA probes targeted to Legionella pneumophila spp 1. A. DNA probe. B. PNA probe. Magnification x 600. Hybridisation using DNA probes resulted in areas of diffuse fluorescence and clumping of cells. Labelling is more uniform when PNA probes are used.*

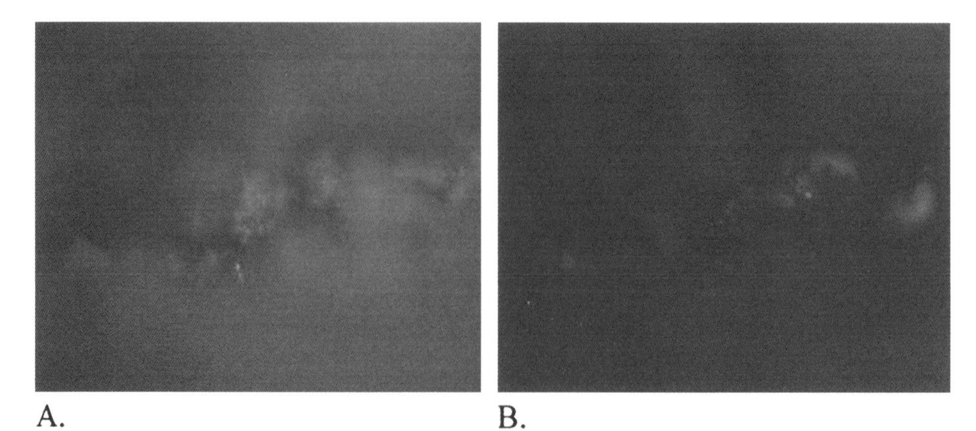

Figure 4. *Hybridisation using a PNA probe specific to L. pneumophila to biofilm on a corroded cast iron pipe. A. EDIC image showing pipe surface. B. Labelled bacteria seen using the TRITC channel. Magnification x 1000. Illustrates clear labelling of L. pneumophila on a pipe surface.*

Figure 5. *Coupon from a lab reactor after spiking of the drinking water biofilm with E. coli O157 and H. pylori. PNA probes targeted to these pathogens have been used in a duplex hybridisation. Cells labelled red are E. coli O157, cells labelled green are H. pylori. Magnification x 1000.*

3 CONCLUSIONS

- Specific PNA probes targeted to the selected bacteria have been successfully developed and validated in extensive slide hybridisation assays.
- Experiments have confirmed the advantages of PNA probes over DNA-based ones for use with complex environmental samples.

- The use of PNA probes in a FISH assay can be applied to the *in situ* detection of these bacteria on highly corroded drinking water pipes when used in conjunction with the EDIC/EF microscope system.
- PNA probes can also be used in duplex and multiplex assays. The careful choice of fluorophores can permit the detection of several species concurrently.
- Ongoing work is validating this technique for the quantification of these bacteria on pipe samples, even when they are present in very low numbers.

References

1 Stender, H., Fiandaca, M. and Hyldig-Nielsen, J.J. (2002). PNA for rapid microbiology. J. Microbiol. Meths., **48**, 1-17.
2 Keevil, C. W. (2003). Rapid detection of biofilms and adherent pathogens using scanning confocal laser microscopy and episcopic differential interference contrast microscopy. Water Sci. Technol. **47,** 105-116.

Acknowledgements

This work has been undertaken as part of a research project which is supported by the European Union within the Fifth Framework Programme, "Energy, environment and sustainable development programme", n° EVK1-2002-00108. There hereby follows a disclaimer stating that the authors are solely responsible for the work, it does not represent the opinion of the Community and the Community is not responsible for any use that might be made of data appearing herein.

DETECTION AND CONFIRMATION OF UNKNOWN CONTAMINANTS IN UNTREATED TAP WATER USING A HYBRID TRIPLE QUADRUPOLE LINEAR ION TRAP LC/MS/MS SYSTEM

Michael T. Baynham[1], David Evans[2], Pat Cummings[2] and Stephen J. Lock[1]

[1]Applied Biosystems, Birchwood Boulevard, Warrington, Cheshire U.K WA3 7QH
E-mail michael.baynham@eur.appliedbiosystems.com
[2]ALcontrol Laboratories, Mill Close, Rotherham, South Yorkshire, UK, S60 1BZ
E-mail pat.cummings@alcontrol.co.uk

1 ABSTRACT

This poster describes the use of a hybrid triple quadrupole linear ion trap LC/MS/MS system with mass spectrometric library searching as a screening tool for the detection and confirmation of trace levels of organic contaminants in tap water without the need for any sample pre-treatment. Two approaches have been investigated:

1) A rapid Multi-Target Screening (MTS) approach for very large numbers of 'target' compounds that have previously been characterised and detailed in a 'target' library.
2) A General Unknown Screen (GUS) whereby sample / control comparisons are made to detect the presence of compounds that are either unique to the sample or compounds present at significantly higher levels than in the control.

Both approaches then utilise novel data dependant hybrid linear ion trap Enhanced Product Ion (EPI) scans to generate library searchable spectra for confirmation of ID at trace levels.

2 INTRODUCTION

Protection of our drinking water resources from contaminants is a major responsibility for both government and water producing bodies. The response taken to a potential drinking water emergency will depend upon both the composition and the nature of the identified contaminant(s). Furthermore, it is essential that there is a high degree of confidence in the correct and rapid identification of the problem before remedial action is taken. To date it has been a necessity to employ the combination of multiple analytical techniques to meet this end.

Our MTS approach utilises a rapid screen for up to 2000 target compounds within 20 minutes. This initial screening employs the highly selective and sensitive triple quadrupole Multiple Reaction Monitoring (MRM) as a 'survey' scan with data dependant hybrid linear ion trap Enhanced Product Ion (EPI) scans. During the initial MRM survey scan, any of the compounds detected above a pre-defined MRM threshold automatically trigger the acquisition of an EPI spectrum which is subsequently searched against entries in the target MTS library.

Conversely, our GUS approach utilises a 'universal' Q3 (full scan) in place of the MRM survey. In this experiment, compounds detected above a set Q3 threshold automatically trigger the generation of an EPI spectrum. Comparison of the Q3 TIC's (Total Ion Chromatograms) of the sample to that of the control reveals compounds that are either unique to the sample, or those that are present at significantly higher concentrations than in the control. Data dependant EPI spectra of these compounds are then submitted to a more comprehensive EPI mass spectrometric library database for confirmation.

It is envisaged that the information obtained by employing either the MRM→EPI or / and Q3→EPI methods will routinely enable rapid identification and confirmation of trace organic contaminants in our drinking water supplies.

3 MATERIALS & METHODS

3.1 HPLC Instrumentation

A Shimadzu Class vp HPLC system (LC10ADvp binary gradient pumping system with SIL-HT autosampler), was used for all HPLC separations with the following conditions.

3.1.1 MRM → EPI Rapid (Multi-Target Screening method)

Column C18 Monolith Guard column
Separation conditions Rapid gradient over 1.5 minutes
Flow rate of 1.40 ml/min (without splitting prior to the MS)
Injection volume = 50 μl
Mobile Phase positive mode
A: Water + 2mM ammonium acetate, B: methanol +0.1% formic acid

Mobile phase negative mode
A: Water, B: methanol +0.1%ammonium hydroxide

3.1.2 Q3 → EPI (General Unknown Screening method)

Column ACE C18 4.6 x 50mm 5mM
Mobile phase and flow rate conditions used are the same as the MRM → EPI Rapid screening method.
Gradient from 25% B to 100% B over 16 minutes.
Injection volume = 50 μl

3.1.3 Mass Spectrometry Conditions – All methods

Mass Spectrometer 4000 Q TRAP® LC/MS/MS System (Applied Biosystems/MDS SCIEX)

Ionisation	Turbo VTM Ion Source (ESI)
Curtain Gas	25
Gas1	50
Gas2	60
CAD Gas	10
Temperature	700 °C
Ionspray voltage	-4500 (neg) +5500 (pos)

4 RESULTS

4.1 Rapid (<2.0 min) initial MTS analysis of samples: MRM → EPI

The LINAC™ collision cell of the 4000 Q TRAP® can simultaneously monitor 296 MRM transitions (contaminants) from single sample injection. This allows for the generation of data dependant EPI spectra in a cycle time of approximately 2.5s without major loss in sensitivity or spectral quality.

Figure 1 shows the data obtained for an injection of 100 ug/litre terbutylazine in both mineral and tap water, using the MRM → EPI rapid MTS approach (53 transitions monitored).

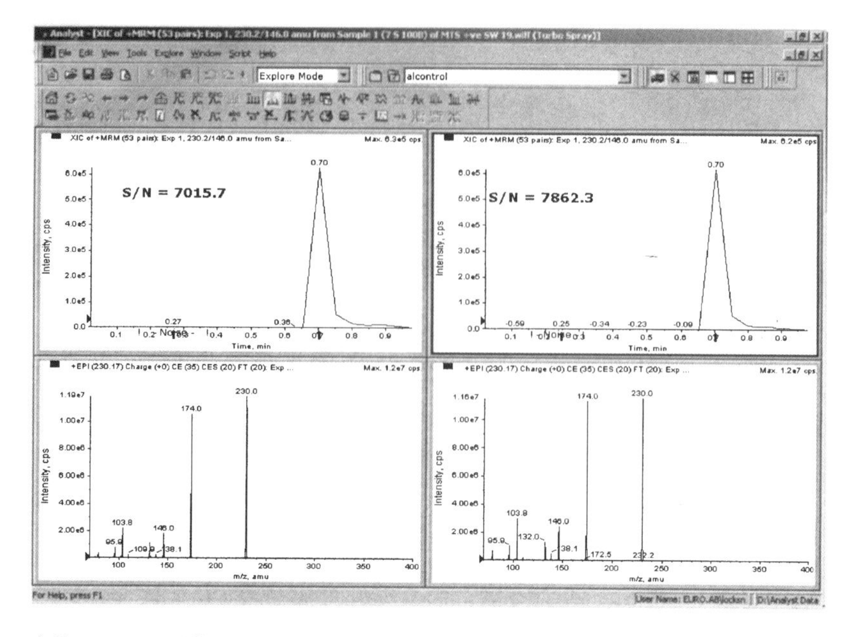

Figure 1 *Positive ion ESI MRM and EPI results from mineral (LEFT TRACE) and tap water (RIGHT TRACE) spiked with 100ppb terbutylazine*

Mineral water typically contains high levels of sodium, which may affect sensitivity due to adduct formation. However, the effect on the S/N is kept to a minimum (see fig.1 above). Figure 2 (below) shows the data obtained for an injection of 100 ug/litre MCPP in both mineral (right trace) and tap water, (left trace) using the (negative ion) MRM to EPI rapid screening approach.

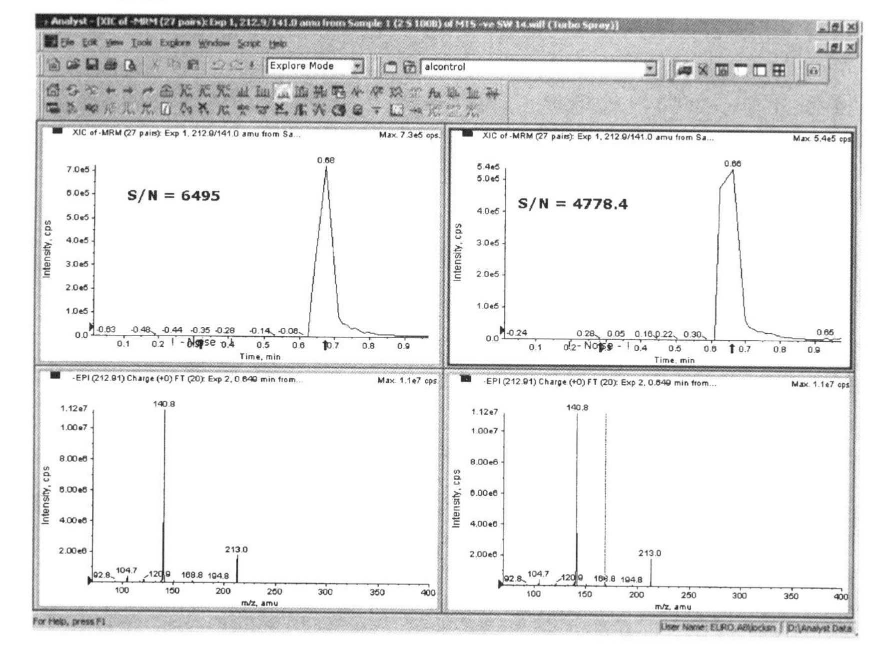

Figure 2 *Negative ion ESI MRM and EPI results from mineral (right trace) and tap water (left trace) spiked with 100 ug/litre MCPP*

4.2 GUS analysis of samples: Q3 → EPI

Figure 3.illustrates the GUS method, showing the comparison of a blank sample control to a sample (A) that has been spiked with 100 ng/litre of isoproturon. A software script makes a detailed comparison of the two Q3 chromatograms and the presence of the spiked compound (m/z=207) is detected in the sample.

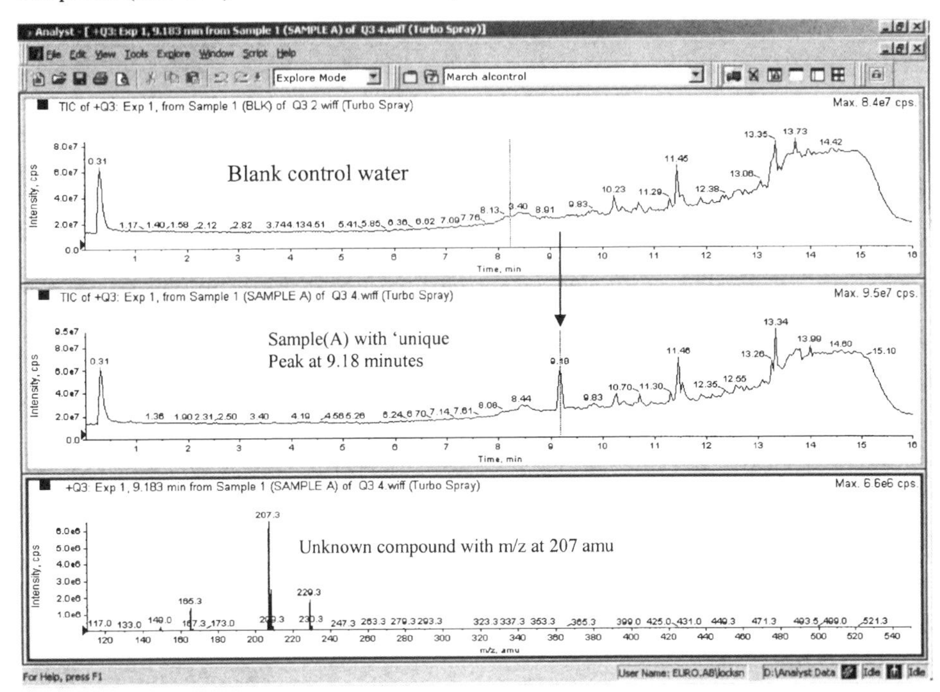

Figure 3 *Control (Blank water) / sample (A) comparison with sample spiked with isoproturon*

Library searching of the EPI spectra (fig.4) obtained for both the blank control and sample (A) at 9.18 minutes shows the presence of isoproturon only in the spiked sample.

Figure 5 shows the data obtained from the sample/control comparison of sample D, which is spiked with an 'unknown' compound. The comparison has indicated the presence of a contaminant at 1.09 min. The Q3 spectrum shows that this compound has a (M+H) mass of 335 amu but the data dependant EPI spectrum is not matched in the EPI library database. This unknown contaminant was later identified to be strychnine by comparing its EPI spectra to that of a strychnine standard.

MTS and GUS methods for a number of compounds spiked into tap water at 10 ug/litre and 100 ug/litre were compared. The comparison of the methods and their relative potentials are shown in Table 1, with sample A – D spiked at 100 ug/litre and samples E – H at 10 ug/litre.

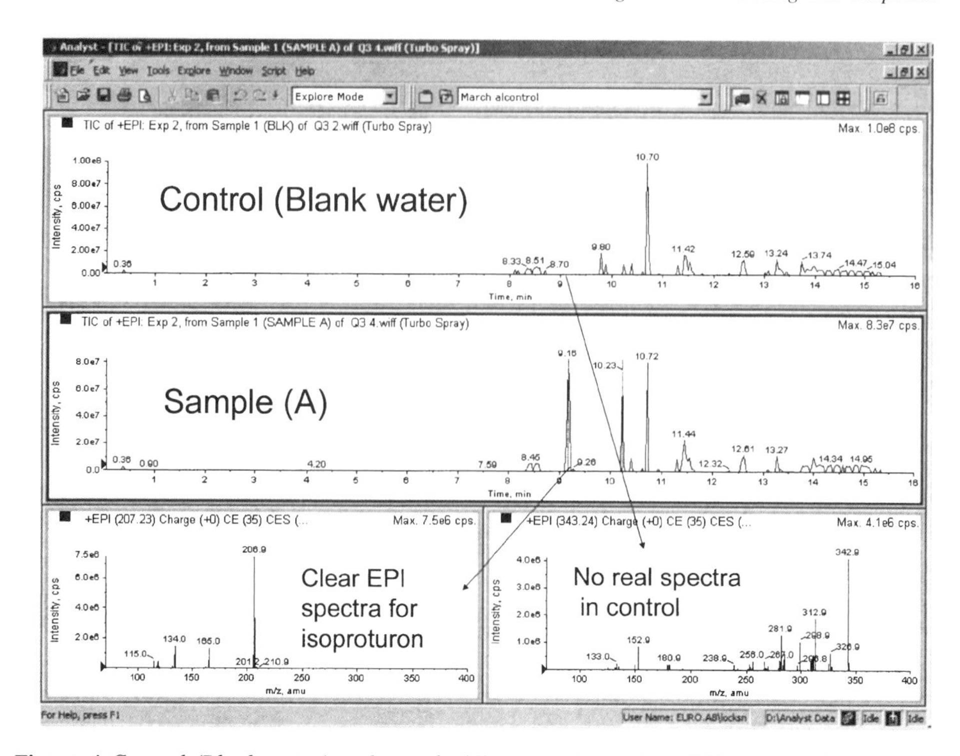

Figure 4 *Control (Blank water) and sample (A) comparison of the EPI spectra obtained for each experiment - isoproturon is clearly detected in the spiked sample (A).*

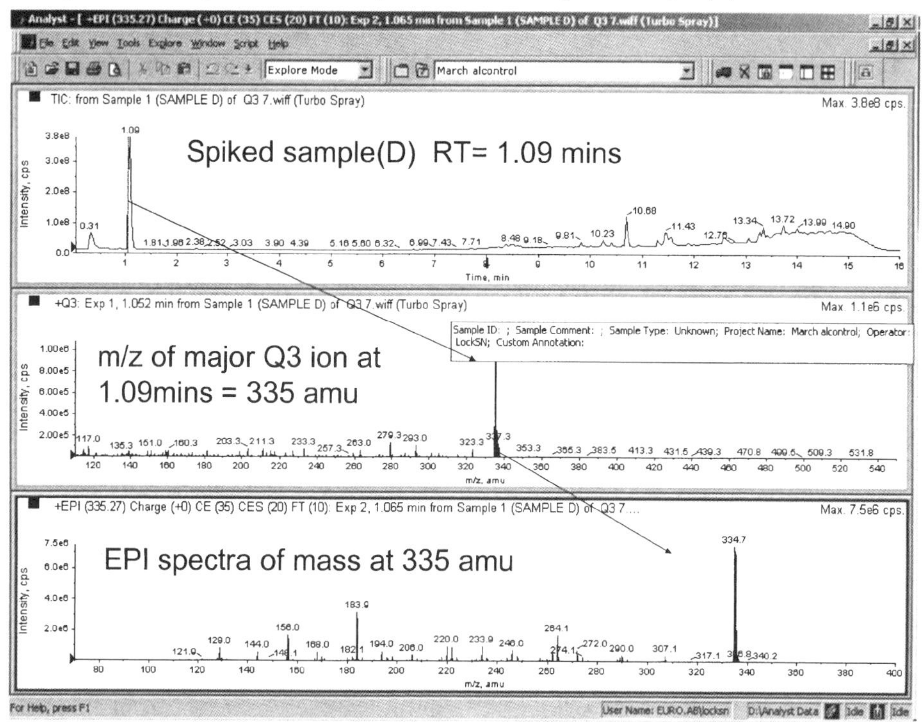

Figure 5 *Sample / Control comparison of sample D – peak at 1.09 min not present in control*

Sample	MRM - EPI Approach Compound detected	Q3 - EPI Approach Compound detected	Retention time (min)
A	Isoproturon	Isoproturon	9.15
B	CHLORPYRIPHOS-ETHYL	CHLORPYRIPHOS-ETHYL	12.05
C	Not in MRM list	m/z 213 amu ?	5.31
D	Not in MRM list	m/z 335.2 amu ?	1.09
E	Isoproturon	Isoproturon	9.15
F	CHLORPYRIPHOS-ETHYL	nothing	
G	Not in MRM list	nothing	
H	Not in MRM list	m/z 335.2 amu ?	1.09
100 ng/litre spike into humic acid	Isoproturon	Isoproturon	9.15

Table 1 *Comparison of screening methods and results obtained*

5 CONCLUSIONS

The MTS approach is the most rapid and sensitive method to screen for and confirm the presence of 'target' organic contaminants in untreated tap water. Using multiple sample injections over 2000 target compounds can be screened for in less than 20 minutes at the sub to low ug/litre level. For compounds not present in the MTS method, the GUS method is the best approach as it does not rely on any assumptions or knowledge as to the identity of the contaminant(s). Here, a sample control comparison will detect unknown contaminants and automatically generate EPI spectra. This spectral information can be searched against a more comprehensive library and the information may be used for identification purposes. However, the GUS approach is the least sensitive of the two methods and requires significantly longer run times. Development of software tools for the automatic generation of a large scale EPI library for water contaminants is ongoing and will further widen the scope of both approaches.

"Mind the Gap" – Facilitated Workshop

Martin Furness

Quality and Environmental Assurance, Severn trent Water

1 INTRODUCTION

The intent of the conference was to consider the "big picture" and delegates were given the opportunity to raise any topics they felt had not been addressed during the event or had perhaps not received enough attention. Interactive discussion was encouraged to identify where the Water Industry could be even better prepared for any water emergencies. Five topics had been signalled in advance by the delegates and two more were raised as the workshop developed. The workshop was free flowing, participative, stimulating and agreed, on the night, to have been of value. The debate was to some extent UK-centric but other nation's views were also discussed.

2 "IS THE COST OF SPECIALIST EQUIPMENT JUSTIFIED?"

If water emergencies are so rare then why invest needlessly?. Whereas delegates supported the view that necessary detection equipment to pre-warn of contamination and protect public health had obvious value there was regret that no universal test was available. Some Companies are utilising technologies with specific detection capabilities. A question was raised about the return on such equipment and could it be used for other purposes. The discussion widened to trying to understand the economic benefit of increased vigilance. Such understanding would go a long way to defining surrogate and pragmatic technologies. A plea was put forward for more support of on line monitoring and querying if it currently was being used to best effect. It was felt that advances in technology will happen anyway in a market motivated by shrinking expenditure revenues.

It was agreed that more awareness of technological developments should be encouraged. Jeni Colebourne, DWI, added that increased responsibility for self-regulation meant the need to undertake appropriate risk assessments to identify specific vulnerabilities and develop resilience?

3 ARE WE READY NOW TO RESPOND TO A TERRORIST THREAT TO THE WATER INDUSTRY?

Did we feel totally prepared for such an event? One comment at the start suggested that the roles and communication channels in bronze, silver and gold commands. Emergency services operate three tiers of command and control for major emergencies headed by the Police. Operational (on site) or Bronze, Tactical (local support) or Silver and Strategic (regional or national support) Gold were not clear for a water emergency. Within the emergency services, this was relatively straightforward. However, with another responsible agency, with its own statutory obligations, this was not self evident. What about reconciling data capture with a cordoned off scene of crime investigation? Specific questions were raised of Water Companies. Do they have triggers for stopping supplies; who makes the decision and how? Reassurances were given to the Group by (whom?). There are clear guidelines in place in all Water Companies but, most importantly, the health risk assessment would be discussed urgently with the local public health doctor. A wider understanding of water quality triggers and the health risk assessment would be beneficial. Reciprocally, it was felt that a broad understanding of operational functions and the scope for errors and omissions would have merit.

The group were reassured that site security was an increasing topic for expenditure and development. One Water Company was trialling hatch lid sensors that would signal attempted or actual entry to the water space.

Lastly, it was conjectured that – "would we recognise a terrorist attack?" The group were informed that through the Water Industry CBRN group a number of scenarios had been investigated. Further industry communication was being considered.

4 ARE SMALL SYSTEMS A GREATER RISK?

Experience in USA and Canada has shown some notorious incidents with small systems and "conventional" pollutants. It was commented that public health aspects of private supplies are currently dealt with in a similar way to food. As part of the general learning it was agreed that specific issues relating to small systems should be included in future conferences or workshops.

5 RESPONSE PLANS – SHOULD WE GET BACK TO BASICS?

Is it better to have simple robust plans for general emergencies and have we all got this anyway? It was accepted that all Water Companies have response plans with which they are comfortable but there is little benchmarking and certainly no sharing of ideas on the decision making processes. For instance, it was felt by one speaker that similar response plans for system intrusion should exist but a review of current practice would need to be instigated to check this. Operational aspects need to be understood in order to elucidate requirements. The group were informed by members of the conference organising committee that this topic would be taken up at the national level.

6 DO WE COLLABORATE ENOUGH?

Do we discuss near misses, for instance. Comparison was made with airlines where there is an international mechanism for sharing close calls. The view of the attendees was that this does not take place as much as it could. There is certainly collaborative work at the research and technology level and engagement by the Regulators is in progress. There was a call for a "lessons learnt" process and regular exercises both nationally and internationally.

7 INCIDENT COMMUNICATION

This was felt to be an under-discussed topic, yet critical to the successful dealing with any emergency. The Industry should share the detail and psychology of internal and external communication processes where they relate to harnessing resources. Public information was important, in proactive involvement in identifying incidents, managing consumer expectations, and managing an event.

8 SAMPLING

Do we know our roles? An opinion that roles are blurred in a water emergency involving CBRN agents was generally supported. When for instance would specialists be called in (bearing in mind it is important to buy as much time as possible in responding to such events)? This may be due to a lack of general understanding of procedures already in place but which have not been sufficiently well promulgated within the industry. Where are the boundaries of a contaminated site? What about forensic aspects? Do microbiological or chemical risks pose a greater threat to a sampler (and would simple precautions and advice be sufficient to avoid chances of exposure)? The group agreed there is an urgent need for more work and guidance in this area particularly regarding sampling strategies (possibly via the current work on the emergency toolbox and availability of appropriate training courses).

The discussion was valuable although limited by time constraints. Delegates were told that the actions identified would be reviewed by the Organising Committee and passed to DEFRA for Water Industry consideration.

A theme started to emerge of more information to aid in the aftermath and recovery of any water emergency.

Subject Index